*Lecture Notes of*
*the Unione Matematica Italiana*

*13*

For further volumes:
http://www.springer.com/series/7172

The Editorial Policy can be found at the back of the volume.

Michel Frémond

# Phase Change in Mechanics

Springer

UMI

Michel Frémond
Università di Roma "Tor Vergata"
Dipartimento di Ingegneria Civile
Via Politecnico 1
00133 Roma
Italy
Michel.Fremond@uniroma2.it

ISSN 1862-9113
ISBN 978-3-642-24608-1 e-ISBN 978-3-642-24609-8
DOI 10.1007/978-3-642-24609-8
Springer Heidelberg Dordrecht London New York

Library of Congress Control Number: 2011945415

Mathematics Subject Classification (2010): 74N25; 74R20; 74M15; 74M20; 80A17; 86A10; 70F35; 76A02; 35Q35; 35Q74



Printed on acid-free paper

Springer is part of Springer Science+Business Media (www.springer.com)

# Preface

A model is a theory which predicts the evolution of a structure. The basic elements of a mechanical predictive theory are described. They involve the choice of the state quantities which characterize the investigated physical phenomenon, the basic equations of mechanics: the equations of motion, the laws of thermodynamics and the constitutive laws. The theory is applied to phase change. Phase change involves microscopic motions which have macroscopic effects. These motions are taken into account in the macroscopic predictive theories. Numerous examples which are used or may be used in engineering are given for different phase change and damage problems. Moreover, when phase change occurs, temperatures may be discontinuous, for instance when warm rain falls on frozen soil. This situation is also investigated.

Some of these problems have been investigated in the framework of the Laboratorio Lagrange with Italian, Tunisian and French scientists, Francesco Ascione, Elena Bonetti, Anna Maria Caucci, Eric Dimnet, Christian Duquennoi, Francesco Freddi, Rym Lassoued, Boumediene Nedjar, Francesca Nerilli and Elisabetta Rocca. All of them are warmly thanked.

These lecture notes are based on a course given at the XXX Scuola Estiva di Fisica Matematica at Ravello in September 2005 and on lectures given at the Università di Roma "Tor Vergata".

The author would like to thank the Scuola Estiva di Fisica Matematica, the Università di Roma "Tor Vergata", the Unione Matematica Italiana, and Professors Mauro Fabrizio, Franco Maceri and Franco Brezzi for the opportunity to give these lectures and to publish them.

Roma, Italy *Michel Frémond*

# Contents

# Chapter 1
# Introduction

A model is a theory which predicts the evolution of a structure. It does not explain the underlying physical phenomenon, it only predicts what is going to happen: *physics explains, mechanics predicts the motion*. A model is schematic and limited to some aspects of the phenomenon which occurs. In this point of view, the scientist or the engineer has an important and major role. He chooses the quantities which are to be predicted. These quantities are the *state quantities*: they characterize the equilibrium, i.e., the structure is at an equilibrium if the state quantities are constant with respect to the time. Let us stress that the notion of equilibrium is subjective.

In the sequel we consider phase change occurring at the engineering level, i.e., at the macroscopic level: solid–liquid phase changes, damage phenomenon, phase change involving temperature discontinuities,... Many other phase change phenomena are relevant to this theory: solid–solid phase changes as in shape memory alloys, phase changes in porous medium, vapor-liquid phase changes,...

The classical ice–water phase change involves microscopic motions which have macroscopic effects. In shape-memory alloys the solid–solid phase change involves also microscopic motions which are responsible for phase twinning. When damaging structures, microscopic motions break the links which are responsible for the cohesion of the material. We think that *all these microscopic motions have to be taken into account in the macroscopic predictive theories*.

The basic idea we have developed is *to account for the power of the microscopic motions in the power of the interior forces*. Thus we modify the expression of the power of the interior forces and assume that it depends on the volume fraction rate, micro-voids volume fractions rate, damage rate,... which are clearly related to the microscopic motions. Furthermore we assume that it depends also on the gradient of these rates to account for *local microscopic interactions*. The consequences of this assumption give the basic equations of motions, one for the macroscopic motions and another one for the microscopic motions. They may be applied in numerous predictive theories: phase change problems, damage, shape memory alloys,... Moreover when phase change occurs, temperatures may be discontinuous, for instance when warm rain falls on frozen soil.

M. Frémond, *Phase Change in Mechanics*, Lecture Notes of the Unione Matematica Italiana 13, DOI 10.1007/978-3-642-24609-8_1,

The notes are organized as follows: a scheme for predictive theories is given in a rather general framework using state quantities related to applications: after choosing the state quantities (Chap. 2), the equations which relate them are given: the balance laws (Sects. 3.1, 3.2 and 3.2.4) and the constitutive laws (Sect. 3.3). Their status is slightly different. The balance laws have a large scope even if they may be carefully modified to adapt to the situation under consideration. For instance, the classical equations of motion are not intangible and they may be completed to deal with enhanced description of motion as mentioned above, [112]. On the contrary the constitutive laws are very specific and in general apply only to the model under consideration. Then examples are described: numerous solid–liquid phase changes (Chap. 4), solid–solid phase changes in shape memory alloys (Chap. 5), damage theories (Chap. 6), contact with adhesion (Chap. 7), the coupling of volume damage and surface adhesion (Chap. 8) and phase change with temperature discontinuities (Chap. 9). In the damage and adhesion theories, we investigate what occurs when the distinction between macroscopic and microscopic motions, which is the very base of the two equations of motion, is not so obvious. We show that effects of vanishing macroscopic motions are taken into account by the equation describing the microscopic motion. Thus the two types of equations of motion may be connected and may be not totally independent. The phase changes produced by the thermal effects of collisions are investigated in Chaps. 10 and 11. It is predicted the behaviour of warm rain which falls on a frozen ground. The previous phase change temperatures are fixed, for instance the ice water phase change occurs at 0°C. But there are phase change temperatures which depend on a physical quantity: for instance, the pressure. In Chap. 12, we investigate liquid vapor phase change, for instance boiling water. This phase change depends on one parameter: when there is phase change occurring on a free surface which separates liquid from vapor, the state quantities satisfy a relationship, for instance, the temperature is a function either of the pressure or of the specific volume. But it is also possible that liquid and vapor coexist in a volume. This is the case in mist, fog and clouds. The air, vapor, liquid mixture is investigated in Chap. 13. The set of state quantities may depend on the temperature: when the temperature is lower than a critical temperature $T_c$, there are two phases vapor and liquid with two volume fractions which are state quantities and when the temperature is larger than the critical temperature, there is only one phase, the gas with only one volume fraction. In the first Appendix A some useful properties of convex analysis are given. In the second Appendix B the small perturbation assumption is investigated.

# Chapter 2
# The State Quantities and the Quantities Describing the Evolution

## 2.1 The State Quantities

As already said, the state quantities are the quantities chosen by the scientist or the engineer to describe a phenomenon he wants to predict. The state quantities define the equilibrium: when the state quantities remain constant with respect to the time, the phenomenon is at an equilibrium. The notion of equilibrium is subjective and depends on the sophistication of the modelling, i.e., on the choice of the state variables.

In order to be specific and apply the results to phase change problems, we choose the state quantities to be:

- $T$, the absolute temperature which is clearly important.
- $\beta$, the volume fraction of a phase of a two phase material. For instance, the liquid water volume fraction in ice–water phase change. This quantity is important to make more precise the description of the mixture of the different phases.
- $\text{grad}\beta$, the gradient of the volume fraction to account for local interactions and to investigate sophisticated phase change phenomena.
- $\varepsilon$, the small deformation which is chosen because it may be useful to take into account the deformation of the materials. For the sake of simplicity the small perturbation assumption (see Appendix B) is made when needed.

Thus the set of the states quantities we have chosen is

$$E = (T, \varepsilon, \beta, \text{grad}\beta).$$

For the sake of simplicity, we assume the material density $\rho$ is constant with respect to time in any of its phases (see Sect. 4.6 in case densities differ in the material phases). The case it is not constant when dealing with clouds is investigated in Chap. 13.

M. Frémond, *Phase Change in Mechanics*, Lecture Notes of the Unione Matematica Italiana 13, DOI 10.1007/978-3-642-24609-8_2,

## 2.2 The Quantities Which Describe the Evolution and the Thermal Heterogeneity

Besides the equilibrium, the engineer or the scientist has to describes evolutions. The velocities of the state quantities are important but it may be useful to have other quantities depending on the present and the past to describe the evolution. This is the case of the gradient of the temperature, gradT, which accounts for the thermal heterogeneities. The set of quantities describing the evolution we use is

$$\delta E = \left( \frac{\mathrm{d}\varepsilon}{\mathrm{d}t}, \frac{\mathrm{d}\beta}{\mathrm{d}t}, \mathrm{grad}\frac{\mathrm{d}\beta}{\mathrm{dt}}, \mathrm{gradT} \right).$$

# Chapter 3
# The Basic Laws of Mechanics

The laws we need to build models for phase change problems are the principle of virtual power to get the equations of motion and the laws of thermodynamics to deal with thermal actions. For the sake of simplicity, we assume the small perturbation assumption (see Appendix B) after deriving the equations of motion. Within this assumption, the mass balance which describes the density evolution has not a major role, except for phase change with voids (Sect. 4.5), phase change with different densities (Sect. 4.6).

## 3.1 The Principle of Virtual Power

The equations of motion are derived with the principle of virtual power (a power is a duality pairing in mathematical terms). The power of the interior forces is chosen to depend on the regular strain rates $\mathsf{D}(\mathbf{U})$ ($\mathbf{U}$ is the macroscopic actual velocity and $D_{i,j}(\mathbf{U}) = 1/2(U_{i,j} + U_{j,i})$), and also on $\mathrm{d}\beta/\mathrm{d}t$ and $\mathrm{grad}(\mathrm{d}\beta/\mathrm{dt})$, where $\beta$ stands for some internal quantity, for instance a phase volume fraction or micro-voids volume fraction. These latter quantities are clearly related to the microscopic motions which intervene in the evolution of the material micro-structure, for instance in the evolution of micro-voids. The gradient of the volume fraction is introduced to take into account local interactions or the influence of a material point on its neighbourhood. By stretch of language, derivative $\mathrm{d}\beta/\mathrm{d}t$ is referred as microscopic velocity.

The virtual power of the interior forces we choose, [112], is

$$\mathscr{P}_{int}(\mathbf{V},\gamma) = -\int_{\Omega} \sigma : \mathsf{D}(\mathbf{V}) d\Omega - \int_{\Omega} (B\gamma + \mathbf{H}\cdot \mathrm{grad}\gamma)\, d\Omega, \tag{3.1}$$

where $\mathbf{V}, \gamma$ are macroscopic and microscopic virtual velocities and $\Omega$ is the domain with boundary $\partial\Omega$, occupied by the material. Two new non classical interior

M. Frémond, *Phase Change in Mechanics*, Lecture Notes of the Unione Matematica Italiana 13, DOI 10.1007/978-3-642-24609-8_3,

forces appear, $B$, the interior microscopic work, and $\mathbf{H}$, the microscopic work flux vector. The tensor $\sigma$ is the stress tensor which is symmetric in this setting (see Remark 3.3 below). Let us note that the expression of the volume density of virtual power gives the quantities which are to be related by constitutive laws and to be measured in experiments. In some case, the generalized forces and the generalized strain rates to be related are not so easy to identify and in this perspective the choice of the power of the internal forces is an important element of the predictive theory, [112].

*Remark 3.1.* The scalar product of $3 \times 3$ matrices is

$$\sigma : \mathsf{D} = \sigma_{ij} D_{ij},$$

see Appendix A.

*Remark 3.2.* The power $\mathscr{P}_{int}(\mathbf{V}, \gamma)$ has to be null for any rigid body motion, [60, 131, 132, 197]. A rigid body motion is a motion such that the distance of the material points remains constant, i.e., a rigid body translation or a rigid body rotation. At the microscopic level, in such a motion there is no phase change which requires deformation of the microscopic structure. Thus volume fraction $\beta$ remains constant, $d\beta/dt = 0$. It results that rigid body motions are defined by $\mathsf{D}(\mathbf{V}) = 0$, $\gamma = 0$ and that $\mathscr{P}_{int}(\mathbf{V}, \gamma) = 0$ for such motions.

The virtual power of the exterior forces is

$$\mathscr{P}_{ext}(\mathbf{V}, \gamma) = \int_{\Omega} \mathbf{f} \cdot \mathbf{V} d\Omega + \int_{\partial\Omega} \mathbf{g} \cdot \mathbf{V} d\Gamma + \int_{\Omega} A \gamma d\Omega + \int_{\partial\Omega} a \gamma d\Gamma,$$

where $\mathbf{f}$ is the volume exterior force, $\mathbf{g}$ the surface exterior force, $A$ and $a$ are respectively the volume and surface exterior sources of microscopic work. Sources of microscopic work can be produced by chemical, electrical, or radiative actions which break the links inside a material, steel or concrete for instance, without macroscopic deformations, [21, 24, 112, 113, 120, 121, 175]. The virtual power of the acceleration forces is

$$\mathscr{P}_{acc}(\mathbf{V}, \gamma) = \int_{\Omega} \rho \frac{\mathrm{d}\mathbf{U}}{\mathrm{d}t} \cdot \mathbf{V} d\Omega + \int_{\Omega} \rho_0 \frac{\mathrm{d}^2 \beta}{\mathrm{d}t^2} \gamma d\Omega,$$

where $\rho$ is the density of the solid and $\rho_0$ is proportional to the density of the microscopic links.

The principle of virtual power

$$\forall \mathbf{V}, \gamma,\ \mathscr{P}_{acc}(\mathbf{V}, \gamma) = \mathscr{P}_{int}(\mathbf{V}, \gamma) + \mathscr{P}_{ext}(\mathbf{V}, \gamma),$$

gives easily the equations of motion

$$\rho \frac{d\mathbf{U}}{dt} - \mathrm{div}\sigma = \mathbf{f},\ in\ \Omega,\ \sigma\mathbf{N} = \mathbf{g},\ on\ \partial\Omega,$$

$$\rho_0 \frac{d^2\beta}{dt^2} - \mathrm{div}\mathbf{H} + \mathrm{B} = A,\ in\ \Omega,\ \mathbf{H}\cdot\mathbf{N} = a,\ on\ \partial\Omega. \tag{3.2}$$

The second equation of motion (3.2) is new. It accounts for the microscopic motions. Both the partial differential equation and the boundary condition have a precise physical meaning: they describe how work is provided to the structure without macroscopic motion.

In the sequel, we consider mostly quasi-static evolutions. Results involving the accelerations are reported in [49, 123].

*Remark 3.3.* The power of the internal forces is null for any rigid body translation velocity because of the Galilean relativity. It is null for any rigid body angular velocity because of the angular momentum balance assuming no external torque and resulting in this setting in the symmetry of the stress tensor. Note that those two properties are quite different. This presentation is very attractive due to its simplicity and its large scope. Its only weakness is to forbids external torques but in small deformation engineering it is rare that applications involve torques.

A more general setting is

$$\mathscr{P}_{int}(\mathbf{V},\gamma) = -\int_\Omega \sigma : \mathrm{grad}\mathbf{V} d\Omega - \int_\Omega (\mathrm{B}\gamma + \mathbf{H}\cdot\mathrm{grad}\gamma)\, d\Omega,$$

$$\mathscr{P}_{ext}(\mathbf{V},\gamma) = \int_\Omega \frac{1}{2}\mathsf{m} : \omega(\mathbf{V}) d\Omega + \int_\Omega \mathbf{f}\cdot\mathbf{V} d\Omega + \int_{\partial\Omega} \mathbf{g}\cdot\mathbf{V} d\Gamma + \int_\Omega A\gamma d\Omega + \int_{\partial\Omega} a\gamma d\Gamma,$$

where

$$\omega(\mathbf{V}) = \frac{1}{2}(\mathrm{grad}\mathbf{V} - (\mathrm{grad}\mathbf{V})^{\mathrm{T}}),$$

the virtual angular velocity is the antisymmetric part of the virtual velocity gradient and antisymmetric matrix $\mathsf{m}$ is an external torque (matrix $(\mathrm{grad}\mathbf{V})^{\mathrm{T}}$ is the transposed matrix of matrix $(\mathrm{grad}\mathbf{V})$). The principle of virtual power gives equations of motion (3.2) where stress $\sigma$ is not symmetric and relationship

$$\sigma - \sigma^T = \mathsf{m},$$

which is the angular momentum balance. It results antisymmetric part of the stress is given by the equation of motion and its symmetric part is going to be given by a constitutive law. In case there is no external torque, $\mathsf{m} = 0$, we recover the presentation we have chosen with power of the internal forces given by (3.1) with symmetric stress $\sigma$.

## 3.2 The Laws of Thermodynamics

Besides the description of motion, the models involve thermal and energetic phenomena. Thus the laws of thermodynamics intervene: the first law, often called the energy balance, together with the second law, see [112] for a detailed presentation.

### 3.2.1 The Energy Balance

The energy balance coupled to the principle of virtual power gives the following balance law, [112]

$$\frac{\mathrm{d}e}{\mathrm{d}t} + \mathrm{div}T\mathbf{Q} = \sigma : \mathsf{D}(\mathbf{U}) + B\frac{\mathrm{d}\beta}{\mathrm{d}t} + \mathbf{H} \cdot \mathrm{grad}\frac{\mathrm{d}\beta}{\mathrm{dt}} + \mathrm{RT},\ \mathrm{in}\ \Omega,$$
$$-T\mathbf{Q} \cdot \mathbf{N} = T\pi,\ on\ \partial\Omega, \tag{3.3}$$

where $T$ is the absolute temperature, $e$ is the volumic internal energy, $\mathbf{Q}$ is the entropy flux vector ($T\mathbf{Q} = \mathbf{q}$ is the heat flux vector), $R$ the volumic exterior rate of entropy production ($TR = r$ is the volumic exterior rate of heat production) and $\pi$ the surfacic rate of entropy provided to the material ($T\pi = \varpi$ is the surfacic rate of heat provided to the material).

By denoting

$$F^{meca} = (\sigma, B, \mathbf{H}, 0),$$

energy balance (3.3) reads

$$\frac{\mathrm{d}e}{\mathrm{d}t} + \mathrm{div}T\mathbf{Q} = F^{meca} \cdot \delta E + RT,\ in\ \Omega,$$
$$-T\mathbf{Q} \cdot \mathbf{N} = T\pi,\ on\ \partial\Omega. \tag{3.4}$$

*Remark 3.4.* The opposite of the power of the interior forces

$$F^{meca} \cdot \delta E = \sigma : \mathsf{D}(\mathbf{U}) + B\frac{\mathrm{d}\beta}{\mathrm{d}t} + \mathbf{H} \cdot \mathrm{grad}\frac{\mathrm{d}\beta}{\mathrm{dt}},$$

is often neglected within the small perturbation assumption (see Appendix B). The energy balance becomes

$$\frac{\mathrm{d}e}{\mathrm{d}t} + \mathrm{div}\mathbf{q} = r,\ in\ \Omega,$$
$$-\mathbf{q} \cdot \mathbf{N} = \varpi,\ on\ \partial\Omega.$$

It is the classical energy balance.

### 3.2.2 The Second Law of Thermodynamics

It is

$$T > 0, \tag{3.5}$$

$$\frac{\mathrm{d}s}{\mathrm{d}t} + \mathrm{div}\mathbf{Q} \geq R, \; in \; \Omega, \tag{3.6}$$

where $s$ is the entropy.

### 3.2.3 The Free Energy

Let us recall a physical property: the internal energy is a convex function of entropy, [132]. By using convex analysis (see Appendix A), we may define the dual function of the internal energy, the free energy

$$-\Psi(T,\varepsilon,\beta,\mathrm{grad}\beta) = \sup_{\mathrm{x}\in\mathbb{R}} \{\mathrm{xT} \quad \mathrm{e}(\mathrm{x},\varepsilon,\beta,\mathrm{grad}\beta)\},$$

giving

$$e(s,\varepsilon,\beta,\mathrm{grad}\beta) = \Psi(\mathrm{T},\varepsilon,\beta,\mathrm{grad}\beta) + \mathrm{Ts}, \tag{3.7}$$

and

$$s = -\frac{\partial \Psi}{\partial T}(T,\varepsilon,\beta,\mathrm{grad}\beta), \; \mathrm{T} = \frac{\partial \mathrm{e}}{\partial \mathrm{s}}(\mathrm{s},\varepsilon,\beta,\mathrm{grad}\beta). \tag{3.8}$$

Function $T \rightarrow \Psi(T,\varepsilon,\beta,\mathrm{grad}\beta)$ is concave. This concavity property is a physical property because it results from convexity of function $s \rightarrow e(s,\varepsilon,\beta,\mathrm{grad}\beta)$, [132]. In the sequel, we choose the free energy as the basic physical potential. In case we need internal energy, we define it by the previous relationship, (3.7).

### 3.2.4 Equivalent Laws of Thermodynamics

The laws of thermodynamics (3.3), (3.5), (3.6) involve two partial differential equations. They have an equivalent formulation involving only one partial differential equation, the entropy balance law, and two inequalities. In order to derive this balance law, the internal forces are split between dissipative interior forces indexed by $^d$ and non dissipative interior forces indexed by $^{nd}$

$$\sigma = \sigma^d + \sigma^{nd},\ B = B^d + B^{nd},\ \mathbf{H} = \mathbf{H}^d + \mathbf{H}^{nd}. \tag{3.9}$$

The non dissipative interior forces are defined with the free energy, $\Psi(E) = \Psi(T, \varepsilon, \beta, \mathrm{grad}\beta)$, which gathers all the static properties of the material

$$s = -\frac{\partial \Psi}{\partial T},$$

$$\sigma^{nd} = \frac{\partial \Psi}{\partial \varepsilon},\ B^{nd} = \frac{\partial \Psi}{\partial \beta},\ \mathbf{H}^{nd} = \frac{\partial \Psi}{\partial (\mathrm{grad}\beta)}. \tag{3.10}$$

By replacing internal energy in energy balance equation (3.3) by its value given by (3.7) and using relationship (3.5), we get equations equivalent to (3.3), (3.5) and (3.6), [41]

$$\frac{\mathrm{d}s}{\mathrm{d}t} + \mathrm{div}\mathbf{Q} = R + \frac{1}{T}\left(\sigma^d : \mathsf{D}(\mathbf{U}) + B^d \frac{\mathrm{d}\beta}{\mathrm{d}t} + \mathbf{H}^d \cdot \mathrm{grad}\frac{\mathrm{d}\beta}{\mathrm{d}t} - \mathbf{Q} \cdot \mathrm{gradT}\right),\ in\ \Omega,$$

$$-\mathbf{Q} \cdot \mathbf{N} = \pi,\ on\ \partial\Omega,$$

$$T > 0,$$

$$\mathscr{D} = \frac{1}{T}\left(\sigma^d : \mathsf{D}(\mathbf{U}) + B^d \frac{\mathrm{d}\beta}{\mathrm{d}t} + \mathbf{H}^d \cdot \mathrm{grad}\frac{\mathrm{d}\beta}{\mathrm{d}t} - \mathbf{Q} \cdot \mathrm{gradT}\right) \geq 0. \tag{3.11}$$

Let us note that only one partial differential equation is involved in this equivalent formulation.

### 3.2.5 The Entropy Balance Equivalent to the Energy Balance

The first two relationships of (3.11) are the entropy balance

$$\frac{\mathrm{d}s}{\mathrm{d}t} + \mathrm{div}\mathbf{Q} = R + \mathscr{D},\ in\ \Omega,$$

$$-\mathbf{Q} \cdot \mathbf{N} = \pi,\ on\ \partial\Omega, \tag{3.12}$$

which is equivalent to the energy balance. The entropy balance shows that the variation of the entropy results from an exterior source, $R$ and from an interior one, $\mathscr{D}$. In terms of physics, this interior entropy source may be experimented when rubbing one's hands. Rubbing produces heat but it never produces cold. This property is $T\mathscr{D} = \mathbf{R}^d \cdot \mathbf{U} \geq 0$, where $\mathbf{R}^d$ is the contact reaction and $\mathbf{U}$ is the hands sliding velocity.

### 3.2.6 An Equivalent Second Law of Thermodynamics

The two other relationships of (3.11) are

$$T > 0,$$

$$\mathscr{D} = \frac{1}{T}\left(\sigma^d : \mathsf{D}(\mathbf{U}) + B^d\frac{\mathrm{d}\beta}{\mathrm{d}t} + \mathbf{H}^d \cdot \mathrm{grad}\frac{\mathrm{d}\beta}{\mathrm{dt}} - \mathbf{Q}\cdot \mathrm{gradT}\right) \geq 0. \tag{3.13}$$

The last relationship means that the interior production of entropy is non negative or that the dissipation is non negative.

In some cases, these equations are more productive in terms of mechanics and mathematics than the classical energy balance equations, [41]. For instance, it may be straightforward to prove that the temperature is positive.

We assume small perturbation (see Appendix B). It results that we may neglect the dissipation

$$\mathscr{D} = \frac{1}{T}\left(\sigma^d : \mathsf{D}(\mathbf{U}) + B^d\frac{\mathrm{d}\beta}{\mathrm{d}t} + \mathbf{H}^d \cdot \mathrm{grad}\frac{\mathrm{d}\beta}{\mathrm{dt}} - \mathbf{Q}\cdot \mathrm{gradT}\right),$$

in (3.12) which becomes

$$\frac{\mathrm{d}s}{\mathrm{d}t} + \mathrm{div}\mathbf{Q} = R, \ in \ \Omega,$$

$$-\mathbf{Q}\cdot\mathbf{N} = \pi, \ on \ \partial\Omega. \tag{3.14}$$

New results on the case where the dissipation is not neglected may be found in [38, 39].

## 3.3 The Constitutive Laws

They are relationships (3.9), (3.10) and relationships defining the dissipative interior forces which depend on the velocities $\mathrm{d}\varepsilon/\mathrm{d}t, \mathrm{d}\beta/\mathrm{d}t, \mathrm{grad}(\mathrm{d}\beta/\mathrm{dt})$, on the thermal heterogeneity gradT and possibly on other quantities $\chi$ which depend on the history of the material. The dissipative interior forces are defined by derivatives

$$\sigma^d = \frac{\partial \Phi}{\partial(\mathrm{d}\varepsilon/\mathrm{d}t)},$$

$$B^d = \frac{\partial \Phi}{\partial(\mathrm{d}\beta/\mathrm{d}t)},$$

$$\mathbf{H}^d = \frac{\partial \Phi}{\partial(\mathrm{grad}(\mathrm{d}\beta/\mathrm{dt}))},$$
$$\mathbf{Q} = -\frac{\partial \Phi}{\partial(\mathrm{gradT})}, \tag{3.15}$$

where

$$\Phi\left(\frac{\mathrm{d}\varepsilon}{\mathrm{d}t}, \frac{\mathrm{d}\beta}{\mathrm{d}t}, \mathrm{grad}\frac{\mathrm{d}\beta}{\mathrm{d}t}, \mathrm{grad}T, \chi\right) = \Phi(\delta E, \chi),$$

is a pseudopotential of dissipation introduced by Jean Jacques Moreau [130, 139, 171]: it is a function of $\delta E$ which is convex with respect to $\delta E$, non negative, with value 0 for $\delta E = 0$.

In a more general setting, relationships (3.15) are

$$\left(\sigma^d, B^d, \mathbf{H}^d, -\mathbf{Q}\right) \in \partial\Phi\left(\frac{\mathrm{d}\varepsilon}{\mathrm{d}t}, \frac{\mathrm{d}\beta}{\mathrm{d}t}, \mathrm{grad}\frac{\mathrm{d}\beta}{\mathrm{dt}}, \mathrm{gradT}, \chi\right) = \partial\Phi(\delta E, \chi), \tag{3.16}$$

where $\partial\Phi$ is the subdifferential set of pseudopotential of dissipation $\Phi$ with respect to $\delta E$, [112]. These constitutive laws ensure that relationship (3.13) is satisfied:

**Theorem 3.1.** *If $T > 0$ and $\Phi(\delta E, \chi)$ is a pseudopotential of dissipation, i.e., function $\delta E \to \Phi(\delta E, \chi)$ is convex, non negative, with value 0 at the origin, $\Phi(0, \chi) = 0$, then constitutive laws (3.16) are such that second law relationship (3.13) is satisfied.*

*Proof.* Due to definition of subdifferential (see Appendix A), we have

$$\left(0 - \sigma^d : \frac{\mathrm{d}\varepsilon}{\mathrm{d}t}\right) + (0 - B^d)\frac{\mathrm{d}\beta}{\mathrm{d}t} + (0 - \mathbf{H}^d)\cdot \mathrm{grad}\frac{\mathrm{d}\beta}{\mathrm{dt}} + (0 - (-\mathbf{Q}))\cdot \mathrm{gradT}$$
$$+\Phi\left(\frac{\mathrm{d}\varepsilon}{\mathrm{d}t}, \frac{\mathrm{d}\beta}{\mathrm{d}t}, \mathrm{grad}\frac{\mathrm{d}\beta}{\mathrm{dt}}, \mathrm{gradT}, \mathrm{E}\right) \le \Phi(0, \mathrm{E}).$$

Because $\Phi(0, E) = 0$, we have

$$\Phi\left(\frac{\mathrm{d}\varepsilon}{\mathrm{d}t}, \frac{\mathrm{d}\beta}{\mathrm{d}t}, \mathrm{grad}\frac{\mathrm{d}\beta}{\mathrm{dt}}, \mathrm{gradT}, \mathrm{E}\right) \le \sigma^{\mathrm{d}} : \frac{\mathrm{d}\varepsilon}{\mathrm{d}t} + B^{\mathrm{d}}\frac{\mathrm{d}\beta}{\mathrm{d}t} + \mathbf{H}^{\mathrm{d}}\cdot \mathrm{grad}\frac{\mathrm{d}\beta}{\mathrm{dt}} - \mathbf{Q}\cdot \mathrm{gradT},$$

which gives relationship (3.13) because $\Phi(\delta E, \chi) \ge 0$ and $T > 0$. □

This way to define the dissipative forces is not the more general but its scope is very large and sufficient in many situations. Experiments are of paramount importance for the choices of $\Psi$ and $\Phi$. The derivatives of these functions are the quantities which can be measured with experiments. There are some empirical rules related to physical properties to choose the two functions.

The choice of the state quantities followed by the choice of $\Psi$ and $\Phi$ are the crucial and delicate steps in modelling. The imagination and the creative ability, even the artistic ability, of the engineer or scientist may appear when establishing the closed forms of functions $\Phi$ and $\Psi$, resulting in elegant and useful predictive theories.

*Remark 3.5.* The main practical problem which is not described by a pseudopotential of dissipation is the Coulomb friction law. New results on constitutive laws where gradT intervene are given in [177].

*Remark 3.6.* The mass balance is not taken into account. We have assumed for the sake of simplicity that the density is constant. If the mass balance is taken into account it result a pressure in the stress constitutive law as may be been seen in a smooth evolution described in Sect. 4.5 and in Chap. 13).

### 3.3.1 An Extra State Quantity

A state quantity $\eta$ of $E$ different from $T$ may be such that its velocity $d\eta/dt$ does not intervene in the power of the internal forces. This case is investigated in [112]. The new equation related to $\eta$ is

$$Z^d + Z^{nd} = 0,$$

with

$$Z^{nd} = \frac{\partial \Psi}{\partial \eta}, \; Z^{nd} = \frac{\partial \Phi}{\partial (d\eta/dt)},$$

in case $\Psi$ and $\Phi$ are smooth functions. In case they are not, the partial derivatives are replaced by the components of the subgradients of $\Psi$ and $\Phi$. In these notes, there are examples in the theory of shape memory alloys, see Sect. 5.5.2, where $\eta = \delta$, the degree of education, and in the predictive theory of clouds in Chap. 13.

## 3.4 Internal Constraints

The quantities of $\delta E$ which describe the evolution, are elements of linear spaces $\delta\mathscr{V}$, because they are velocities and gradT is an element of a linear space, $\delta E \in \delta\mathscr{V}$. Their actual values are often not the entire linear space but a subset $\delta\mathscr{C} \subset \delta\mathscr{V}$. It is said that the evolution quantities satisfy *internal constraints*. For example, the material may be incompressible

$$\mathrm{div}\mathbf{U} = \mathrm{tr}\left(\frac{\mathrm{d}\varepsilon}{\mathrm{dt}}\right) = 0,$$

or a mass balance has to be satisfied

$$\frac{\partial}{\partial t}(\rho_1\beta_1 + \rho_2\beta_2) + \rho \text{div}\mathbf{U} = 0, \tag{3.17}$$

or an evolution is irreversible

$$\frac{d\beta}{dt} \leq 0, \tag{3.18}$$

or an non interpenetration condition has to be satisfied

$$\mathbf{U} \cdot \mathbf{N} \geq 0,$$

where **U** is the velocity of a point sliding on a solid with exterior normal vector **N**. These properties are physical properties and they are not mathematical properties. Thus they have to be taken into account by the two functions which gather the mechanical properties, either $\Psi$ or $\Phi$. It is obvious that

$$\delta E \in \delta\mathscr{C}, \tag{3.19}$$

has to be taken into account with $\Phi$. We use a very productive and general way to satisfy (3.19) having the pseudopotential of dissipation to involve the indicator function of $\delta\mathscr{C}$, $I_{\delta\mathscr{C}}$

$$\Phi(\delta E, \chi) = \Phi(\delta E, \chi) + I_{\delta\mathscr{C}}(\delta E).$$

A difficulty appears immediately due to the non differentiability of function $\Phi$. It is possible to overcome it by defining generalized derivatives. In many cases the set $\delta\mathscr{C}$ is convex (this is the case for the examples) allowing to use subgradients as generalized derivatives. A short presentation of convex analysis is given in Appendix A.

There are also internal constraints on the state quantities $E$. We apply the same idea. If the actual internal quantities have to satisfy

$$E \in \mathscr{C} \subset \mathscr{V},$$

where $\mathscr{C}$ is a subset of space $\mathscr{V}$, we choose the free energy satisfying

$$\Psi(E) = \Psi(E) + I_{\mathscr{C}}(E).$$

In many cases, space $\mathscr{V}$ is linear but this is not always the case, [112]. In many practical situations, $\mathscr{C}$ is a convex subset of linear space $\mathscr{V}$ (for instance, if the small perturbation assumption is assumed), allowing to have subgradients as generalized derivatives. Examples of internal constraints on the state quantities are

$$\beta \in [0, 1], \tag{3.20}$$

implying that $\beta$ is a volume fraction,

$$(\beta_1, \beta_2) \in K,$$

the volume fractions $(\beta_1, \beta_2)$ of a mixture satisfy some relationship. If $K$ is

$$K = \{(\beta_1, \beta_2) \,|\, 0 \leq \beta_1 \leq 1,\ 0 \leq \beta_2 \leq 1,\ \beta_1 + \beta_2 \in [0,1]\},$$

voids may appear in the mixture (see Sect. 4.5). The temperature $T$ satisfy the internal constraint

$$T \geq 0. \tag{3.21}$$

For the sake of simplicity, we assume that (3.21) is always satisfied and we do not take it into account. The case $T = 0$ is investigated in [112]. We prove that solutions of most of the problems we solve below, satisfy this internal constraint if the initial temperature is positive.

### 3.4.1 Constitutive Laws in Case There Are Internal Constraints on the State Quantities

For the sake of simplicity, we assume internal constraint (3.20) is satisfied by $\beta$ and that

$$tr\varepsilon = 0.$$

The free energy becomes

$$\Psi(T, \varepsilon, \beta, \mathrm{grad}\beta) + \mathrm{I}(\beta) + \mathrm{I}_0(\mathrm{tr}\varepsilon),$$

where $I$ and $I_0$ are the indicator functions of segment $[0,1]$ and of the origin 0 of $\mathbb{R}$, see Appendix A. We define the non dissipative reaction forces

$$B^{reac} \in \partial I(\beta),\ -p^{reac} \in \partial I_0(tr\varepsilon),\ \sigma^{reac} = -p^{reac}\mathbf{1},$$

where $\partial I(\beta)$ and $\partial I_0(tr\varepsilon)$ are the subdifferential sets of the indicator functions $I$ and $I_0$, see Appendix A. These constitutive laws imply that the subdifferential sets $\partial I(\beta)$ and $\partial I_0(tr\varepsilon)$ are not empty. Thus due to Theorem A.2 of Appendix A that $I(\beta) = 0$ and $I_0(tr\varepsilon) = 0$. Thus $\beta \in [0,1]$ and $tr\varepsilon = 0$.

We replace definitions (3.9) by

$$\sigma = \sigma^d + \sigma^{reac} + \sigma^{nd},\ B = B^d + B^{reac} + B^{nd},\ \mathbf{H} = \mathbf{H}^d + \mathbf{H}^{nd}. \tag{3.22}$$

It results entropy balance (3.12) where we keep definition (3.10), becomes

$$\frac{ds}{dt}+\mathrm{div}\mathbf{Q}=R+\frac{1}{T}\left(\sigma^d:\mathsf{D}(\mathbf{U})+B^d\frac{d\beta}{dt}+\mathbf{H}^d\cdot\mathrm{grad}\frac{d\beta}{dt}-\mathbf{Q}\cdot\mathrm{gradT}\right)$$
$$+\frac{1}{T}\left(\sigma^{reac}:\mathsf{D}(\mathbf{U})+B^{reac}\frac{d\beta}{dt}\right),\ in\ \Omega,$$
$$-\mathbf{Q}\cdot\mathbf{N}=\pi,\ on\ \partial\Omega. \tag{3.23}$$

We let

$$\mathscr{D}=\frac{1}{T}\left(\sigma^d:\mathsf{D}(\mathbf{U})+B^d\frac{d\beta}{dt}+\mathbf{H}^d\cdot\mathrm{grad}\frac{d\beta}{dt}-\mathbf{Q}\cdot\mathrm{gradT}\right),$$
$$\mathscr{D}^{reac}=\frac{1}{T}\left(\sigma^{reac}:\mathsf{D}(\mathbf{U})+B^{reac}\frac{d\beta}{dt}\right).$$

Quantity $\mathscr{D}$ is the dissipation and $\mathscr{D}^{reac}$ is the reactions dissipation. The entropy balance is

$$\frac{ds}{dt}+\mathrm{div}\mathbf{Q}=R+\mathscr{D}+\mathscr{D}^{reac},\ in\ \Omega,$$
$$-\mathbf{Q}\cdot\mathbf{N}=\pi,\ on\ \partial\Omega,$$

where $R$ is the exterior entropy source, $\mathscr{D}$ is the interior dissipative source and $\mathscr{D}^{reac}$ is the non dissipative reaction source.

The definition of the dissipative forces are still given by (3.16) if there are internal constraints on the velocities of $\delta E$, for instance (3.17) or (3.18).

$$\left(\sigma^d,B^d,\mathbf{H}^d,-\mathbf{Q}\right)\in\partial\Phi\left(\frac{d\varepsilon}{dt},\frac{d\beta}{dt},\mathrm{grad}\frac{d\beta}{dt},\mathrm{gradT},\chi\right)=\partial\Phi(\delta E,\chi).$$

*Remark 3.7.* The reactions $\sigma^{reac}$ and $B^{reac}$ are subgradients of the free energy thus they may be said non dissipative as the smooth derivatives $\sigma^{nd}$, $B^{nd}$ and $\mathbf{H}^{nd}$ which do not intervene in the entropy production. But these reactions may have a non null power which may intervene in the entropy production. We choose to keep the non dissipative qualifying adjective because the power of these reactions is null in any smooth evolution as shown in the following section.

#### 3.4.1.1 The Power of the Non Dissipative Reactions Forces

Let us note that

$$\sigma^{reac}:\mathsf{D}(\mathbf{U})=-p^{reac}tr(\mathsf{D}(\mathbf{U}))=-p^{reac}\frac{d}{dt}(tr\varepsilon)=0.$$

This property is general: in case the constraint is to belong to a linear subspace, the power of the non dissipative reaction force is null.

In relationship (3.23), the velocities, for instance $\mathrm{d}\beta/\mathrm{d}t$ may be discontinuous with respect to time. Based on the causality principle [112], we choose to have the left derivative which depends on the past, in the formula

$$\frac{\mathrm{d}^l\beta}{\mathrm{d}t} = \lim_{\Delta t \to 0, \Delta t > 0} \frac{\beta(t) - \beta(t - \Delta t)}{\Delta t}. \tag{3.24}$$

Let us investigate the power of reaction $B^{reac}$:

- In case $0 < \beta(t, \mathbf{x}) < 1$, we have $0 = B^{reac}(t, \mathbf{x}) \in \partial I(\beta(t, \mathbf{x}))$ and

$$B^{reac}(t, \mathbf{x}) \frac{\mathrm{d}^l\beta}{\mathrm{d}t}(t, \mathbf{x}) = 0.$$

- In case $\beta(\tau, \mathbf{x}) = 1$ in some interval $[t_0, t]$ with $t_0 < t$, we have $(d^l\beta/dt)(t, \mathbf{x}) = 0$ and

$$B^{reac}(t, \mathbf{x}) \frac{\mathrm{d}^l\beta}{\mathrm{d}t}(t, \mathbf{x}) = 0.$$

  We have the same result in case $\beta(\tau, \mathbf{x}) = 0$ in some interval $[t_0, t]$ with $t_0 < t$.
- In case $\beta(t, \mathbf{x}) = 1$ and $\beta(\tau, \mathbf{x}) < 1$ in some interval $[t_0, t]$ and $(d^l\beta/dt)(t, \mathbf{x}) > 0$, we have $B^{reac}(t, \mathbf{x}) \in \partial I(1) = \mathbb{R}^+$. If $B^{reac}(t, \mathbf{x}) > 0$, we have

$$B^{reac}(t, \mathbf{x}) \frac{\mathrm{d}^l\beta}{\mathrm{d}t}(t, \mathbf{x}) > 0.$$

  If $B^{reac}(t, \mathbf{x}) = 0$, we have

$$B^{reac}(t, \mathbf{x}) \frac{\mathrm{d}^l\beta}{\mathrm{d}t}(t, \mathbf{x}) = 0.$$

  Let us note that $B^{reac}(t, \mathbf{x}) = 0$, if function $\tau \to B^{reac}(\tau, \mathbf{x})$ is continuous at time $t$ because $B^{reac}(\tau, \mathbf{x}) = 0$ for $t_0 \le \tau < t$. Analogous result hold in case $\beta(t, \mathbf{x}) = 0$ and $\beta(\tau, \mathbf{x}) > 0$ in some interval $[t_0, t]$.

We conclude that the power of reaction force $B^{reac}$ is null if it is a continuous function of time. In this situation we say that the *constraints are workless* (les liaisons sont parfaites in French). Let us note that the reaction power is always non negative

**Theorem 3.2.** *If temperature is positive and if the velocities are the left derivatives, we have*

$$\mathscr{D}^{reac} \ge 0.$$

*If the left derivatives are equal to the right derivatives, i.e., the state quantities the derivatives of which are involved in $\mathscr{D}^{reac}$ are differentiable, we have*

$$\mathscr{D}^{reac} = 0.$$

*Proof.* Proof is based on convex analysis, see [112]. □

*Remark 3.8.* A precise derivation of this point of view is given in [112] together with examples of situations where $\delta\mathscr{V}$ is not a linear space and with an investigation of the internal constraint (3.21) on the temperature. In [112], it is also proved that when the velocities are discontinuous with respect to time, the left derivatives, are the quantities which appear in the basic laws of mechanics in agreement with the causality principle. Let us note that if the right derivative is chosen, it is not possible to prove that the second law is satisfied. In the sequel, the time derivatives are left derivatives. For the sake of simplicity, we write $\mathrm{d}\beta/\mathrm{d}t$ instead of $\mathrm{d}^l\beta/\mathrm{d}t$. Right derivatives appear in collision theories but they sum up the very sophisticated phenomena which occur in a collision. With this point of view, they depend on the past in agreement with the causality principle, [114].

*Remark 3.9.* The power of the reaction is always non negative. We say that the evolution is smooth in case it is workless, i.e., it is null. This is the case, if either the reaction is a continuous function of time or the generalized velocity of deformation is continuous with respect to time. In our example, this condition is equivalent to the differentiability of the state quantity $\beta$. In terms of mathematics, it is often sufficient that the properties are satisfied almost everywhere with respect to $\mathbf{x}$ and $t$.

In the sequel, we consider evolutions which are smooth, i.e., evolutions such that

$$\mathscr{D}^{reac} = 0. \tag{3.25}$$

The case where the evolutions are not smooth is investigated when dealing with discontinuities, in Sects. 3.5 and 3.6 of this chapter and in Chaps. 9, 10 and 11 dedicated to discontinuities of temperature $T$ and volume fraction $\beta$ and to discontinuities of velocity $\mathbf{U}$ occurring in collisions.

*Remark 3.10.* As already said, relationship (3.25) is chosen to define a smooth evolution. This is the case in rigid bodies mechanics where a workless constraint is defined by the fact its power is null. When solving a problem, one may verify a posteriori that the non dissipative reactions are workless, i.e., $\mathscr{D}^{reac} = 0$. It may also occur that it is impossible to solve the problem, i.e., to find a smooth solution. In this situation, a non smooth solution satisfying $\mathscr{D}^{reac} > 0$, may be found. The reaction dissipation is positive, i.e., the reactions are not workless (non parfaites in French). This is the case in non smooth evolutions, for instance in collisions as described in Chaps. 10, 11 and in [114].

### 3.4.2 Some Notations

In case there are many state quantities and velocities as in Chap. 13, it is useful to introduce some vectors gathering the dissipative internal forces $F^d$, the non dissipative reactions to the internal constraints on the state quantities $F^{reac}$ and the non dissipative internal forces $F^{nd}$

$$\begin{aligned}
E &= (T,\varepsilon,\beta,\mathrm{grad}\beta,\eta),\ \delta E = \left(\frac{\mathrm{d}\varepsilon}{\mathrm{d}t},\frac{\mathrm{d}\beta}{\mathrm{d}t},\mathrm{grad}\frac{\mathrm{d}\beta}{\mathrm{dt}},\frac{\mathrm{d}\eta}{\mathrm{dt}},\mathrm{gradT}\right),\\
F &= (\sigma,B,\mathbf{H},0,-\mathbf{Q}),\ F = F^d + F^{reac} + F^{nd},\\
F^d &= \left(\sigma^d,B^d,\mathbf{H},Z^d,-\mathbf{Q}\right),\ F^{reac} = (-p^{reac}\mathbf{1},B^{reac},0,0,0),\\
F^{nd} &= \left(\sigma^{nd},B^{nd},\mathbf{H},Z^{nd},0\right).
\end{aligned} \tag{3.26}$$

We have

$$\begin{aligned}
\mathscr{D} &= \frac{1}{T}\left(F^d\cdot\delta E\right),\\
\mathscr{D}^{reac} &= \frac{1}{T}(F^{reac}\cdot\delta E).
\end{aligned}$$

Let us recall that a smooth evolution is characterized by $\mathscr{D}^{reac} = 0$.

## 3.5 The Basic Laws of Mechanics on a Discontinuity Surface

The quantities which appear in the preceding laws have to satisfy some mathematical smoothness properties in order to define the derivatives. There are cases where these quantities are discontinuous and the results are no longer valid. Of course, the basic laws apply in these circumstances but they have to be adapted to take the discontinuities into account, [111, 112]. The discontinuities can be on a surface, which is often called a free boundary, when solving a problem in $\mathbb{R}^3$ (on a line when in $\mathbb{R}^2$, on a point when in $\mathbb{R}$).

We assume that the specific volume $1/\rho$ is one of the state quantities $E$. We no longer assume small perturbations. Of course, results also make sense within the small perturbation assumption. When velocities $\mathbf{U}$, $\dot{\beta} = \mathrm{d}\beta/\mathrm{d}t$, temperature $T$, entropy $s = \rho\widehat{s}$, internal energy $e = \rho\widehat{e}$, entropy flux vector $\mathbf{Q}$ or interior forces $\sigma$ and $\mathbf{H}$ are discontinuous with respect to $\mathbf{x},t$ on a surface $S_{x,t}$ of $\mathbb{R}^3\times\mathbb{R}$, the computations which give the basic relationships are no longer valid.

In this section, notation $[A] = A_2 - A_1$ denotes the discontinuity of the quantity $A$, $m$ is the mass flux ($m = \rho v$, where $v = \mathbf{V}_{rel}\cdot\mathbf{N}$ is the material normal velocity with respect to the surface, (see formula (3.31) below) and $Q$ is the entropy received through contact actions by the two domains which are on the two sides of the discontinuity surface $\Gamma = S_{x,t} \cap (\mathbb{R}^3 \times \{t\})$. The indices 1 and 2 refer to the two sides of the discontinuity surface, the normal vector $\mathbf{N}$ is directed from side 1 toward side 2.

*Remark 3.11.* In this section, $\Gamma$ has dimension 2. When $\Gamma = S_{x,t} \cap (\mathbb{R}^3 \times \{t\})$ has dimension 3, i.e., when the domain where there is a discontinuity at time $t$ has dimension 3 and the discontinuity is with respect to time, another analysis is needed. This situation is investigated in the following section.

The second law of thermodynamics on the discontinuity surface $\Gamma$ with unit normal vector $\mathbf{N}$ in $\mathbb{R}^3$, is

$$m[\widehat{s}] \geq [Q], \tag{3.27}$$

where $\widehat{s}$ is the specific entropy and $Q = -\mathbf{Q}.\overrightarrow{N}$ is the entropy flow through surface $\Gamma$, $TQ$ is the heat flow at temperature $T$. The inequality depends only on the difference $[Q] = Q_2 - Q_1$, the entropy received by the surface. The distribution of $[Q]$ between the received entropies from the two sides, $-Q_1$ from side 1 and $Q_2$ from side 2, results from the constitutive laws which are established in the sequel.

We denote $\mathbf{T} = \sigma\mathbf{N}$ the force applied on the discontinuity surface $S_x$ with normal vector $\mathbf{N}$. Different average values are defined by

$$\underline{T} = \frac{T_1+T_2}{2}, \underline{Q} = \frac{Q_1+Q_2}{2}, \underline{\mathbf{U}} = \frac{\mathbf{U}_1+\mathbf{U}_2}{2}, \underline{\mathbf{T}} = \frac{\mathbf{T}_1+\mathbf{T}_2}{2}, \underline{A} = \frac{A_1+A_2}{2},$$

where $\underline{A}$ is the average of a quantity $A$ on the discontinuity surface. Let us recall that the temperatures $T$ are positive because they are the traces on the discontinuity surface of the volume temperatures which are positive. We assume that the average temperature $\underline{T}$ is positive. The particular case where $\underline{T} = 0$ is investigated in [112].

The equation of motion are

$$m[\mathbf{U}] = [\mathbf{T}], \tag{3.28}$$

and

$$[\mathbf{H} \cdot \mathbf{N}] = 0.$$

The energy balance is

$$m\left[\widehat{e} + \frac{1}{2}\mathbf{U}^2\right] - [\mathbf{T} \cdot \mathbf{U}] - [\mathbf{H} \cdot \mathbf{N}\dot{\beta}] = [TQ] = \underline{T}[Q] + \underline{Q}[T], \tag{3.29}$$

where we recall that $\widehat{e}$ is the specific internal energy. Relationship (3.28) gives

$$[\mathbf{T} \cdot \mathbf{U}] = \underline{\mathbf{T}} \cdot [\mathbf{U}] + \underline{\mathbf{U}} \cdot [\mathbf{T}] = \underline{\mathbf{T}}.[\mathbf{U}] + m\underline{\mathbf{U}}.[\mathbf{U}] = \underline{\mathbf{T}}.[\mathbf{U}] + \frac{m}{2}[\mathbf{U}^2]. \tag{3.30}$$

Note that due to the mass balance $[\rho(\mathbf{U}-\mathbf{W})]\cdot\mathbf{N}=0$,

$$[U_N]=[\mathbf{U}]\cdot\mathbf{N}=[\mathbf{V}_{rel}]\cdot\mathbf{N}=[V_N]=[v]=\left[\frac{\rho v}{\rho}\right]=m\left[\frac{1}{\rho}\right], \tag{3.31}$$

where, $U_N=\mathbf{U}\cdot\mathbf{N}$, $\mathbf{V}_{rel}=\mathbf{U}-\mathbf{W}$ is the velocity with respect to the discontinuity surface, whose velocity is $\mathbf{W}$, $v=V_N=\mathbf{V}_{rel}\cdot\mathbf{N}$, the relative normal velocity. Let us also define $\underline{\mathbf{V}}=(\mathbf{V}_{rel2}+\mathbf{V}_{rel1})/2$, the average relative velocity. With the relationships (3.30) and (3.31), the energy balance (3.29) gives

$$m\left([\widehat{e}]-\underline{T}_N\left[\frac{1}{\rho}\right]\right)-\underline{\mathbf{T}}_T\cdot[\mathbf{U}_T]-\mathbf{H}\cdot\mathbf{N}[\dot{\beta}]-\underline{T}[Q]-\underline{Q}[T]=0, \tag{3.32}$$

where $\underline{T}_N=\underline{\mathbf{T}}\cdot\mathbf{N}$, $\underline{\mathbf{T}}_T=\underline{\mathbf{T}}-\underline{T}_N\mathbf{N}$ and $\mathbf{U}_T=\mathbf{U}-U_N\mathbf{N}$. In relationship (3.32), the internal energy $\widehat{e}$ is replaced by $\widehat{\Psi}+T\widehat{s}$, where $\widehat{\Psi}$ is the specific free energy to get

$$m\left([\widehat{\Psi}+T\widehat{s}]-\underline{T}_N\left[\frac{1}{\rho}\right]\right)-\underline{\mathbf{T}}_T\cdot[\mathbf{U}_T]-\mathbf{H}\cdot\mathbf{N}[\dot{\beta}]-T[Q]-\underline{Q}[T]=0,$$

or

$$\underline{T}\{[m[\widehat{s}]-[Q]\}-\left\{-([\widehat{\Psi}]+\underline{\widehat{s}}[T]-\underline{T}_N\left[\frac{1}{\rho}\right]\right\}m+\underline{\mathbf{T}}_T\cdot[\mathbf{U}_T]+\mathbf{H}\cdot\mathbf{N}[\dot{\beta}]+\underline{Q}[T]\}=0. \tag{3.33}$$

Quantity $-([\widehat{\Psi}]+\underline{\widehat{s}}[T]-\underline{T}_N[1/\rho])$ is perfectly defined. Actually, if $\widehat{\Psi}$ is replaced by $\widehat{\Psi}+CT+A$, ($A$ and $C$ are constants, let us recall that the free energy is defined up to an affine function of the temperature), $[\widehat{\Psi}]+\underline{\widehat{s}}[T]$ is replaced by $([\widehat{\Psi}]+C[T])+(\underline{\widehat{s}}-C)[T]=[\widehat{\Psi}]+\underline{\widehat{s}}[T]$. In order to use equality (3.33), the state quantities and the actual or virtual evolution of the material are to be defined. We decide to characterize the state of the material by the quantities $(\varepsilon,1/\rho,\beta,\mathrm{grad}\beta,\mathrm{T})$ on the two sides of the discontinuity surface

$$E=\left(\varepsilon_1,\frac{1}{\rho_1},\beta_1,\mathrm{grad}\beta_1,\mathrm{T}_1,\varepsilon_2,\frac{1}{\rho_2},\beta_2,\mathrm{grad}\beta_2,\mathrm{T}_2\right),$$

and its evolution by the velocities and discontinuities

$$\Delta\tilde{E}=\left(V_{N2}=\mathbf{V}_{rel2}\cdot\mathbf{N},V_{N1}=\mathbf{V}_{rel1}\cdot\mathbf{N},[T],[\mathbf{V}_T],[\dot{\beta}]\right),$$

which describe the way the state of the material evolves when crossing the discontinuity surface and the way the discontinuity line moves. Note that this set is equivalent to $((V_{N2}+V_{N1})/2,[T],[\mathbf{V}],[\dot{\beta}])$ or to $((V_{N2}+V_{N1})/2,[T],[\mathbf{U}],[\dot{\beta}])$ since $[\mathbf{V}]=[\mathbf{U}]$. This last choice is coherent with the choice of $\varepsilon(\mathbf{U})$, grad$\beta$ and gradT as quantities describing the evolution when there is no discontinuity. Actually $[\mathbf{U}]$,

$[\dot{\beta}]$ and $[T]$ are the non-smooth part of the space derivatives which appear in $\delta E$, $(\varepsilon(\mathbf{U}), \text{grad}\dot{\beta}, \text{gradT})$. The average normal velocity is added in order to describe the evolution of the discontinuity line. One can say that the constitutive laws depend on $(\varepsilon(\mathbf{U}), \text{grad}\dot{\beta}, \text{gradT})$ in the distributive sense.

*Remark 3.12.* More details are given in [112]. Following [112], the choice of $\Delta\tilde{E}$ is in agreement with the characterizations of the evolution by left derivatives. The quantities $\Delta\tilde{E}$ are, of course, known, because the left derivatives are known in the smooth situation, as soon as the history of the material is known. The equations on a discontinuity surface give the initial velocities which depend on the past for the future evolution.

The inverse of the densities $\rho$, the specific volumes, are state quantities. Thus we can choose $\rho_2 V_{N2}$ and $\rho_1 V_{N1}$ as elements equivalent to $V_{N2}$ and $V_{N1}$, since the set $(E, \Delta\tilde{E})$ gives the same information as the set $\{E, (\rho_2 V_{N2}, \rho_1 V_{N1}, [T], [\mathbf{V}_T])\}$. Assuming the mass balance (3.31) is satisfied, we find that knowing $(E, \Delta\tilde{E})$ is equivalent to knowing $(E, \Delta E)$ with

$$\Delta E = (m, [T], [\mathbf{U}_T], [\dot{\beta}]).$$

Thus we decide to characterize the evolution of the material by $\Delta E$.

*Remark 3.13.* The quantity $[\rho]$ is not chosen as an element of $\Delta\tilde{E}$ or $\Delta E$ because the mass balance gives

$$[\rho V_N] = [\rho]\underline{V}_N + \underline{\rho}[V_N] = 0.$$

Thus $[\rho]$ is a function of $\underline{\rho}[V_N]$ which is an element of a set equivalent to $\Delta\tilde{E}$) .

*Remark 3.14.* In the smooth situation, vector $\text{gradT}(\mathbf{x},\text{t})$ is known as soon as the function $T(\mathbf{x},t)$ is known. Nevertheless we have chosen $T \in E$ and $\text{gradT} \in \delta\text{E}$. On the discontinuity surface, to know $T_1(\mathbf{x},t)$ and $T_2(\mathbf{x},t)$ is equivalent to know function $T(\mathbf{x},t)$ in the smooth situation, and to know $[T(\mathbf{x},t)]$ is equivalent to know $\text{gradT}(\mathbf{x},\text{t})$ in the smooth situation. Thus the choices of $T_1(\mathbf{x},t) \in E$, $T_2(\mathbf{x},t) \in E$ and $[T(\mathbf{x},t)] \in \Delta E$ are consistent.

Vector

$$F^{dd} = \left\{ -\left( [\widehat{\Psi}] + \underline{\widehat{s}}[T] - \underline{T}_N \left[\frac{1}{\rho}\right] \right), \underline{Q}, \underline{\mathbf{T}}_T, \mathbf{H}\cdot\mathbf{N} \right\},$$

depends on the volume constitutive laws, i.e., on $E$. Relationship (3.33) becomes

$$\underline{T}\{[m\widehat{s}] - [Q]\} - F^{dd} \cdot \Delta E = 0,$$

The constitutive laws on the discontinuity surface are relationships between $F^{dd}$, $\Delta E$ and $E$ such that

$$F^{dd} \cdot \Delta E \geq 0,$$

which is equivalent to the second law (3.27). Constitutive law for $\underline{Q}$ gives the distribution of the received entropy $[Q]$ between the two sides of the discontinuity surface or between the temperatures $T_1$ and $T_2$. In the same way, constitutive law for the force $\underline{\mathbf{T}}_T$ gives the distribution of the force $[\mathbf{T}]$ between the two sides of discontinuity surface.

The constitutive laws are defined by a surface pseudopotential of dissipation $\Phi(\Delta E, E) = \Phi(m, [T], [\mathbf{U}_T], [\dot{\beta}], E)$ where $E$ is a parameter. In case this potential is a smooth function, constitutive laws are

$$-([\widehat{\Psi}] + \underline{\widehat{s}}[T] - \underline{T}_N[\frac{1}{\rho}]) = \frac{\partial \Phi}{\partial m},$$

$$\underline{Q} = \frac{\partial \Phi}{\partial [T]}, \tag{3.34}$$

$$\underline{\mathbf{T}}_T = \frac{\partial \Phi}{\partial [\mathbf{U}_T]}, \tag{3.35}$$

$$\mathbf{H} \cdot \mathbf{N} = \frac{\partial \Phi}{\partial [\dot{\beta}]}. \tag{3.36}$$

In Sect. 4.1 and in Chaps. 9 and 12 down below, examples are given.

In the classical situation, there are three constitutive laws on a discontinuity surface. Law (3.36) with

$$\Phi(m, [T], [\mathbf{U}_T]) = I_0([T]) + \varphi([\mathbf{U}_T]),$$

gives $[T] = 0$ and

$$-\underline{T}_N = -\frac{[\widehat{\Psi}]}{[1/\rho]},$$

which is the discontinuous analogue to $p = -\partial\widehat{\Psi}/\partial(1/\rho)$, which gives the pressure in a fluid where there is no dissipation with respect to $\partial\rho/\partial t$ or to div$\mathbf{U}$ due to the mass balance, (see for instance formula (13.46) below). Law (3.34) is analogous to law (3.15) which gives Fourier law. Law (3.35) is analogous to law (3.15) which gives the dissipative stresses depending on the strain rates. The tangential force $\underline{\mathbf{T}}_T$ is only dissipative. There is indeed no contribution of the free energy to its value.

Note that the constitutive laws are actually causal: they characterize completely a material if its past is known, i.e., if its state $E$, its evolution $\Delta E$ depending on its history are known.

Note that $[\widehat{\Psi}]$, $[T]$, $[1/\rho]$ and $m$, $\underline{Q}$ and $[T]$, $\underline{\mathbf{T}}_T$ and $[\mathbf{U}_T]$, $\mathbf{H}\cdot\mathbf{N}$ and $[\dot{\beta}]$ change sign when the orientation of the normal vector $\mathbf{N}$ is changed whereas $\underline{\widehat{s}}$, $\underline{T}_N$, $\underline{T}$, and $[Q]$ do not change sign. We assume that the pseudo-potential $\Phi$ depends on the orientation and satisfies

$$\Phi(m, [T], [\mathbf{U}_T], [\dot{\beta}], -\mathbf{N}) = \Phi(-m, -[T], -[\mathbf{U}_T], -[\dot{\beta}], \mathbf{N}),$$

vector $\mathbf{N}$ being understood as a state quantity.

## 3.6 The Basic Laws of Mechanics in a Discontinuity Volume

Discontinuities can be at a time $t$ in the whole volume. This is the case for collisions which are investigated in Chap. 11. We recall very rapidly the collision theory detailed in [112, 114]. The collisions being dissipative phenomena, they produce heat which intervene in the thermal evolution and may produce phase changes.

In this section, we denote for function $t \to A(t)$

$$\underline{A} = \frac{A^+ + A^-}{2}, \ [A] = A^+ - A^-.$$

where

$$A^+ = \lim_{\Delta t \to 0, \Delta t > 0} A(t + \Delta t), \ A^- = \lim_{\Delta t \to 0, \Delta t > 0} A(t - \Delta t),$$

are the values of quantity $A$ after and before the discontinuity. Note that in an other context, values $A^+$ and $A^-$ have been denoted right and left values, $A^r$ and $A^l$, see Sect. 3.4.1.1 and formula (3.24).

### *3.6.1 The Principle of Virtual Power and the Equations of Motion*

The virtual work of the interior forces, the interior percussion forces, is

$$\mathcal{T}_{int}(\mathbf{V}, \gamma) = -\int_\Omega \Sigma : \mathsf{D}\left(\frac{\mathbf{V}^+ + \mathbf{V}^-}{2}\right) d\Omega - \int_\Omega B_p [\gamma] + \mathbf{H}_p \cdot \mathrm{grad}\,[\gamma]\, \mathrm{d}\Omega,$$

where $\mathbf{V}$, $\gamma$ are virtual velocities: $\mathbf{V}^-$, $\gamma^-$ are the velocity before the collision and $\mathbf{V}^+$, $\gamma^+$ are the velocity after. It is denoted $[X] = X^+ - X^+$, the discontinuity of quantity $X$.

It appears percussion stresses $\Sigma$ and percussion work $B_p$ and work flux vector $\mathbf{H}_p$. The virtual work of the acceleration forces is

$$\mathcal{T}_{acc}(\mathbf{V}, \gamma) = \int_\Omega \rho\, [\mathbf{U}] \cdot \frac{\mathbf{V}^+ + \mathbf{V}^-}{2} d\Omega.$$

*Remark 3.15.* The mass balance is

$$[\rho] = 0.$$

Thus the densities are constant in the equations of motion. The incompressibility in collision of fluids is investigated in [85, 112, 115].

The virtual work of the exterior forces is

$$\mathcal{T}_{ext}(\mathbf{V},\gamma) = \int_{\Omega} \mathbf{F}_p \cdot \frac{\mathbf{V}^+ + \mathbf{V}^-}{2} d\Omega + \int_{\partial\Omega} \mathbf{G}_p \cdot \frac{\mathbf{V}^+ + \mathbf{V}^-}{2} d\Gamma$$

$$+ \int_{\Omega} A_p [\gamma] d\Omega_1 + \int_{\partial\Omega} a_p [\gamma] d\Gamma.$$

We assume that the surface exterior percussions $\mathbf{G}_p$ are applied to the whole boundary of the solid. The $\mathbf{F}_p$ are the volume exterior percussions. The $A_p$ and $a_p$ are the volume and surface percussion work provided by the exterior by electrical, radiative,... actions. The equations of motion results from the principle of virtual work

$$\forall \mathbf{V}, \forall \gamma, \ \mathcal{T}_{acc}(\mathbf{V},\gamma) = \mathcal{T}_{int}(\mathbf{V},\gamma) + \mathcal{T}_{ext}(\mathbf{V},\gamma).$$

Different choices of the virtual velocities $\mathbf{V}$ and $\gamma$ give

$$\rho [\mathbf{U}] = \mathrm{div}\Sigma + \mathbf{F}_\mathrm{p}, \ \mathrm{in}\ \Omega, \ \Sigma\mathbf{N} = \mathbf{G}_\mathrm{p}, \ \mathrm{on}\ \partial\Omega,$$

$$-B_p + \mathrm{div}\mathbf{H}_\mathrm{p} + \mathrm{A}_\mathrm{p} = 0, \ in\ \Omega, \ \mathbf{H}_p\mathbf{N} = a_p, \ on\ \partial\Omega.$$

### 3.6.2 The Laws of Thermodynamics

The first law is

$$[\mathcal{E}] + [\mathcal{K}] = \mathcal{T}_{ext}(\mathbf{U},\beta) + \mathcal{C},$$

where $\mathcal{K}$ is the kinetic energy, and $\mathcal{C}$ is the thermal impulse received by the structure. With the principle of virtual work where the velocities are the actual velocitics, the first law gives

$$[\mathcal{E}] = -\mathcal{T}_{int} + \mathcal{C}.$$

The temperature may be discontinuous: we denote $T^-$ the temperature before the collision and $T^+$ the temperature after the collision. We assume that heat is received either at temperature $T^-$ or at temperature $T^+$

$$\mathcal{C} = \int_{\partial\Omega} -\left(T^+\mathbf{Q}_p^+ + T^-\mathbf{Q}_p^-\right) \cdot \mathbf{N} d\Gamma + \int_{\Omega} T^+\mathcal{B}^+ + T^-\mathcal{B}^- d\Gamma,$$

where $\mathbf{Q}_p$ is the impulsive entropy flux vector and $\mathcal{B}$ the impulsive entropy source. With

$$\mathcal{E} = \int_{\Omega} e d\Omega,$$

we get

$$[e] = \Sigma : \mathsf{D}\left(\frac{\mathbf{U}^+ + \mathbf{U}^-}{2}\right) + B_p[\beta] + \mathbf{H}_p \cdot \mathrm{grad}[\beta]$$
$$- \mathrm{div}(\mathrm{T}^+\mathbf{Q}_\mathrm{p}^+ + \mathrm{T}^-\mathbf{Q}_\mathrm{p}^-) + \mathrm{T}^+\mathscr{B}^+ + \mathrm{T}^-\mathscr{B}^-.$$

By using the Helmholtz relationship, $e = \Psi + Ts$, we have

$$[\Psi] + \underline{s}[T] + \underline{T}[s] = \Sigma : \mathsf{D}\left(\frac{\mathbf{U}^+ + \mathbf{U}^-}{2}\right) + B_p[\beta] + \mathbf{H}_p \cdot \mathrm{grad}[\beta]$$
$$- \mathrm{div}(\underline{\mathrm{T}}\Sigma(\mathbf{Q}_\mathrm{p}) + [\mathrm{T}]\Delta(\mathbf{Q}_\mathrm{p})) + \underline{\mathrm{T}}\Sigma(\mathscr{B}) + [\mathrm{T}]\Delta(\mathscr{B}), \quad (3.37)$$

where a sum

$$T^+\mathscr{B}^+ + T^-\mathscr{B}^- = \Sigma(T\mathscr{B}),$$

is split in an other sum

$$\Sigma(T\mathscr{B}) = \underline{T}\Sigma(\mathscr{B}) + [T]\Delta(\mathscr{B}),$$

with

$$\Sigma(\mathscr{B}) = \mathscr{B}^+ + \mathscr{B}^-,\ \Delta(\mathscr{B}) = \frac{\mathscr{B}^+ - \mathscr{B}^-}{2},$$

and

$$\underline{T} = \frac{T^+ + T^-}{2},\ [T] = T^+ - T^-.$$

The second law is

$$[\mathscr{S}] = \int_\Omega [s]\, d\Omega \geq -\int_\Gamma (\mathbf{Q}_p^+ + \mathbf{Q}_p^-) \cdot \mathbf{N} d\Gamma + \int_\Omega \mathscr{B}^+ + \mathscr{B}^- d\Omega,$$

which gives

$$[s] \geq -\mathrm{div}\Sigma(\mathbf{Q}_\mathrm{p}) + \Sigma(\mathscr{B}). \quad (3.38)$$

Combining relationships (3.37) and (3.38), we get

$$[\Psi] + \underline{s}[T] + \mathrm{div}\left([\mathrm{T}]\Delta(\mathbf{Q}_\mathrm{p})\right) - [\mathrm{T}]\Delta(\mathscr{B})$$
$$\leq \Sigma : \mathsf{D}(\frac{\mathbf{U}^+ + \mathbf{U}^-}{2}) + B_p[\beta] + \mathbf{H}_p \cdot \mathrm{grad}[\beta] - \mathrm{grad}\underline{\mathrm{T}} \cdot \Sigma(\mathbf{Q}_\mathrm{p}). \quad (3.39)$$

Let us note that the right hand side is a scalar product between internal forces and related evolution quantities whereas it is not the case of the left hand side. Let us try to relate $[\Psi]$ to a scalar product. We have

$$\begin{aligned}[\Psi] &= \Psi(T^+,\beta^+,\mathrm{grad}\beta^+) - \Psi(\mathrm{T}^-,\beta^-,\mathrm{grad}\beta^-) \\ &= \Psi(T^+,\beta^+,\mathrm{grad}\beta^+) - \Psi(\mathrm{T}^+,\beta^-,\mathrm{grad}\beta^-) \\ &\quad + \Psi(T^+,\beta^-,\mathrm{grad}\beta^-) - \Psi(\mathrm{T}^-,\beta^-,\mathrm{grad}\beta^-).\end{aligned}$$

Because the free energy is a concave function of temperature $T$, we have

$$\Psi(T^+,\beta^-,\mathrm{grad}\beta^-) - \Psi(\mathrm{T}^-,\beta^-,\mathrm{grad}\beta^-) \leq -\mathrm{s}^{\mathrm{fe}}\,[\mathrm{T}], \tag{3.40}$$

with

$$-s^{fe} \in \hat{\partial}\Psi_T(T^-,\beta^-,\mathrm{grad}\beta^-),$$

where $\hat{\partial}\Psi_T$ is the set of the uppergradients of the concave function

$$T \to \Psi_T(T,\beta^-,\mathrm{grad}\beta^-) = \Psi(\mathrm{T},\beta^-,\mathrm{grad}\beta\ \ ),$$

(see Appendix A). We assume $\Psi$ is a convex function of $(\beta, \mathrm{grad}\,\beta)$. Thus we have

$$\Psi(T^+,\beta^+,\mathrm{grad}\beta^+) - \Psi(\mathrm{T}^+,\beta^-,\mathrm{grad}\beta^-) \leq \mathrm{B}^{\mathrm{fe}}\,[\beta] + \mathbf{H}^{\mathrm{fe}}\cdot\mathrm{grad}\,[\beta],$$

with

$$\left(B^{fe},\mathbf{H}^{fe}\right) \in \partial\Psi_{\beta,\mathrm{grad}\beta}(T^+,\beta^+,\mathrm{grad}\beta^+). \tag{3.41}$$

The internal forces $\left(B^{fe},\mathbf{H}^{fe}\right)$ depend on the future state $(T^+,\beta^+,\mathrm{grad}\beta^+)$, in agreement with our idea that the constitutive laws sum up what occurs during the discontinuity of the state quantities. It results

$$\begin{aligned}&[\Psi] + \underline{s}[T] + \mathrm{div}\,[\mathrm{T}]\Delta(\mathbf{Q}_\mathrm{p}) - [\mathrm{T}]\Delta(\mathscr{B}) \\ &\leq B^{fe}\,[\beta] + \mathbf{H}^{fe}\cdot\mathrm{grad}\,[\beta] - \mathrm{s}^{\mathrm{fe}}\,[\mathrm{T}] + \underline{\mathrm{s}}\,[\mathrm{T}] + \mathrm{div}\,[\mathrm{T}]\Delta(\mathbf{Q}_\mathrm{p}) - [\mathrm{T}]\Delta(\mathscr{B}) \\ &= B^{fe}\,[\beta] + \mathbf{H}^{fe}\cdot\mathrm{grad}\,[\beta] + (-\mathrm{s}^{\mathrm{fe}} + \underline{\mathrm{s}} + \mathrm{div}\Delta(\mathbf{Q}_\mathrm{p}) - \Delta(\mathscr{B}))\,[\mathrm{T}] + \Delta(\mathbf{Q}_\mathrm{p})\cdot\mathrm{grad}[\mathrm{T}].\end{aligned}$$

As usual, we assume no dissipation with respect to $[T]$ and have

$$\Delta(\mathbf{Q}_p) = 0,$$

$$-s^{fe} + \underline{s} + \mathrm{div}\Delta(\mathbf{Q}_\mathrm{p}) - \Delta(\mathscr{B}) = 0.$$

Thus

$$\Delta(\mathscr{B})+S=0, \tag{3.42}$$

with $S = s^{fe} - \underline{s}$. This relationship splits either the received heat, $T^+\mathscr{B}^+ + T^-\mathscr{B}^-$, or the received entropy, $\mathscr{B}^+ + \mathscr{B}^-$, between the two temperatures, $T^+$ and $T^-$. Let us note that this relationship depends on the future state via the average entropy $\bar{s}$.

We assume that the interior forces satisfy the inequality

$$0 \leq \Sigma : \mathsf{D}\left(\frac{\mathbf{U}^+ + \mathbf{U}^-}{2}\right) + (B_p - B^{fe})[\beta] + (\mathbf{H}_p - \mathbf{H}^{fe}) \cdot \mathrm{grad}[\beta] - 2\mathrm{grad}\underline{\mathrm{T}} \cdot \mathbf{Q}_\mathrm{p}. \tag{3.43}$$

*Remark 3.16.* The discontinuity $[\Psi]$ may be split in a different manner

$$\begin{aligned}
[\Psi] &= \Psi(T^+,\beta^+,\mathrm{grad}\beta^+) - \Psi(\mathrm{T}^-,\beta^-,\mathrm{grad}\beta^-)\\
&= \Psi(T^+,\beta^+,\mathrm{grad}\beta^+) - \Psi(\mathrm{T}^-,\beta^+,\mathrm{grad}\beta^+)\\
&\quad + \Psi(T^-,\beta^+,\mathrm{grad}\beta^+) - \Psi(\mathrm{T}^-,\beta^-,\mathrm{grad}\beta^-).
\end{aligned}$$

We get

$$\Psi(T^+,\beta^+,\mathrm{grad}\beta^+) - \Psi(\mathrm{T}^-,\beta^+,\mathrm{grad}\beta^+) \leq -\hat{\mathrm{s}}^{\mathrm{fe}}[\mathrm{T}],$$

with

$$-\hat{s}^{fe} \in \hat{\partial}\Psi_T(T^-,\beta^+,\mathrm{grad}\beta^+).$$

If $\Psi$ it is a convex function of $(\beta, \mathrm{grad}\,\beta)$

$$\Psi(T^-,\beta^+,\mathrm{grad}\beta^+) - \Psi(\mathrm{T}^-,\beta^-,\mathrm{grad}\beta^-)) \leq \hat{\mathrm{B}}^{\mathrm{fe}}[\beta] + \hat{\mathrm{H}}^{\mathrm{fe}} \cdot \mathrm{grad}[\beta],$$

with

$$\left(\hat{B}^{fe},\hat{H}^{fe}\right) \in \partial\Psi_{\beta,\mathrm{grad}\beta}(T^-,\beta^-,\mathrm{grad}\beta^-).$$

The internal forces $\left(\hat{B}^{fe},\hat{H}^{fe}\right)$ depend entirely on the past state $(T^-,\beta^-,\mathrm{grad}\beta^-)$. Because we think that the constitutive laws sum up what occurs during the discontinuity, it is compulsory that the internal forces depend on the future state $(T^+,\beta^+,\mathrm{grad}\,\beta^+)$. Thus this splitting of the free energy does not seem as good as the one we have chosen.

We may choose a pseudopotential of dissipation

$$\Phi(\Delta E^\pm,E^\pm,\chi) = \Phi\left(\mathsf{D}\left(\frac{\mathbf{U}^+ + \mathbf{U}^-}{2}\right),[\beta],\mathrm{grad}[\beta],\mathrm{grad}\underline{\mathrm{T}},\mathrm{E}^\pm,\chi\right),$$

and the constitutive laws

$$\left(\Sigma, (B_p - B^{fe}), (\mathbf{H}_p - \mathbf{H}^{fe}), -2\mathbf{Q}_p\right)$$
$$\in \partial\Phi\left(\mathsf{D}\left(\frac{\mathbf{U}^+ + \mathbf{U}^-}{2}\right), [\beta], \mathrm{grad}\,[\beta], \mathrm{grad}\underline{\mathrm{T}}, \mathrm{E}^{\pm}, \chi\right). \tag{3.44}$$

**Theorem 3.3.** *If constitutive laws (3.40), (3.41), (3.42) and (3.44) are satisfied, then the second law is satisfied.*

*Proof.* If relationship (3.44) is satisfied, inequality (3.43) is satisfied. Then it is easy to prove that the inequality (3.39) which is equivalent to the second law is satisfied.

*Remark 3.17.* In this chapter, there are discontinuities either with respect to time or with respect to space. It is possible that both discontinuities are concomitant. This is the case when a solid colliding an obstacle is fractured. This non smooth situation is investigated in [114] and [100].

# Chapter 4
# Solid–Liquid Phase Change

Besides the classical ice–water phase change, the solid–liquid phase change exhibits many different behaviours: diffuse with respect to time and space phase change, phase change with voids and bubbles, phase change with different densities, dissipative phase change, irreversible phase change, phase change with thermal memory,... The different predictive theories result from the choices of the free energies and pseudopotentials of dissipation. Examples are given: soil freezing and the insulation of an house in winter, liquid gas underground storage, fossil permafrost.

## 4.1 The Stefan Problem

The classical ice–water phase change or the Stefan problem [2, 34, 96, 149, 200], introduces as state quantities the temperature $T$ and the liquid or water volume fraction $\beta$, [3, 133]. The free energy and pseudopotential are [78, 112]

$$\Psi(T,\beta) = -CT\ln T - \beta\frac{L}{T_0}(T-T_0) + I(\beta),$$

$$\Phi(\delta E,\chi) = \Phi(\mathrm{gradT},\mathrm{T}) = \frac{\lambda}{2\mathrm{T}}(\mathrm{gradT})^2, \tag{4.1}$$

with $\delta E = \mathrm{gradT}$ and $\chi = T$. Let us comment the different terms of $\Psi$ and $\Phi$:

- $-CT\ln T$ is responsible for heat storage ($C$ is the heat capacity).
- $-\beta(L/T_0)(T-T_0)$ is a linear function with respect to $\beta$: its derivative being constant with respect to $\beta$ defines a threshold (we prove below that when $(T-T_0)<0$ the solid phase, the ice, is present and when $(T-T_0)>0$, the liquid phase, the water, is present). The quantity $L$ is the latent heat at phase change temperature $T_0$.

M. Frémond, *Phase Change in Mechanics*, Lecture Notes of the Unione Matematica Italiana 13, DOI 10.1007/978-3-642-24609-8_4,

- Indicator function $I$ of the interval $[0,1]$, (see Appendix A), takes into account the internal constraint on the liquid water volume fraction

$$0 \leq \beta \leq 1. \tag{4.2}$$

- In the classical ice–water phase change, it is assumed that no macroscopic motion occurs. Thus the deformation $\varepsilon$ is not a state quantity. We assume that there is no local interaction, thus grad$\beta$ does not intervene.
- The pseudopotential of dissipation describes the thermal conduction: $\lambda$ is the thermal conductivity.

The constitutive laws are

$$\sigma = 0\,,\; B \in -\frac{L}{T_0}(T - T_0) + \partial I(\beta)\,,\; \mathbf{H} = 0,$$

$$\mathbf{Q} = -\frac{\lambda}{T}\text{gradT},$$

$$s = C(1 + \ln T) + \frac{L}{T_0}\beta,$$

where $\partial I$ is the subdifferential set of function $I$, (see Appendix A).

On the free boundary between ice and water, the constitutive laws are those given in Sect. 3.5. Let us recall that the mass flux through the free boundary is $m = -\rho W$ where $W = \overrightarrow{W}.\overrightarrow{N}$ is the normal velocity of the free boundary. In the sequel, it is denoted $[A] = A_2 - A_1$, the discontinuity of quantity $A$ on the free boundary (the indices 1 and 2 refer to the two sides of the discontinuity surface). The pseudopotential of dissipation has to imply the well known experimental results: the temperature is continuous on the free boundary and the ice–water phase change is perfectly reversible or non dissipative. Thus we choose the pseudopotential

$$\Phi(m, [T]) = I_0([T]),$$

where $I_0$ is the indicator function of the origin of $\mathbb{R}$ (see Appendix A). It results the constitutive laws from (3.36)

$$[T] = 0,$$

$$-[\Psi] = \frac{\partial \Phi}{\partial m} = 0,$$

which imply the continuity of the free energy on the free boundary $\Gamma(t)$. This relationship implies that $\beta(T - T_0)$ is continuous. Thus the temperature on the free boundary is $T_0$ because $[\beta] \neq 0$: the phase change occurs only at this temperature which is, of course, called the phase change temperature.

### 4.1.1 The Equations

They result from the equations of motion, the entropy balance and the constitutive laws. We have assumed that the ice–water system does not move. Thus there is only the equation of microscopic motion. Moreover we assume from now on, that there is no volume exterior source of exterior work $A = 0$ (in the sequel we are going to assume when needed, no surface exterior source of work $a = 0$). Outside the free boundary $\Gamma(t)$ we have

$$C\frac{\partial \ln T}{\partial t} + \frac{L}{T_0}\frac{\partial \beta}{\partial t} - \lambda \Delta \ln T = R\,,\ in\ \Omega\backslash\Gamma(t), \tag{4.3}$$

$$-\frac{L}{T_0}(T - T_0) + \partial I(\beta) \ni 0\,,\ in\ \Omega\backslash\Gamma(t)\,, \tag{4.4}$$

$$\lambda\frac{\partial \ln T}{\partial N} = \pi\,,\ on\ \partial\Omega, \tag{4.5}$$

$$T(\mathbf{x},0) = T^0(\mathbf{x})\,,\ \beta(\mathbf{x},0) = \beta^0(\mathbf{x})\,,\ on\ \Omega, \tag{4.6}$$

where $R$ and $\pi$ are the volume and surface entropy sources. On free boundary $\Gamma(t)$, we have the constitutive laws

$$[T] = [\ln T] = 0\,,\ T = T_0, \tag{4.7}$$

and the entropy balance

$$m[s] - [Q] = 0,$$

or

$$\rho W\frac{L}{T_0}[\beta] = \left[\frac{\lambda}{T}\frac{\partial T}{\partial N}\right] = \left[\lambda\frac{\partial \ln T}{\partial N}\right], \tag{4.8}$$

or

$$\rho W L[\beta] = \left[\lambda\frac{\partial T}{\partial N}\right],$$

because $T = T_0$. This relationship is the classical energy balance on the free boundary of the Stefan problem.

The functions $T^0$ and $\beta^0$ are the initial temperature and liquid water content which have to satisfy an obvious compatibility condition. Equation (4.4) results from the equation of microscopic motion. It gives the value of the temperature difference, $T - T_0$, versus the water volume fraction. It is easy to read this relationship:

1. If $T - T_0 < 0$, then $\beta = 0$ and there is ice.
2. If $T - T_0 > 0$, then $\beta = 1$ and there is liquid water.
3. If $T - T_0 = 0$, then $0 \leq \beta \leq 1$ and there is a mixture of ice and liquid water.

*Remark 4.1.* It is reasonable that the microscopic motions do not intervene on the free boundary because the microscopic equation of motion (4.4) does not involve derivatives with respect to space. The macroscopic equation of motion which contains space derivative does not intervene because we have assumed that the macroscopic velocities **U** are zero.

*Remark 4.2.* If the term related to $\beta$ in the free energy

$$-\beta \frac{L}{T_0}(T - T_0) + I(\beta),$$

is replaced by a quadratic function

$$\frac{L}{T_0}\left\{-\beta(T - T_0) + \frac{\beta^2}{2}(T_1 - T_0)\right\} + I(\beta),$$

with $T_1 \geq T_0$, the relationship (4.4) becomes

$$-\frac{L}{T_0}\{(T - T_0) - \beta(T_1 - T_0)\} + \partial I(\beta) \ni 0\,,\ in\ \Omega.$$

The phase change occurs for $T_0 \leq T \leq T_1$ with

$$\beta = \frac{T - T_0}{T_1 - T_0},$$

and for $T > T_1$ there is liquid: $\beta = 1$, for $T < T_0$ there is solid: $\beta = 0$. When the temperature is increasing phase change begins at $T = T_0$ and ends at $T = T_1$. This type of phenomenon occurs in shape memory alloys (see the references on shape memory alloys).

### 4.1.2 A Way to Solve the Equations Where the Free Boundary Is a By-Product: The Freezing Index

Let us note that (4.3) and (4.4) in the distributive sense are valid outside free boundary $\Gamma(t)$ and on the free boundary where they are equivalent to (4.7) and to (4.8). We may integrate (4.3) or the entropy balance (3.14) with respect to time and get

$$s(t) - s(0) + \int_0^t \mathrm{div}\mathbf{Q}\mathrm{dt} = \int_0^t R dt \;,\; in\; \Omega,$$
$$-\int_0^t \mathbf{Q}\cdot\mathbf{N} dt = \int_0^t \pi dt \;,\; on\; \partial\Omega.$$

It is possible to prove that

$$\int_0^t \mathrm{div}\mathbf{Q}\mathrm{dt} = \mathrm{div}\int_0^t \mathbf{Q}\mathrm{dt}.$$

Assuming the Fourier law, it is also possible to prove that

$$\int_0^t \mathbf{Q} dt = -\lambda\int_0^t \mathrm{grad}\ln\mathrm{Tdt} = -\lambda\,\mathrm{grad}\left(\int_0^t \ln\mathrm{Tdt}\right).$$

Thus we have

$$s(t) - s(0) - \lambda\Delta\left(\int_0^t \ln T dt\right) = \int_0^t R dt \;,\; in\; \Omega,$$
$$\lambda\frac{\partial\left(\int_0^t \ln T dt\right)}{\partial N} = \int_0^t \pi dt \;,\; on\; \partial\Omega. \tag{4.9}$$

By letting

$$u(\mathbf{x},t) = \int_0^t \ln T(\mathbf{x},t) dt,$$

we get

$$s(\mathbf{x},t) = C(1+\ln T) + \frac{L}{T_0}\beta = C(1+\frac{\partial u}{\partial t}) + \frac{L}{T_0}\beta.$$

The equation of motion (4.4)

$$T - T_0 \in \partial I(\beta),$$

is equivalent to

$$\exp\left(\frac{\partial u}{\partial t}\right) - T_0 \in \partial I(\beta)\,. \tag{4.10}$$

Equation (4.10) may be solved and the water volume fraction may be eliminated to give

$$C(1+\frac{\partial u}{\partial t})+\frac{L}{T_0}\partial I^*(\exp(\frac{\partial u}{\partial t})-T_0)-\lambda\Delta u=s(0)+\int_0^t Rdt\ ,\ in\ \Omega,$$

$$\lambda\frac{\partial u}{\partial N}=\int_0^t \pi dt\ ,\ on\ \partial\Omega, \qquad (4.11)$$

where $I^*$ is the dual function of $I$ (see Appendix A or [112, 172]). This partial differential equation completed by the boundary condition (4.9) and initial condition $u(\mathbf{x},0)=0$ may be solved, [41]. There exists a variational formulation because $\ln T$ and $u$ being continuous on the free boundary, gradu involves no surface measure and has only density with respect to the Lebesgue measure. For the numerical methods, it is also possible to use the energy balance instead of the entropy balance. The resulting equation for the freezing index defined by

$$w(\mathbf{x},t)=\int_0^t T(\mathbf{x},t)dt,$$

is

$$C\frac{\partial w}{\partial t}+L\partial I^*(\frac{\partial w}{\partial t}-T_0)-\lambda\Delta w=e(0)+\int_0^t rdt\ ,\ in\ \Omega,$$

$$\lambda\frac{\partial w}{\partial N}=\int_0^t \varpi dt\ ,\ on\ \partial\Omega, \qquad (4.12)$$

where $r=TR$ and $\varpi=T\pi$ are assumed to be known functions. The main numerical advantage of (4.11) and (4.12) is that they do not involve the free boundary which is a by-product given by the level set either $\partial u/\partial t=\ln T_0$ or $\partial w/\partial t=T_0$, [32, 112]. Approximations useful for numerics are given in [51, 55, 146].

## 4.2 Examples

The computer program CESARGEL, [141], solving (4.12) may be used for engineering purposes. Figure 4.1 shows the temperature inside an insulated house in winter, [89].

Figure 4.2 shows the extension of the permafrost during the last 120,000 years in some place in France [150]. Time in year is the abscissa and depth in meter is the ordinate. Applications to soil freezing are given in [33, 128, 135, 198].

## 4.3 Unfrozen Water in the Frozen Zone

When freezing a water saturated porous medium, experiments show that the water freezes progressively. There exists unfrozen liquid water in the frozen zone, i.e., the zone where $T<T_0$, [112]. The predictive theory which takes this phenomenon

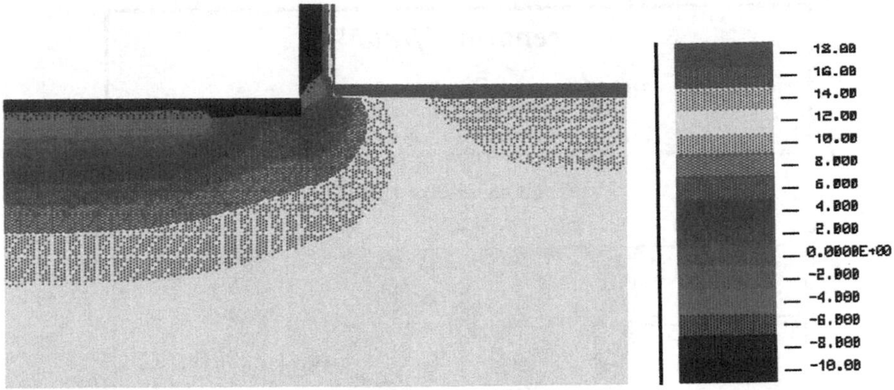


**Fig. 4.1** The temperature in winter of the wall of a warm house and the temperature in the surrounding ground. The temperatures are computed with the freezing index $w(x,t)$. The temperature scale in C is on the right: red color is warm and blue is cold

into account introduces a function $h(\beta)$ in the free energy. Its derivative is directly related to the unfrozen water content (see (4.14) below). The free energy and pseudopotential are

$$\Psi(T,\beta) = -CT\ln T - \beta\frac{L}{T_0}(T-T_0) + h(\beta),$$

$$\Phi(\mathrm{gradT},\mathrm{T}) = \frac{\lambda}{2T}(\mathrm{gradT})^2.$$

Convex function $h$ takes into account the internal constraint (4.2) by being equal to $+\infty$ outside interval $[0,1]$. The constitutive laws are

$$\sigma = 0\,,\; B \in -\frac{L}{T_0}(T-T_0) + \partial h(\beta)\,,\; \mathbf{H} = 0,$$

$$\mathbf{Q} = -\frac{\lambda}{T}\mathrm{gradT},$$

$$s = C(1+\ln T) + \frac{L}{T_0}\beta.$$

Subdifferential set $\partial h(\beta)$ is

$$\partial h(\beta) = \begin{cases} \frac{dh}{d\beta}(0) + \partial I(0),\; if\; \beta = 0, \\ \frac{dh}{d\beta}(\beta),\; if\; 0 < \beta < 1, \\ \frac{dh}{d\beta}(1) + \partial I(1),\; if\; \beta = 1, \end{cases} \tag{4.13}$$

with $\partial I(0) = \mathbb{R}^-$ and $\partial I(0) = \mathbb{R}^+$.

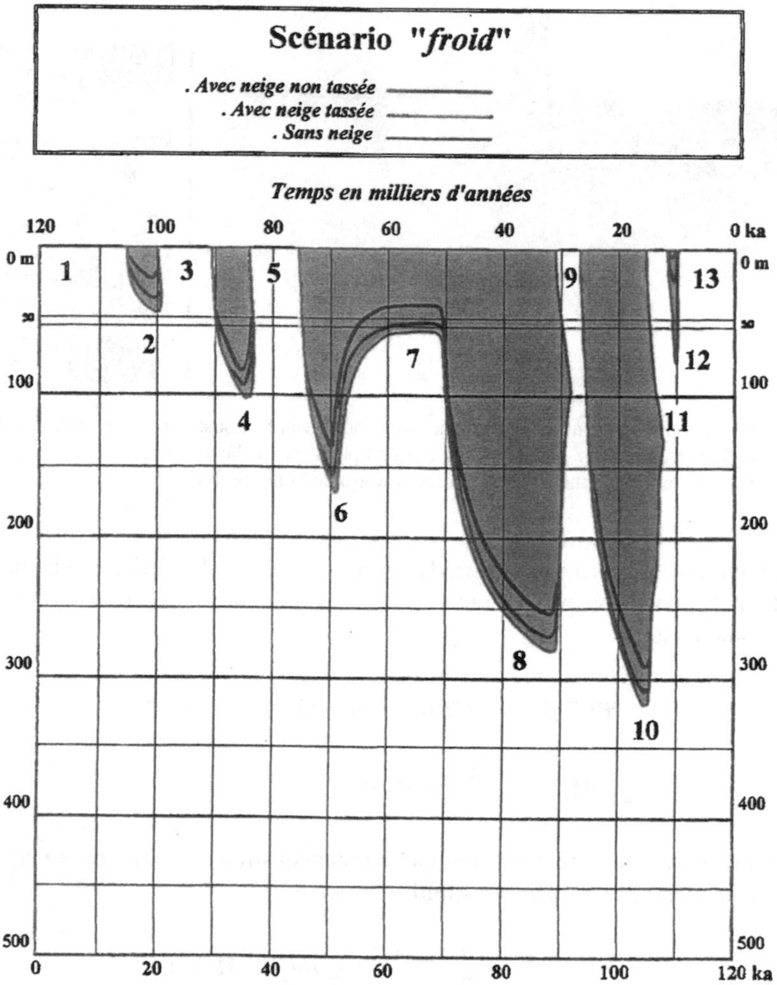


**Fig. 4.2** The extension of permafrost in the last 120,000 years at some place in France. The $0°C$ curves in the time-depth plane. The results depend on the snow cover on the ground (three scenarii have been investigated). The blue domain is the time-depth domain where the temperature is negative. One may note than during one of the cold periods the soil was frozen up to 300 m

## *4.3.1 The Equations*

The equations resulting from the equations of motion, balance of energy and constitutive laws are outside the free boundary $\Gamma(t)$

$$C\frac{\partial \ln T}{\partial t} + \frac{L}{T_0}\frac{\partial \beta}{\partial t} - \lambda \Delta \ln T = R\,,\ in\ \Omega\backslash\Gamma(t),$$

$$-\frac{L}{T_0}(T - T_0) + \partial h(\beta) \ni 0\,,\ in\ \Omega\backslash\Gamma(t), \tag{4.14}$$

$$\lambda \frac{\partial \ln T}{\partial N} = \pi \text{ , } on\ \partial\Omega,$$

$$T(\mathbf{x},0) = T^0(\mathbf{x}) \text{ , } \beta(\mathbf{x},0) = \beta^0(\mathbf{x}) \text{ , } in\ \Omega,$$

and on the free boundary $\Gamma(t)$ (which exists if $h(\beta)$ is constant on some interval) where

$$[\Psi] = 0,$$

which implies that $T - T_0 = 0$,

$$\rho W \frac{L}{T_0}[\beta] = \left[\frac{\lambda}{T}\frac{\partial T}{\partial N}\right] = \left[\lambda \frac{\partial \ln T}{\partial N}\right].$$

Equation (4.14) which results from the equation of microscopic motion, gives the temperature versus the unfrozen water content. Thus function $h$ may be measured with experiments, [112]. Measurement of unfrozen water content versus temperature, [87, 88], gives function $(dh/d\beta)(\beta)$ depending on $T$ (relationship (4.14)).

### 4.3.2 An Example

An example in Fig. 4.3 shows the temperature in the ground surrounding a decommissioned natural gas storage. The CESARGEL computer program, [141], involves

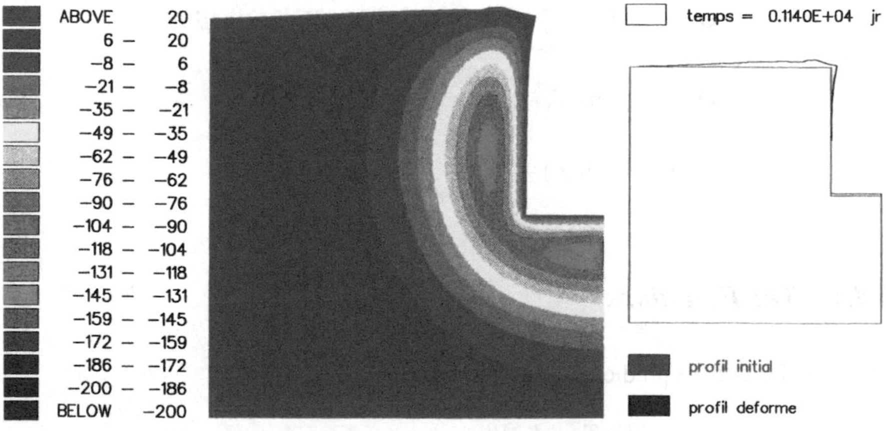


**Fig. 4.3** The temperature in the ground around a decommissioned liquefied natural gas storage, 3 years after the removing of the liquefied gas. Part of the soil is still frozen. The soil has heaved due to the cryogenic suction inducing water motion, [33]. The temperature scale in C is given on the left: red color is warm and blue is cold

function $h(\beta)$. Moreover, a computation of the frost heave induced by the water motion is shown. Due to the cryogenic suction induced by freezing, a displacement of the ground up to 0.5 m occurs, [33, 112]. One may remark that ground freezing and thawing are slow phenomena, the Fig. 4.3 shows the temperature 11,340 days, more than 3 years, after the removing of the natural liquefied gas.

## 4.4 Phase Change with a Freezing Fringe

In some phase changes, the phenomena are spread over a small zone around the surface where the temperature is equal to the phase change temperature $T_0$. This freezing fringe results from spatial interactions. Thus the gradient of $\beta$ intervene and we choose, [112]

$$\Psi(T,\beta,\text{grad}\beta) = -CT\ln T - \beta\frac{L}{T_0}(T-T_0) + I(\beta) + \frac{k}{2}(\text{grad}\beta)^2,$$

$$\Phi(\text{gradT},\text{T}) = \frac{\lambda}{2T}(\text{gradT})^2.$$

Parameter $k$ measures the intensity of the local interactions: when $k$ is large the freezing fringe is widely spread, when $k$ is small, the freezing fringe is narrow and tends to be a surface in dimension three, a line in dimension two. The term involving grad$\beta$ is related to interface properties which influence the phase change phenomenon. For instance, a free energy involving quantity grad$\beta$ takes into account surface tension effects. The constitutive laws are

$$\sigma = 0\,,\ B \in -\frac{L}{T_0}(T-T_0) + \partial I(\beta)\,,\ \mathbf{H} = k\text{grad}\beta,$$

$$\mathbf{Q} = -\frac{\lambda}{T}\text{gradT},$$

$$s = C(1+\ln T) + \frac{L}{T_0}\beta.$$

### 4.4.1 *The Equations*

The equations of the predictive theory are

$$C\frac{\partial \ln T}{\partial t} + \frac{L}{T_0}\frac{\partial \beta}{\partial t} - \lambda\Delta\ln T = R\,,\ in\ \Omega,$$

$$-k\Delta\beta - \frac{L}{T_0}(T-T_0) + \partial I(\beta) \ni 0\,,\ in\ \Omega,$$

$$\lambda \frac{\partial \ln T}{\partial N} = \pi \, , \; k \frac{\partial \beta}{\partial N} = 0 \, , \; on \; \partial \Omega,$$

$$T(\mathbf{x},0) = T^0(\mathbf{x}) \, , \; \beta(\mathbf{x},0) = \beta^0(\mathbf{x}) \, , \; in \; \Omega.$$

In this model there is no free boundary. The phase change is spread in the freezing fringe. The indicator function $I(\beta)$ may be replaced by a function $h(\beta)$.

## 4.5 Phase Change with Voids and Bubbles

Bubbles or voids may appear in phase change. In frozen ice, it may be seen small bubbles. In cast iron voids or bubbles are present. In this situation there are two phase volume fractions $\beta_1$ and $\beta_2$, $\beta = (\beta_1, \beta_2)$, [124]. The two phases do not fill the whole volume. As for an example, we assume that $\beta_1$ is the liquid volume fraction, water for instance, and $\beta_2$ is the solid volume fraction, ice for instance. The voids volume fraction is $v = 1 - \beta_1 - \beta_2$. We assume, for the sake of simplicity, the same constant density $\rho$ (i.e., the material is incompressible) and the same velocity $\partial \mathbf{u}/\partial t$ for the two phases ($\mathbf{u}$ is the small displacement). The mass balance is

$$\frac{\partial}{\partial t}(\beta_1 + \beta_2) + (\beta_1 + \beta_2)\,\mathrm{div}\mathbf{U} = 0. \tag{4.15}$$

The equations of motion for the microscopic motions responsible of the evolution of $\beta_1$ and $\beta_2$ and for the macroscopic motion resulting from the body and surface exterior forces $\mathbf{f}$ and $\mathbf{g}$ are

$$-\begin{pmatrix} B_1 \\ B_2 \end{pmatrix} + \mathrm{div}\begin{pmatrix} \mathbf{H}_1 \\ \mathbf{H}_0 \end{pmatrix} = 0 \, , \; \text{in } \Omega, \; \begin{pmatrix} \mathbf{H}_1 \\ \mathbf{H}_0 \end{pmatrix} \cdot \mathbf{N} = 0 \, , \; \text{on } \partial\Omega,$$

$$\mathrm{div}\sigma + \mathbf{f} = 0 \, , \; \text{in } \Omega \, , \; \sigma \cdot \mathbf{N} = \mathbf{g} \, , \; \text{on } \partial\Omega,$$

where works $B_1, B_2$, work flux vectors $\mathbf{H}_1$, $\mathbf{H}_2$ and stress $\sigma$ are the interior forces. We assume that the evolution is quasistatic and that there is no exterior source of work producing the phase change.

The volumic free energy we choose is

$$\Psi(T, \varepsilon, \beta, \mathrm{grad}\beta) = \sum_{\mathrm{j}=1}^{2} \beta_{\mathrm{j}} \Psi_{\mathrm{j}} + \mathrm{I}_{\mathrm{K}}(\beta) + \frac{\mathrm{k}}{2} |\mathrm{grad}\beta|^2,$$

where $\Psi_j$ denotes the free energy of phase $_j$, $I_K$ is the indicator function (see Appendix A) of the convex set

$$K = \{(\beta_1, \beta_2) \,|\, 0 \le \beta_1 \le 1, \, 0 \le \beta_2 \le 1, \, \beta_1 + \beta_2 \in [0,1]\}.$$

The term $I_K(\beta)+(k/2)\,|\mathrm{grad}\beta|^2$ is an *interaction or mixture free energy*. The effect of $I_K(\beta)$ is to guarantee that the fractions $\beta_1$ and $\beta_2$ take admissible physical values.

Let us note that even if the free energy of the voids phase is 0, the voids phase has physical properties due to the mixture free energy which depends on the gradients of $\beta_1$ and of $\beta_2$. The gradients are related to the interfaces properties: $\mathrm{grad}\beta_1$ describes properties of the voids–liquid interface and $\mathrm{grad}\beta_2$ describes properties of the voids–solid interface, for instance surface tension. In this setting, the voids have a role in the phase change and make it different from a phase change without voids. The model is simple and schematic but it may be upgraded by introducing sophisticated interaction free energy depending on $\beta$ and on $\mathrm{grad}\beta$.

Moreover, for the free energies of the phases, we consider the following simplified expressions

$$\Psi_1(T,\varepsilon) = -CT\ln T - \frac{L}{T_0}(T-T_0) + \mathscr{L}(T),$$

$$\Psi_2(T,\varepsilon) = -CT\ln T + \frac{1}{2}\{\lambda_e(tr\varepsilon)^2 + 2\mu_e\varepsilon:\varepsilon\} + \mathscr{L}(T),$$

where $\lambda_e$ and $\mu_e$ are the elasticity Lamé parameters. The linear function $\mathscr{L}(T)$ is defined by

$$\mathscr{L}(T) = CT_0\ln T_0 + C(1+\ln T_0)(T-T_0).$$

It has been chosen in such a way that within the small perturbation assumption when $T-T_0=\theta$, with $\theta$ small compared to $T_0$, the quantity $-CT\ln T+\mathscr{L}(T)$ is equivalent to its leading term

$$-CT\ln T + \mathscr{L}(T) \simeq \frac{C}{2T_0}(T-T_0)^2. \tag{4.16}$$

*Remark 4.3.* The free energy of a material is defined up to a linear function of the temperature. The free energies $\Psi_1$ and $\Psi_2$ are the free energies of the phases of the same material. Then they are defined up to any linear function of $T$ but, of course, this linear function has to the same for $\Psi_1$ and $\Psi_2$.

Thus the volumic free energy is

$$\Psi(T,\varepsilon,\beta,\mathrm{grad}\beta) = (\beta_1+\beta_2)(-\mathrm{CT}\ln\mathrm{T} + \mathscr{L}(\mathrm{T}))$$
$$-\frac{\beta_1 L}{T_0}(T-T_0) + \frac{\beta_2}{2}\{\lambda_e(tr\varepsilon)^2 + 2\mu_e\varepsilon:\varepsilon\} + I_K(\beta) + \frac{k}{2}|\mathrm{grad}\beta|^2.$$

For the pseudopotential of dissipation with quantity $\chi = (T,\beta_1,\beta_2)$, we choose

$$\Phi(\mathrm{gradT},\dot{\varepsilon},\frac{\partial\beta}{\partial \mathrm{t}},\mathrm{T},\beta_1,\beta_2)$$

$$=\frac{(\beta_1+\beta_2)\lambda}{2T}(\mathrm{gradT})^2+\frac{\mathrm{c}}{2}\left(\frac{\partial\beta}{\partial \mathrm{t}}\right)^2+\frac{\beta_1}{2}\{\lambda_{\mathrm{v}}(\mathrm{tr}\dot{\varepsilon})^2+2\mu_{\mathrm{v}}\dot{\varepsilon}:\dot{\varepsilon}\}$$

$$+I_0(\frac{\partial}{\partial t}(\beta_1+\beta_2)+(\beta_1+\beta_2)\,\mathrm{div}\mathbf{U}), \tag{4.17}$$

where $\dot{\varepsilon}=\partial\varepsilon/\partial t$ and $I_0$ is the indicator function of the origin. The viscosity parameters of the fluid are $\lambda_v$ and $\mu_v$. The last term in (4.17) is zero if mass balance (4.15) is satisfied and it is $+\infty$ otherwise. In other words we may say that the presence of the last term in (4.17) is due to the fact that mass balance (4.15) is an internal constraint between velocities and it must be included in the expression of the pseudopotential of dissipation $\Phi$. Let us recall that the pseudopotential accounts for the properties of the velocities. We will see that this term is related to the *pressure* in the system. Parameter $c$ is the viscosity parameter of the microscopic motion. The constitutive laws are

$$\sigma=\beta_1\left(\lambda_v(tr\dot{\varepsilon})1+2\mu_v\dot{\varepsilon}\right)+\beta_2\left(\lambda_e(tr\varepsilon)1+2\mu_e\varepsilon\right)-(\beta_1+\beta_2)\,p\,1,$$

$$\begin{pmatrix}B_1\\B_2\end{pmatrix}\in\begin{pmatrix}-\frac{L}{T_0}(T-T_0)\\ \frac{1}{2}\{\lambda_e(tr\varepsilon)^2+2\mu_e\varepsilon:\varepsilon\}\end{pmatrix}+\begin{pmatrix}-CT\ln T+\mathscr{L}(T)\\-CT\ln T+\mathscr{L}(T)\end{pmatrix}$$

$$+\partial I_K(\beta)+c\frac{\partial\beta}{\partial t}-p\begin{pmatrix}1\\1\end{pmatrix},$$

$$\mathbf{H}_i=k\,\mathrm{grad}\beta_{\mathrm{i}}\,,\ \mathbf{Q}=-\frac{\lambda\,(\beta_1+\beta_2)}{\mathrm{T}}\mathrm{gradT},$$

$$s=(\beta_1+\beta_2)\,C\ln\frac{T}{T_0}+\frac{L}{T_0}\beta_1,$$

where 1 is the identity matrix, $p$ is the pressure defined by

$$-p\in\partial I_0(\frac{\partial}{\partial t}(\beta_1+\beta_2)+(\beta_1+\beta_2)\mathrm{div}\mathbf{U}).$$

This constitutive law implies that the mass balance is satisfied because the subdifferential set

$$\partial I_0\left(\frac{\partial}{\partial t}(\beta_1+\beta_2)+(\beta_1+\beta_2)\,\mathrm{div}\mathbf{U}\right),$$

is not empty. The stress $\sigma$ is an elastic stress in the solid part and a viscous stress in the fluid.

The small perturbation assumption for $T$ and $\beta_1+\beta_2$ are: $\theta = T - T_0$ is small compare to $T$ and $(\beta_1+\beta_2)-(\beta_1(0)+\beta_2(0))$ is small compare to $(\beta_1(0)+\beta_2(0))$. For the sake of simplicity, we assume there is no voids at the initial time: $\beta_1(0)+\beta_2(0)=1$. With the small perturbation assumption, the constitutive laws are

$$\sigma = \beta_1\left(\lambda_v(tr\dot{\varepsilon})1+2\mu_v\dot{\varepsilon}\right)+\beta_2\left(\lambda_e(tr\varepsilon)1+2\mu_e\varepsilon\right)-p1,$$

$$\begin{pmatrix} B_1 \\ B_2 \end{pmatrix} \in \begin{pmatrix} -\frac{L}{T_0}\theta \\ 0 \end{pmatrix} + \partial I_K(\beta)+c\frac{\partial\beta}{\partial t}-p\begin{pmatrix} 1 \\ 1 \end{pmatrix},$$

$$\mathbf{H}_i = k\mathrm{grad}\beta_\mathrm{i}\ ,\ \mathbf{Q} = -\frac{\lambda}{\mathrm{T}}\mathrm{gradT},$$

$$s = C(1+\frac{\theta}{T_0})+\frac{L}{T_0}\beta_1,$$

where the deformations $\varepsilon$ and temperatures $\theta$ at power 2 have been neglected in agreement with the small deformation assumption (see Appendix B).

*Remark 4.4.* An other way to have simple constitutive laws is to assume that the voids are filled with vapor with heat capacity $C$ and thermal conductivity $\lambda$. The functions $\Psi$ and $\Phi$ become

$$\Psi(T,\varepsilon,\beta,\mathrm{grad}\beta) = -CT\ln T - \frac{\beta_1 L}{T_0}(T-T_0)+\frac{\beta_2}{2}\{\lambda_e(tr\varepsilon)^2+2\mu_e\varepsilon:\varepsilon\}$$
$$+I_K(\beta)+\frac{k}{2}|\mathrm{grad}\beta|^2,$$

and

$$\Phi(\mathrm{gradT},\dot{\varepsilon},\frac{\partial\beta}{\partial \mathrm{t}},\mathrm{T},\beta_1,\beta_2)$$
$$= \frac{\lambda}{2T}(\mathrm{gradT})^2+\frac{\mathrm{c}}{2}\left(\frac{\mathrm{d}\beta}{\mathrm{dt}}\right)^2+\frac{\beta_1}{2}\{\lambda_\mathrm{v}(\mathrm{tr}\dot{\varepsilon})^2+2\mu_\mathrm{v}\dot{\varepsilon}:\dot{\varepsilon}\}$$
$$+I_0(\frac{\partial}{\partial t}(\beta_1+\beta_2)+(\beta_1+\beta_2)\,\mathrm{div}\mathbf{U}).$$

They give the following constitutive laws

$$\sigma = \beta_1\left(\lambda_v(tr\dot{\varepsilon})1+2\mu_v\dot{\varepsilon}\right)+\beta_2\left(\lambda_e(tr\varepsilon)1+2\mu_e\varepsilon\right)-(\beta_1+\beta_2)\,p1,$$

$$B \in \begin{pmatrix} -\frac{L}{T_0}(T-T_0) \\ \frac{1}{2}\{\lambda_e(tr\varepsilon)^2+2\mu_e\varepsilon:\varepsilon\} \end{pmatrix} + \partial I_K(\beta)+c\frac{\partial\beta}{\partial t}-p\begin{pmatrix} 1 \\ 1 \end{pmatrix},$$

$$\mathbf{H}_i = k\mathrm{grad}\beta_{\mathrm{i}}\ ,\ \mathbf{Q} = -\frac{\lambda}{\mathrm{T}}\mathrm{gradT},$$

$$s = C(1+\ln T) + \frac{L}{T_0}\beta_1.$$

## 4.5.1 The Equations

The equations of the predictive theory are

$$\frac{C}{T_0}\frac{\partial \theta}{\partial t} + \frac{L}{T_0}\frac{\partial \beta_1}{\partial t} - \frac{\lambda}{T_0}\Delta\theta = R\ ,\ in\ \Omega,$$

$$c\frac{\partial \beta}{\partial t} - k\Delta\beta + \partial I_K(\beta) + \begin{pmatrix} -\frac{L}{T_0}\theta \\ 0 \end{pmatrix} - p\begin{pmatrix} 1 \\ 1 \end{pmatrix} \ni 0\ ,\ in\ \Omega,$$

$$\mathrm{div}\left(\beta_1\{\lambda_{\mathrm{v}}(\mathrm{tr}\varepsilon(\frac{\partial \mathbf{u}}{\partial \mathrm{t}}))1 + 2\mu_{\mathrm{v}}\varepsilon(\frac{\partial \mathbf{u}}{\partial \mathrm{t}})\} + \beta_2\{\lambda_{\mathrm{e}}\mathrm{tr}\varepsilon(\mathbf{u})1 + 2\mu_{\mathrm{e}}\varepsilon(\mathbf{u})\}\right)$$

$$-\mathrm{gradp} + \mathbf{f} = 0\ ,\ \mathrm{in}\ \Omega,$$

$$\frac{\partial}{\partial t}(\beta_1 + \beta_2) + \mathrm{div}\frac{\partial \mathbf{u}}{\partial \mathrm{t}} = 0\ ,\ \mathrm{in}\ \Omega,$$

$$\lambda\frac{\partial \theta}{\partial N} = T_0\pi\ ,\ k\frac{\partial \beta}{\partial N} = 0\ ,\ \sigma\mathbf{N} = \mathbf{g}\ ,\ on\ \partial\Omega,$$

$$\theta(\mathbf{x},0) = \theta^0(\mathbf{x})\ ,\ \beta(\mathbf{x},0) = \beta^0(\mathbf{x})\ ,\ \mathbf{u}(\mathbf{x},0) = \mathbf{u}^0(\mathbf{x})\ ,\ in\ \Omega. \tag{4.18}$$

Let us note that the pressure intervene in both equations for macroscopic and microscopic motions which are coupled. Mathematical results may be found in [124].

## 4.5.2 The Case Where There Is Only Liquid: The Cavitation Phenomenon

Let us assume that there are only water and voids, $0 < \beta_1 < 1$, $\beta_2 = 0$ and interpret equation (4.18) assuming an homogeneous state. We have

$$\partial I_K(\beta_1, 0) = \begin{pmatrix} 0 \\ -Q \end{pmatrix},$$

with $Q \geq 0$. Equation (4.18) is

$$c\begin{pmatrix} \frac{\partial \beta_1}{\partial t} \\ 0 \end{pmatrix} + \begin{pmatrix} 0 \\ -Q \end{pmatrix} + \begin{pmatrix} -\frac{L}{T_0}\theta \\ 0 \end{pmatrix} - p\begin{pmatrix} 1 \\ 1 \end{pmatrix} \ni 0.$$

It results

$$p = -Q < 0,$$
$$c\frac{\partial \beta_1}{\partial t} = p + \frac{L}{T_0}\theta.$$

The pressure is negative in the water. If $p + (L/T_0)\theta < 0$,

$$\frac{\partial \beta_1}{\partial t} < 0.$$

The voids volume fraction increases and from the mass balance we get

$$\mathrm{div}\mathbf{U} > 0.$$

This is the cavitation phenomenon which results from the pressure becoming negative.

Let us also note that if

$$p + \frac{L}{T_0}\theta \geq 0\,,\; \theta \geq 0,$$

we may have only liquid at an equilibrium: $\beta_1 = 1$ and $(\partial \beta_1/\partial t) = 0$. Thus large temperature and large pressure make the liquid to be the only phase present at an equilibrium.

### 4.5.3 *The Case There Is Only Solid: Soil Mechanics*

In case there is only solid and voids, $\beta_1 = 0$, $0 < \beta_2 < 1$, (4.18) becomes

$$c\begin{pmatrix} 0 \\ \frac{\partial \beta_2}{\partial t} \end{pmatrix} + \begin{pmatrix} -P \\ 0 \end{pmatrix} + \begin{pmatrix} -\frac{L}{T_0}\theta \\ 0 \end{pmatrix} - p\begin{pmatrix} 1 \\ 1 \end{pmatrix} \ni 0.$$

We get that the voids volume fraction either increases or decreases if the pressure is either low or large. This is a property of soils mechanics, [162].

Let us also note that if

$$p \geq 0\,,\; \theta \leq 0,$$

we may have only solid at an equilibrium: $\beta_2 = 1$ and $(\partial\beta_2/\partial t) = 0$. Thus non negative pressure and non positive temperature make the solid to be the only phase present at an equilibrium.

We conclude that the pressure $p$ is the main practical tool to govern the voids volume fraction and that the temperature $T$ is the main practical tool to govern the phase change.

### 4.5.4 Numerical Examples

We investigate the evolution of a one-dimensional continuous body occupying domain $(-h/2, h/2)$. The numerical results are due to Francesco Ascione, [12]. We denote $w(z,t)$, the vertical displacement; $\theta(z,t) = T(z,t) - T_0$, the difference between the absolute temperature and phase change temperature (one may think of the Celsius temperature); $\beta_i(z,t)$ the phase volume fractions, with $i = 1$ for liquid phase volume fraction and $i = 2$ for solid phase volume fraction; $p(z,t)$, the pressure. The exterior forces applied to the structure are $F$, the vertical body force and $G$, the surface traction on the top of the structure; $R(z,t)$, is the volume entropy source and $\pi_{top} = \pi(z = h/2, t)$, $\pi_{bottom} = \pi(z = -h/2, t)$ are the surface entropy flows. Quantities $R(z,t)T_0$ and $\pi(z = \pm h/2, t)T_0$ are the volume and surface heat sources. We assume the solid is fixed on its bottom face.

We investigate the influence of the physical parameters which intervene in the phase change phenomenon: in example 1, the influence of the external load when cavitation occurs; in examples 2 and 3, the effect of microscopic motion parameter $k$ which accounts for spatial interactions; in examples 4 and 5, the influence of the other microscopic motion parameter, $c$ which accounts for the dissipative effects; finally, in example 6, the influence of pressure $p$.

For all the examples we assume the following values of the boundary conditions:

$$\text{at top},\ z = \frac{h}{2},\ \lambda\frac{\partial\theta}{\partial z} = T_0\pi_{top},\ k\frac{\partial\beta_1}{\partial z} = 0,\ k\frac{\partial\beta_2}{\partial z} = 0,\ \sigma_{zz} = G,$$

$$\text{at bottom},\ z = -\frac{h}{2},\ -\lambda\frac{\partial\theta}{\partial z} = T_0\pi_{bottom},\ k\frac{\partial\beta_1}{\partial z} = 0,\ k\frac{\partial\beta_2}{\partial z} = 0,\ w = 0.$$

The initial conditions are for all the examples

$$\theta(z,0) = 0,\ \beta_1(z,0) = 0,\ \beta_2(z,0) = 1,\ w(z,0) = 0,$$

except in the example 1 (cavitation) where the volume fraction $\beta_1$ and $\beta_2$ have inverted values. The thickness is $h = 1\,\text{m}$, the surface entropy fluxes are $\pi_{top} = 0.5\ \text{W}/(\text{m}^2\ \text{K})$, $\pi_{bottom} = -0.05\ \text{W}/(\text{m}^2\ \text{K})$, phase change temperature is $T_0 = 273.15$ K, the other mechanical and thermal parameters have usual values, elasticity parameters $\lambda_e = 10.4\ 10^9$ Pa, $\mu_e = 3.9\ 10^9$ Pa, viscosity parameters

$\lambda_v = 1.00\ 10^9$ Pa s, $\mu_v = 1.827\ 10^9$ Pa s, latent heat $3.35\ 10^8\ \mathrm{J/m^3}$, heat capacity $4.186\ 10^6\ \mathrm{J/(kg\ K)}$ and thermal conductivity $2.23\ \mathrm{W/(K\,m)}$. The top tension $G$, phase change viscosity $c$ and local influence parameter $k$ change in the different examples. Note that the thermal action mostly heats the structure which is frozen at initial time (except for the cavitation experiment where it is liquid water).

### 4.5.5 *Example 1: The Cavitation Phenomenon*

We investigate the cavitation phenomenon: an increase of voids volume fraction and a consequent decrease of the water volume fraction due to a low pressure resulting from a tension applied at the top of the structure. The parameters are, respectively: $G = 1.01\ 10^5$Pa ($G$ is a tension), $k = 10^7$ J/m, $c = 10^7 (\mathrm{J\ s})/\mathrm{m}^3$.

The numerical results are reported in Fig. 4.4a–e where it is shown the evolution of $\theta$, $v$, $\beta_1$, $w$ and $p$.

As expected solid volume fraction remains null and the voids volume fraction increases due to the low pressure. The voids and water volume fractions at the two ends of the structure at final time are: at bottom $z = -h/2$, $v = 0.278283$, $\beta_1 = 0.721362$, and at top $z = h/2$, $v = 0.411954$, $\beta_1 = 0.588632$.

### 4.5.6 *Example 2: The Non Dissipative Spatial Interaction Is Important*

To emphasize the effect of the interaction parameter $k$, we choose an extreme large value for this parameter, almost equal to $10^{20}$ J/m. The other two variable data are $c = 10^7\ (\mathrm{Js})/\mathrm{m}^3$ and $G = -1.01\ 10^5$ Pa. Volume fractions $\beta_1$ and $\beta_2$ are shown in Fig. 4.5a, b.

Functions $z \rightarrow \beta_i(z)$ become constant. Note that at the both ends of the structure the two volume fractions have the same values: at bottom $z = -h/2$, $\beta_1 = 0.003632$ and $\beta_2 = 0.9963$ and at top $z = h/2$, $\beta_1 = 0.003728$ and $\beta_2 = 0.996296$.

A large spatial interaction parameter makes the evolution of the $\beta_i$'s very dependent of the neighbourhood. On the contrary a very low interaction parameter makes the evolution decoupled and independent of the neighbourhood, as shown in the following section.

### 4.5.7 *Example 3: No Spatial Interaction*

In opposition to previous example 2, parameter $k$ is now almost null, $k = 10^{-20}$ J m. The other data are those of the previous example. It results that the evolution of the

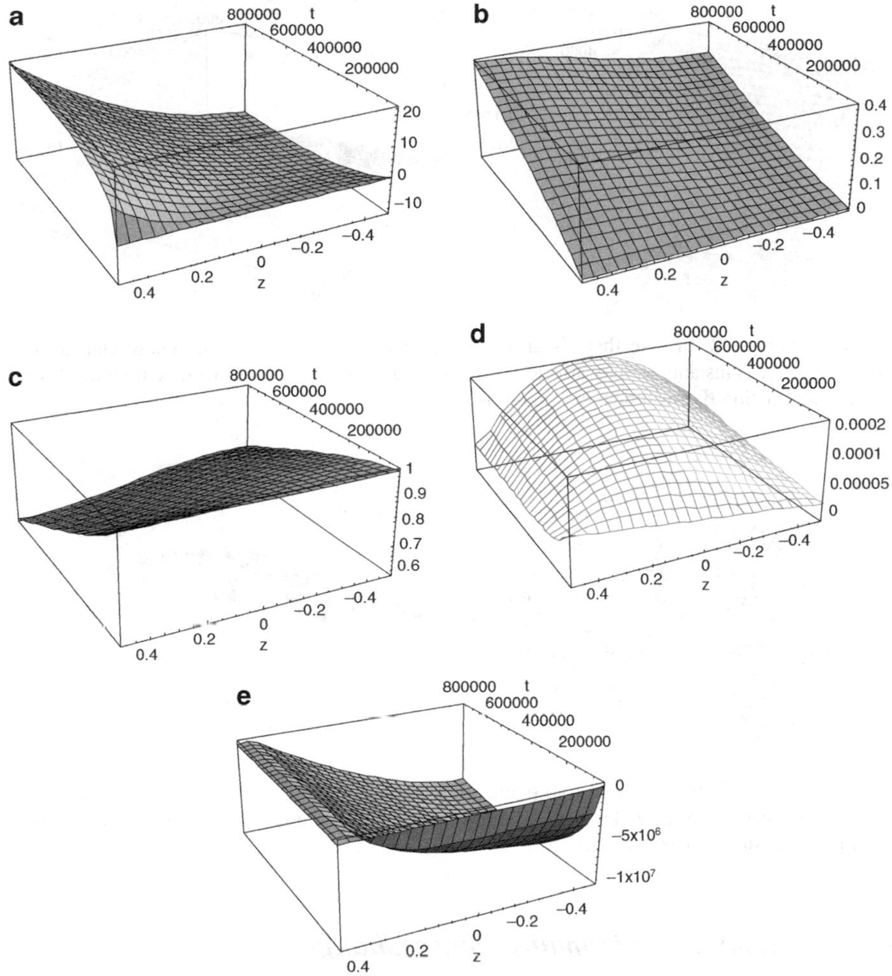


**Fig. 4.4** Example 1. The cavitation phenomenon is due to low pressure $p$ shown in (**e**), resulting in creation of voids (voids volume fraction is shown in (**b**)). Temperature $\theta$, water volume fraction $\beta_1$, displacement $w$ are shown in (**a**), (**c**) and (**d**). Note that the vertical scale of figures giving $\beta_i$ versus $z$ and $t$ are not the same in the figures of this section

$\beta_i$'s are independent of their neighbourhood. Because temperature $\theta$ governing the evolution of the $\beta_i$'s, is smooth with respect to space, the $\beta_i$'s are also smooth with respect to space. Water and ice volume fractions are shown in Fig. 4.6a, b.

At the two ends of the structure, the volume fractions have independent values: at bottom $z = -h/2$, $\beta_1 = 0.278283$ and $\beta_2 = 0.721362$, and at top $z = h/2$, $\beta_1 = 0.411954$ and $\beta_2 = 0.588632$.

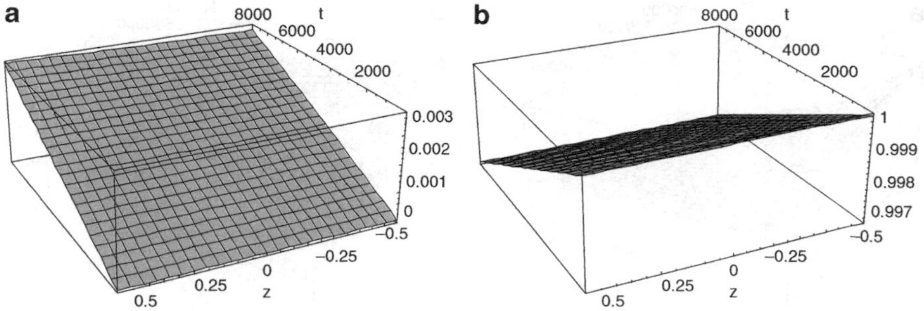


**Fig. 4.5** Example 2. In case there is an important spatial interaction in the phase change, the whole structure is instantaneously affected by the boundary actions. Water volume fraction $\beta_1$ and ice volume fraction $\beta_2$ are shown in (**a**) and (**b**)

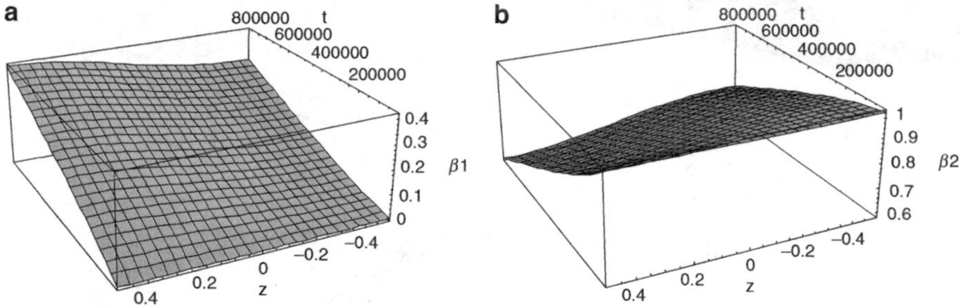


**Fig. 4.6** Example 3. In case there is no spatial interaction in the phase change, the whole structure is not instantaneously affected by the boundary actions. Water volume fraction $\beta_1$ and ice volume fraction $\beta_2$ are shown in (**a**) and (**b**)

### 4.5.8 Example 4: Dissipative Phase Change

In these next two sections we investigate the effect of the dissipative microscopic motion parameter $c$. It controls the velocity of the phase change. For this purpose, we assume firstly it has a very large value, $c = 10^{20}$ J s$/$m$^3$. The other two variable data, $k$ and $G$ have the following values, respectively: $k = 10^7$ J$/$m and $G = -1.01\ 10^5$ Pa. Volume fractions $\beta_1$ and $\beta_2$ are shown in Fig. 4.7a, b.

As expected, functions $t \to \beta_i(t)$ are constant. For instance, we have: at bottom $z = -h/2$, $\beta_1 = 0.00154885$, $\beta_2 = 0.998433$, and at top $z = h/2$, $\beta_1 = 0.00711434$, $\beta_2 = 0.992941$.

### 4.5.9 Example 5: Non Dissipative Phase Change

Contrary to the previous case, we choose $c$ to be almost null with value, $c = 10^{-20}$ (J $s$)$/$m$^3$, and keep the values of $k$ and $G$ as in example 4. In agreement with

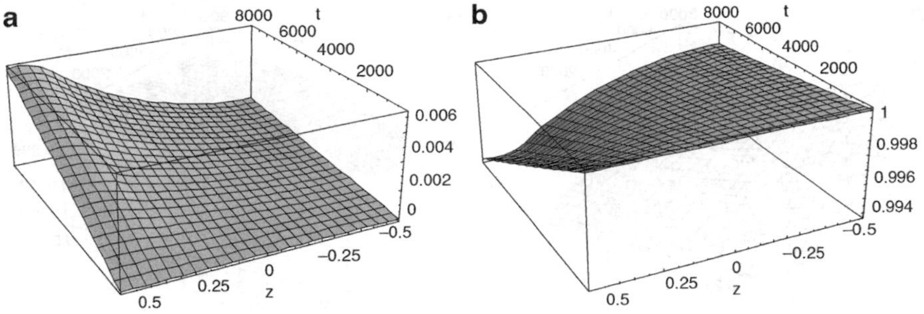


**Fig. 4.7** Example 4. A dissipative phase change is slow. Water volume fraction $\beta_1$ and ice volume fraction $\beta_2$ are shown in (**a**) and (**b**)

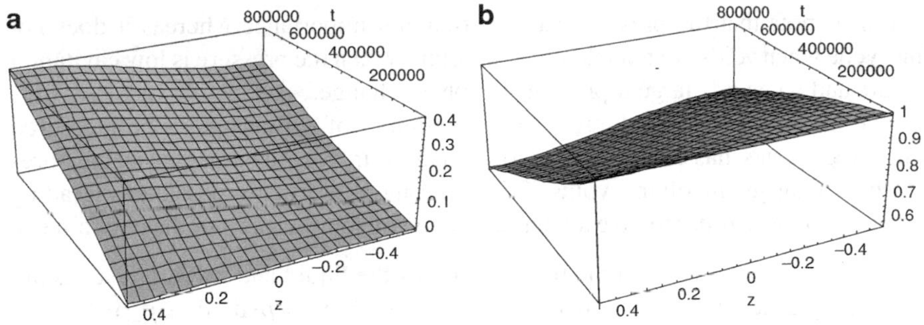


**Fig. 4.8** Example 5. Phase change is rapid in case it is not dissipative. Water volume fraction $\beta_1$ and ice volume fraction $\beta_2$ are shown in (**a**) and (**b**)

the predictive theory, phase change is almost instantaneous as seen in Fig. 4.8a, b. The corresponding volume fractions $\beta_1$ and $\beta_2$ are: at bottom $z = -h/2$, $\beta_1 = 0.278283$, $\beta_2 = 0.721362$, and at top $z = h/2$, $\beta_1 = 0.411954$, $\beta_2 = 0.588632$.

As $c$ decreases, the velocity of the phase change increases.

### *4.5.10 Example 6: Effect of the Pressure*

In this last example, we investigate the effect of the pressure $p$ on the phase change phenomenon. We assume parameters $k$ and $c$ have the following values, respectively: $10^7$ J/m and $10^7$ (J s)/m$^3$. We apply a very large compression at the top of the structure with value $G = -1.01\ 10^5$ Pa.

In agreement with the predictive theory, very high pressure prevents ice to be transformed into water even if temperatures are positive, as shown in Fig. 4.9a, b. The volume fractions have the following values: at bottom $z = -h/2$, $\beta_1 = 0.001500$, $\beta_2 = 0.998433$, and at top $z = h/2$, $\beta_1 = 0.007114$, $\beta_2 = 0.992941$.

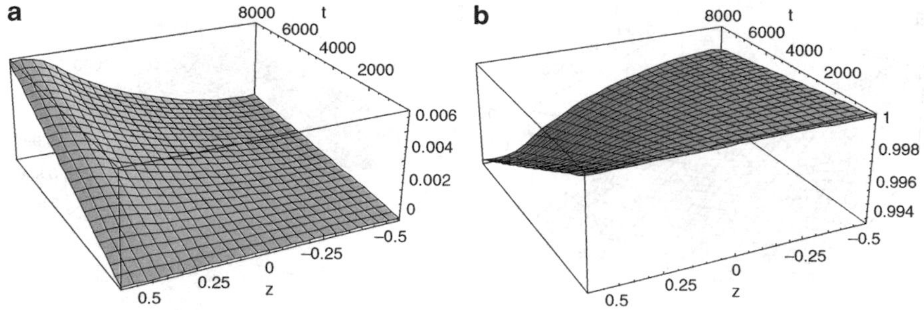


**Fig. 4.9** Example 6. High pressure prevents phase change. Water volume fraction $\beta_1$ and ice volume fraction $\beta_2$ are shown in (**a**) and (**b**)

Let us note that the pressure has a paramount importance whereas it does not intervene when voids are not taken into account. When the pressure is low cavitation occurs and when it is large it prevents the phase change.

The predictive theory which spares the number of parameters to be measured with experiments, has a scope large enough to account for the basic physical features of phase changes involving voids. Of course there are many possible upgrading sophistications in order to give a finer prediction of some aspects of the phenomena.

*Remark 4.5.* In case we assume the densities of the liquid and solid to be constant but slightly different, the mass balance becomes with $\rho^0 = \rho_1\beta_1(0) + \rho_2\beta_2(0)$

$$\frac{\partial}{\partial t}(\rho_1\beta_1 + \rho_2\beta_2) + \rho^0 \mathrm{div}\mathbf{U} = 0.$$

In case we have no voids and a mixture of ice and water, $\beta_1 + \beta_2 = 1$, at an equilibrium, we have

$$\partial I_K(\beta_1, \beta_2) = \begin{pmatrix} P \\ P \end{pmatrix},$$

with $P \geq 0$. It results that

$$\begin{pmatrix} P \\ P \end{pmatrix} + \begin{pmatrix} -\frac{L}{T_0}\theta \\ 0 \end{pmatrix} - p\begin{pmatrix} \rho_1 \\ \rho_2 \end{pmatrix} \ni 0,$$

or

$$-\frac{L}{T_0}\theta - p(\rho_1 - \rho_2) = 0, \; p \geq 0. \tag{4.19}$$

In this situation the phase change temperature depends on the pressure. It is known that this is actually the case for the ice–water phase change. Because $\rho_1 > \rho_2$ the phase change temperature decreases when the pressure increases. A consequence of this property is that the ice melts at the bottom of a glacier and lubricates the rock ice contact surface allowing the downhill motion of the glacier.

The previous relationship shows that the pressure is 0, when the phase change occurs at temperature 0°C. Thus the stress $-\rho^0 p1$ resulting from the mass balance constraint is the pressure with respect to the atmospheric pressure. This phenomenon is investigated in the following section where the macroscopic equation of motion is not as sophisticated as it is in this section.

## 4.6 Solid–Liquid Phase Change with Different Densities

Let us consider water in a rigid and impermeable container and freeze the container. Because density of ice $\rho_2$ is lower than density of water $\rho_1$ it seems impossible that water transforms into ice: the water remains liquid even at low temperature. Experiments, for instance the freezing of a glass bottle filled with water, show that the water pressure increases up to the rupture of the bottle. When the container is not impermeable, freezing may produce a non homogeneous material, for instance water ice or sorbet.

We investigate this problem, [125]. The theory takes into account the density difference between ice and water but the equation for the macroscopic motion is less sophisticated than that of the previous section: we assume the ice has zero velocity. Water velocity is $\mathbf{U}_1$.

### 4.6.1 The Mass Balance

It is

$$\rho_1 \frac{d\beta_1}{dt} + \rho_2 \frac{\partial \beta_2}{\partial t} + \rho_1 \beta_1 \mathrm{div}\mathbf{U}_1 = 0\,,\ in\ \Omega,$$

$$-\rho_1 \beta_1 \mathbf{U}_1 \cdot \mathbf{N} = \varpi\,,\ on\ \partial\Omega, \tag{4.20}$$

where $\varpi$ is the liquid water intake and $\mathbf{N}$ is the outward normal vector to domain $\Omega$ occupied by the ice water mixture.

### 4.6.2 The Equations of Motion

For the macroscopic motion of the water, it is

$$\rho_1 \frac{d(\beta_1 \mathbf{U}_1)}{dt} = \mathrm{div}\sigma_1 + \mathbf{f}_1\,,\ \text{in}\ \Omega,$$

$$\sigma_1 \mathbf{N} = \mathbf{g}_1\,,\ on\ \partial\Omega, \tag{4.21}$$

where $\sigma_l$ is the water stress, $\mathbf{f}_l$ is the action of the exterior on the liquid water, for instance the friction of the water on the solid ice phase and $\mathbf{g}_l$ is the boundary exterior action on the liquid water.

For the microscopic motions responsible for the ice water phase change, it is

$$-\begin{pmatrix} B_1 \\ B_2 \end{pmatrix} + \text{div}\begin{pmatrix} \mathbf{H}_1 \\ \mathbf{H}_2 \end{pmatrix} = 0\,,\ in\ \Omega,$$

$$\begin{pmatrix} \mathbf{H}_1 \\ \mathbf{H}_2 \end{pmatrix} \cdot \mathbf{N} = 0\,,\ on\ \partial\Omega, \tag{4.22}$$

assuming no exterior source of work.

### 4.6.3 The Entropy Balance

For the sake of simplicity, till the end of this section we assume the small perturbation assumption. In this setting the entropy balance is

$$\begin{aligned} &\frac{\partial s}{\partial t} + \text{div}\mathbf{Q} = \text{R} \\ &+\frac{1}{T}\left\{\sigma_l^d : D(\mathbf{U}_l) + B_1^d \frac{\partial \beta_1}{\partial t} + B_2^d \frac{\partial \beta_2}{\partial t}\right. \\ &\left. +\mathbf{H}_1^d \cdot \text{grad}\frac{\partial \beta_1}{\partial \text{t}} + \mathbf{H}_2^{\text{d}} \cdot \text{grad}\frac{\partial \beta_2}{\partial \text{t}} - \text{gradT}.\mathbf{Q}\right\}\,,\ in\ \Omega, \end{aligned} \tag{4.23}$$

where the internal forces are split into dissipative forces indexed by $^d$ and non dissipative interior forces indexed by $^{nd}$.

### 4.6.4 The Constitutive Laws

The non dissipative forces are defined with the water free energy

$$\begin{aligned} \sigma_l^{nd} &= 0, \\ B_1^{nd} &= \frac{\partial \Psi}{\partial \beta_1}\,,\ \mathbf{H}_1^{nd} = \frac{\partial \Psi}{\partial \text{grad}\beta_1}, \\ B_2^{nd} &= \frac{\partial \Psi}{\partial \beta_2}\,,\ \mathbf{H}_2^{nd} = \frac{\partial \Psi}{\partial \text{grad}\beta_2}. \end{aligned}$$

Free energy $\Psi$ takes into account the internal constraint on the state quantities, for instance

$$(\beta_1,\beta_2)\in K=\{(\gamma_1,\gamma_2)\,|0\le\gamma_1\le 1,\ 0\le\gamma_2\le 1,\ \gamma_1+\gamma_2\in[0,1]\}.$$

We choose

$$\Psi(\theta,\beta_1,\beta_2,\mathrm{grad}\beta_1,\mathrm{grad}\beta_2)=-\beta_1\frac{\mathrm{C}_1}{2\mathrm{T}_0}\theta^2-\beta_2\frac{\mathrm{C}_2}{2\mathrm{T}_0}\theta^2$$
$$-\frac{\beta_1 L}{T_0}\theta+I_K(\beta_1,\beta_2)+\frac{k}{2}(\mathrm{grad}\beta_1)^2+\frac{k}{2}(\mathrm{grad}\beta_2)^2,$$

where $\theta=T-T_0$ is the Celsius temperature assumed to be small. Thus we have within the small perturbation assumption

$$s=\beta_1\frac{C_1}{T_0}\theta+\beta_2\frac{C_2}{T_0}\theta+\frac{\beta_1 L}{T_0},$$

$$\begin{pmatrix}B_1^{nd}\\B_2^{nd}\end{pmatrix}=\begin{pmatrix}-\frac{L}{T_0}\theta\\0\end{pmatrix}+\partial I_K(\beta_1,\beta_2), \tag{4.24}$$

$$\begin{pmatrix}\mathbf{H}_1^{nd}\\\mathbf{H}_2^{nd}\end{pmatrix}=k\begin{pmatrix}\mathrm{grad}\beta_1\\\mathrm{grad}\beta_2\end{pmatrix}. \tag{4.25}$$

The dissipative forces are defined with pseudopotential of dissipation $\Phi$ which takes into account the internal constraints related to the velocities. The mass balance is such a constraint. For instance, we may choose

$$\Phi(\frac{\partial\beta_1}{\partial t},\frac{\partial\beta_2}{\partial t},\mathrm{grad}\frac{\partial\beta_1}{\partial t},\mathrm{grad}\frac{\partial\beta_2}{\partial t},\mathrm{D}(\mathbf{U}_1),\mathrm{grad}\theta)$$
$$=I_0(\rho_1\frac{\partial\beta_1}{\partial t}+\rho_2\frac{\partial\beta_2}{\partial t}+\rho_1\beta_1\mathrm{div}\mathbf{U}_1)$$
$$+\frac{c}{2}\left(\frac{\partial\beta_1}{\partial t}\right)^2+\frac{c}{2}\left(\frac{\partial\beta_2}{\partial t}\right)^2+\frac{\lambda}{2T_0}(\mathrm{grad}\theta)^2,$$

where $D(\mathbf{U}_1)$ is the strain rate. Let us denote $-p$ an element of the subdifferential set $\partial I_0$. Thus the constitutive laws are

$$\sigma_1^d=-p\rho_1\beta_1 1, \tag{4.26}$$

$$B_1^d=c\frac{\partial\beta_1}{\partial t}-p\rho_1,\ \mathbf{H}_1^d=0, \tag{4.27}$$

$$B_2^d=c\frac{\partial\beta_2}{\partial t}-p\rho_2,\ \mathbf{H}_2^d=0. \tag{4.28}$$

Quantity $p\rho_1\beta_1$ is the pressure in the liquid phase.

### 4.6.5 Approximation of the Equation for the Macroscopic Motions

The equation of motion (4.21) and constitutive law (4.26) give

$$\rho_1 \frac{\partial(\beta_1 \mathbf{U}_1)}{\partial t} = -\mathrm{grad}(\mathrm{p}\rho_1\beta_1) + \mathbf{f}_1.$$

We assume the volume exterior force $\mathbf{f}_1$ results mainly from friction on the solid ice phase and that this force is proportional to the relative velocity

$$\mathbf{f}_1 = -\frac{(\rho_1\beta_1)^2}{m}\mathbf{U}_1.$$

Because the force has to be 0 when there is no water, we assume force $\mathbf{f}_1$ to be proportional to the square of the liquid water density. We neglect the acceleration forces and have

$$\frac{(\rho_1\beta_1)^2}{m}\mathbf{U}_1 = -\mathrm{grad}(\mathrm{p}\rho_1\beta_1),$$

which is within the small perturbation theory

$$\rho_1\beta_1\mathbf{U}_1 = -m\mathrm{gradp}. \tag{4.29}$$

Parameter $m/\rho_1$ is the mobility of the water and velocity $\beta_1\mathbf{U}_1$ is the filtration velocity which is measured in experiments. This relationship is in agreement with the Darcy law.

*Remark 4.6.* Force $\mathbf{f}_1$ is the density of force applied to the liquid water with respect to volume $d\Omega$. The actual density of force applied to the volume occupied by the liquid water $\beta_1 d\Omega$ is

$$\tilde{\mathbf{f}}_1 = \frac{\mathbf{f}_1}{\beta_1} = -\frac{\rho_1^2\beta_1}{m}\mathbf{U}_1.$$

Force $\tilde{\mathbf{f}}_1$ is proportional to $\beta_1$ as expected. It is also proportional to the water momentum $(\rho_1\beta_1)\mathbf{U}_1$.

### 4.6.6 The Equations for the Phase Change

They result from mass balance (4.20), equation of motion (4.21), (4.22), entropy balance (4.23) and constitutive laws (4.24), (4.27), (4.25), (4.28) and Darcy law (4.29)

$$\rho_1 \frac{\partial \beta_1}{\partial t} + \rho_2 \frac{\partial \beta_2}{\partial t} - m\Delta p = 0 \, , \, in \, \Omega,$$

$$c \begin{pmatrix} \frac{\partial \beta_1}{\partial t} \\ \frac{\partial \beta_2}{\partial t} \end{pmatrix} - k \begin{pmatrix} \Delta \beta_1 \\ \Delta \beta_2 \end{pmatrix} - p \begin{pmatrix} \rho_1 \\ \rho_2 \end{pmatrix} + \begin{pmatrix} -\frac{L}{T_0}\theta \\ 0 \end{pmatrix} + \partial I_K(\beta_1, \beta_2) \ni 0 \, , \, in \, \Omega,$$

$$\beta_1^0 C_1 \frac{\partial \theta}{\partial t} + \theta^0 C_1 \frac{\partial \beta_1}{\partial t} + \beta_2^0 C_2 \frac{\partial \theta}{\partial t} + \theta^0 C_2 \frac{\partial \beta_2}{\partial t} + L \frac{\partial \beta_1}{\partial t}$$
$$-\lambda \Delta \theta = T_0 R \, , \, in \, \Omega. \tag{4.30}$$

This set of equation is completed by boundary conditions on $\partial\Omega$

$$m\frac{\partial p}{\partial N} + \alpha_p(p - p_{ext}) = 0,$$
$$\frac{\partial \beta_1}{\partial N} = 0 \, , \, \frac{\partial \beta_1}{\partial N} = 0,$$
$$\lambda \frac{\partial \theta}{\partial N} + \alpha_\theta(\theta - \theta_{ext}) = 0,$$

where $\alpha_p$ and $\alpha_\theta$ measure the hydraulic and thermal conductivity of the container boundary ($\alpha_p = 0$ if the boundary is watertight, $\alpha_p = \infty$ if the boundary is connected to a water supply with pressure $p_{ext}$), and initial conditions

$$\beta_1(\mathbf{x},0) = \beta_1^0(\mathbf{x}) \, , \, \beta_2(\mathbf{x},0) = \beta_2^0(\mathbf{x}) \, , \, \theta(\mathbf{x},0) = \theta^0(\mathbf{x}) \, .$$

This set of partial differential equations allows to compute pressure $p(\mathbf{x},t)$, liquid water and ice volume fractions $\beta_1(\mathbf{x},t)$, $\beta_2(\mathbf{x},t)$ and temperature $\theta(\mathbf{x},t)$ depending on the external actions resulting from the exterior pressure $p_{ext}(\mathbf{x},t)$ and exterior temperature $\theta_{ext}(\mathbf{x},t)$ and rate of heat production $T_0R(\mathbf{x},t)$ very often equal to 0. In an engineering situation the governing action is the exterior temperature cooling and heating the container. Mathematical results are reported in [125].

*Remark 4.7.* More sophisticated boundary condition result from more sophisticated physical boundary properties. The boundary of the container can be semi-permeable, in this case we have

$$\rho_1 \beta_1 \mathbf{U}_1 \cdot \mathbf{N} = -\alpha_p(p - p_{ext})^-,$$

or

$$m\frac{\partial p}{\partial N} - \alpha_p(p - p_{ext})^- = 0.$$

This boundary condition means that when the pressure is lower than the exterior pressure $p_{ext}$, water flows inside the container but when the pressure is larger than the exterior pressure, no water flows outside .

The pressure may also be controlled on the boundary

$$m\frac{\partial p}{\partial N} + \partial I_{-}(p - p_{ext}) \ni 0.$$

Water flows outside the container in order to maintain the pressure lower than the outside pressure $p_{ext}$ which may be the atmospheric pressure.

*Remark 4.8.* The energy balance, the last (4.30), may be simplified within the small perturbation assumption (reference temperature $\theta^0$ is small)

$$(\beta_1^0 C_1 + \beta_2^0 C_2)\frac{\partial \theta}{\partial t} + L\frac{\partial \beta_1}{\partial t} - \lambda\Delta\theta = T_0 R\,,\ in\ \Omega.$$

### 4.6.7 An Example: Liquid Water in an Impermeable Container

We have

$$\beta_1^0 = 1\,,\ \beta_2^0 = 0\,,\ \theta^0 > 0.$$

The boundary condition for the pressure is

$$\frac{\partial p}{\partial N} = 0.$$

We investigate what occurs when the container is cooled. In order to look for closed form solutions, we assume all the quantities are homogeneous. Assuming the temperature is known, we focus on (4.30) which describes the phase change. Because pressure is homogeneous, (4.30) is

$$c\begin{pmatrix} \frac{\partial \beta_1}{\partial t} \\ \frac{\partial \beta_2}{\partial t} \end{pmatrix} - p\begin{pmatrix} \rho_1 \\ \rho_2 \end{pmatrix} + \begin{pmatrix} -\frac{L}{T_0}\theta \\ 0 \end{pmatrix} + \partial I_K(\beta_1, \beta_2) \ni 0.$$

When the temperature is positive the solution of this equation is $\beta_1 = 1$ and $\beta_2 = 0$. When the container is cooled and its temperature becomes negative, the phase change which is expected does not occur. Indeed according to the mass balance, to values $\beta_1 = 1$ and $\beta_2 = 0$ and condition $\beta_1 + \beta_2 \leq 1$, we should have

$$\rho_2\frac{\partial \beta_2}{\partial t} + \rho_1\frac{\partial \beta_1}{\partial t} = 0\,,\ \frac{\partial \beta_1}{\partial t} < 0\,,\ \frac{\partial \beta_2}{\partial t} + \frac{\partial \beta_1}{\partial t} \leq 0.$$

But the two first condition imply

$$\frac{\partial \beta_2}{\partial t} + \frac{\partial \beta_1}{\partial t} = (1 - \frac{\rho_1}{\rho_2})\frac{\partial \beta_1}{\partial t} > 0,$$

which contradicts the third condition: thus phase change is impossible. Let us check that the water remains liquid ($\partial\beta_1/\partial t = 0$) even if temperature is negative. We must have

$$-p\begin{pmatrix}\rho_1\\ \rho_2\end{pmatrix}+\begin{pmatrix}-\frac{L}{T_0}\theta\\ 0\end{pmatrix}+\partial I_K(1,0)\ni 0,$$

with

$$\partial I_K(1,0)=\begin{pmatrix}P\\ Q\end{pmatrix},\ P\geq 0\,,\ Q\leq P,$$

or

$$p\rho_1+\frac{L}{T_0}\theta\geq 0\,,$$

$$0\leq p(\rho_1-\rho_2)+\frac{L}{T_0}\theta. \tag{4.31}$$

These two conditions may be satisfied because $\rho_1 - \rho_2 > 0$. The minimum value of the pressure is

$$p=-\frac{L}{(\rho_1-\rho_2)T_0}\theta.$$

Thus when cooling the container, water remains liquid and pressure increases. This is in agreement with experiment: a glass container filled of water explodes when freezing. The explosion occurs when the pressure overcomes the glass resistance.

### 4.6.8 An Other Example: Ice in an Impermeable Container

We have

$$\beta_1^0=0\,,\ \beta_2^0=1\,,\ \theta^0<0,$$

and we investigate what occurs when the container is heated. In order to look for closed form solutions we assume all the quantities are homogeneous. Equation (4.30) which describes the phase change becomes

$$c\begin{pmatrix}\frac{\partial\beta_1}{\partial t}\\ \frac{\partial\beta_2}{\partial t}\end{pmatrix}-p\begin{pmatrix}\rho_1\\ \rho_2\end{pmatrix}+\begin{pmatrix}-\frac{L}{T_0}\theta\\ 0\end{pmatrix}+\partial I_K(\beta_1,\beta_2)\ni 0. \tag{4.32}$$

When temperature is negative, $\beta_1 = 0,\ \beta_2 = 1$, is solution with $p = 0$, for instance

$$\begin{pmatrix} -\frac{L}{T_0}\theta \\ 0 \end{pmatrix} + \partial I_K(0,1) \ni 0.$$

When temperature becomes positive, $\beta_1 = 0,\ \beta_2 = 1$, is no longer solution because

$$\partial I_K(0,1) = \begin{pmatrix} P \\ Q \end{pmatrix}\ ,\ Q \geq 0\ ,\ P \leq Q,$$

implying

$$-p\begin{pmatrix} \rho_1 \\ \rho_2 \end{pmatrix} + \begin{pmatrix} -\frac{L}{T_0}\theta) \\ 0 \end{pmatrix} + \begin{pmatrix} P \\ Q \end{pmatrix} \ni 0,$$

or

$$\begin{aligned} Q &= p\rho_2 \geq 0, \\ Q - P &= p(\rho_2 - \rho_1) - \frac{L}{T_0}\theta \geq 0. \end{aligned} \tag{4.33}$$

The two last inequalities are impossible. Thus the water cannot remain frozen. It thaws according to equation

$$c\begin{pmatrix} \frac{\partial \beta_1}{\partial t} \\ \frac{\partial \beta_2}{\partial t} \end{pmatrix} - p\begin{pmatrix} \rho_1 \\ \rho_2 \end{pmatrix} + \begin{pmatrix} -\frac{L}{T_0}\theta \\ 0 \end{pmatrix} = 0,$$

with

$$\rho_2 \frac{\partial \beta_2}{\partial t} + \rho_1 \frac{\partial \beta_1}{\partial t} = 0,$$

due to the mass balance. The pressure is

$$p = -\frac{\rho_1 L}{(\rho_1^2 + \rho_2^2) T_0}\theta.$$

The solution satisfies

$$\begin{aligned} \frac{\partial \beta_1}{\partial t} &= \frac{\rho_2^2 L}{c(\rho_1^2 + \rho_2^2) T_0}\theta > 0\ ,\ \frac{\partial \beta_2}{\partial t} = \frac{-\rho_2 \rho_1 L}{c(\rho_1^2 + \rho_2^2) T_0}\theta < 0, \\ \frac{\partial \beta_2}{\partial t} + \frac{\partial \beta_1}{\partial t} &= \frac{(\rho_2^2 - \rho_2 \rho_1) L}{c(\rho_1^2 + \rho_2^2) T_0}\theta < 0, \end{aligned}$$

as long as $\beta_1 \leq 1$, $\beta_2 \geq 0$. The evolution stops when $\beta_2$ is 0. The final value of $\beta_1$

$$\beta_1 = \frac{\rho_2}{\rho_1} < 1,$$

shows that there are voids in agreement with the homogeneity assumption. We have

$$-p\begin{pmatrix}\rho_1\\ \rho_2\end{pmatrix} + \begin{pmatrix}-\frac{L}{T_0}\theta\\ 0\end{pmatrix} + \partial I_K(\frac{\rho_2}{\rho_1},0) \ni 0,$$

with

$$\partial I_K(\frac{\rho_2}{\rho_1},0) = \begin{pmatrix}0\\ Q\end{pmatrix}, \ Q \leq 0,$$

giving the pressure

$$p = -\frac{L}{\rho_1 T_0}\theta.$$

### 4.6.9 An Other Example: Freezing of a Water Emulsion in an Impermeable Container

We have

$$0 < \beta_1^0 < 1\,,\ \beta_2^0 = 0\,,\ \theta^0 > 0.$$

The emulsion, a mixture of voids and water, is cooled. Assuming an homogeneous evolution, it may be seen that the water freezes till either $\beta_1 = 0$, (in this case $\beta_1^0 < \rho_2/\rho_1$ and the cooling results in a mixture of ice and voids), or $\beta_1 + \beta_2 = 1$, (in this case $\beta_1^0 \geq \rho_2/\rho_1$ and the cooling results in a mixture of ice and liquid water, $\beta_1 = (\rho_1\beta_1^0 - \rho_2)/(\rho_1 - \rho_2)$, $\beta_2 = (\rho_1 - \rho_1\beta_1^0)/(\rho_1 - \rho_2)$). This mixture is a water ice or sorbet.

### 4.6.10 The Non Homogeneous Evolution

In practice cooling and heating are performed with surface actions. Thus the temperature is non homogeneous. One may expect that the resulting mixture is non homogeneous. In the last example, the ice volume fraction is larger in the neighbourhood of the surface of the container than in the middle where the liquid volume fraction is more important. The evolution is given by solving the set of partial differential equations.

### 4.6.11 The Motion of Glaciers

In case we have an homogeneous mixture of ice and water without voids at an equilibrium,

$$\beta_1 + \beta_2 = 1\,,\ \frac{\partial \beta_1}{\partial t} = \frac{\partial \beta_2}{\partial t} = 0,$$

(4.32) gives

$$-p\begin{pmatrix}\rho_1\\ \rho_2\end{pmatrix} + \begin{pmatrix}-\frac{L}{T_0}\theta\\ 0\end{pmatrix} + \begin{pmatrix}P\\ P\end{pmatrix} = 0,$$

with $P \geq 0$. We get from this equation

$$-p(\rho_1 - \rho_2) - \frac{L}{T_0}\theta,\ p \geq 0. \tag{4.34}$$

This is relationship (4.19) up to the pressure unit, either $p$ with pascal unit, or $\rho p$ with $p$ with meter unit for relationship (4.34). Because $\rho_1 > \rho_2$ the phase change temperature decreases when the pressure increases. A consequence of this property is that the ice melts at the bottom of a glacier and lubricates the rock ice contact surface allowing the downhill motion of the glacier.

## 4.7 Dissipative and Irreversible Phase Change

When the phase change is progressive or when the microscopic motion is viscous, there is dissipation with respect to the phase volume fraction, [112, 127]. The evolution can even be irreversible (for instance when cooking an egg, [127]). The $\Psi$ and $\Phi$ with parameter $\chi = T$ are

$$\Psi(T, \beta, \mathrm{grad}\beta) = -CT\ln T - \beta\frac{L}{T_0}(T - T_0) + I(\beta) + \frac{k}{2}(\mathrm{grad}\beta)^2,$$

$$\Phi(\mathrm{gradT}, \frac{\mathrm{d}\beta}{\mathrm{dt}}, \mathrm{T}) = \frac{\lambda}{2T}(\mathrm{gradT})^2 + \frac{\mathrm{c}}{2}\left(\frac{\mathrm{d}\beta}{\mathrm{dt}}\right)^2 + \mathrm{I}_-(\frac{\mathrm{d}\beta}{\mathrm{dt}}),$$

where $I_-$ is the indicator function of the set of the non positive numbers. It takes into account the irreversibility of the phase change

$$\frac{d\beta}{dt} \leq 0.$$

The constitutive laws are

$$\sigma = 0\,,\ B \in -\frac{L}{T_0}(T-T_0) + c\frac{d\beta}{dt} + \partial I(\beta) + \partial I_-(\frac{d\beta}{dt})\,,\ \mathbf{H} = k\mathrm{grad}\beta,$$

$$\mathbf{Q} = -\frac{\lambda}{T}\mathrm{gradT},$$

$$s = C(1+\ln T) + \frac{L}{T_0}\beta.$$

### 4.7.1 The Equations

The equations of the predictive theory, assuming small perturbations, are

$$C\frac{\partial \ln T}{\partial t} + \frac{L}{T_0}\frac{\partial \beta}{\partial t} - \lambda\Delta \ln T = R\,,\ in\ \Omega,$$

$$c\frac{\partial \beta}{\partial t} - k\Delta\beta - \frac{L}{T_0}(T-T_0) + \partial I(\beta) + \partial I_-(\frac{\partial \beta}{\partial t}) \ni 0\,,\ in\ \Omega,$$

$$\lambda\frac{\partial \ln T}{\partial N} = \pi,\ k\frac{\partial \beta}{\partial N} = 0\,,\ on\ \partial\Omega,$$

$$T(\mathbf{x},0) = T^0(\mathbf{x})\,,\ \beta(\mathbf{x},0) = \beta^0(\mathbf{x})\,,\ in\ \Omega.$$

When the evolution is not irreversible, the subdifferential $\partial I_-(\partial\beta/\partial t)$ is absent from the equations. The phase change occurs at temperatures which are close to $T_0$ as it may be seen in equation

$$c\frac{\partial \beta}{\partial t} - k\Delta\beta - \frac{L}{T_0}(T-T_0) + \partial I(\beta) \ni 0.$$

Other results on dissipative phase change and related problems may be found in [14, 35, 47, 48, 56, 68, 70, 73, 127, 144, 160, 207, 208].

## 4.8 The Phase Volume Fraction Depends on the Temperature

It is possible that volume fractions depend on the temperature. To account for this phenomenon, we choose a the pseudopotential of dissipation $\Phi$ depending on $\chi = (T,\beta)$

$$\Psi(T,\beta,\mathrm{grad}\beta) = -CT\ln T - \beta\frac{L}{T_0}(T-T_0) + I(\beta) + \frac{\bar{k}}{2}\beta^2 + \frac{k}{2}(\mathrm{grad}\beta)^2,$$

$$\Phi(\mathrm{gradT},\frac{\mathrm{d}\beta}{\mathrm{dt}},\mathrm{T},\beta) = \frac{\lambda}{2T}(\mathrm{gradT})^2 + \frac{\mathrm{c}}{2}\left(\frac{\mathrm{d}\beta}{\mathrm{dt}}\right)^2 + \mathrm{K}(\mathrm{T},\beta)\left|\frac{\mathrm{d}\beta}{\mathrm{dt}}\right|,$$

where $K(T,\beta)$ is a non negative function. The physical parameter $\bar{k}$ is positive. In this situation, the pseudopotential of dissipation depends on the parameters $T,\beta$ of $E$. The constitutive laws are

$$\sigma = 0\,,\; B \in -\frac{L}{T_0}(T - T_0) + \bar{k}\beta + c\frac{d\beta}{dt} + \partial I(\beta) + K(T,\beta)sgn(\frac{d\beta}{dt})\,,\; \mathbf{H} = k\mathrm{grad}\beta,$$

where the sign graph is defined by

$$sign(x) = \begin{cases} 1, & \text{if } x > 0, \\ [-1,1], & \text{if } x = 0, \\ -1, & \text{if } x < 0. \end{cases}$$

### *4.8.1 The Equations*

The equations of the predictive theory, assuming small perturbations, are

$$C\frac{\partial \ln T}{\partial t} + \frac{L}{T_0}\frac{\partial \beta}{\partial t} - \lambda \Delta \ln T = R\,,\; in\; \Omega,$$

$$c\frac{\partial \beta}{\partial t} - k\Delta\beta - \frac{L}{T_0}(T - T_0) + \bar{k}\beta + \partial I(\beta) + K(T,\beta)sgn(\frac{d\beta}{dt}) \ni 0\,,\; in\; \Omega,$$

$$\lambda\frac{\partial \ln T}{\partial N} = \pi\,,\; k\frac{\partial \beta}{\partial N} = 0\,,\; on\; \partial\Omega,$$

$$T(\mathbf{x},0) = T^0(\mathbf{x})\,,\; \beta(\mathbf{x},0) = \beta^0(\mathbf{x})\,,\; in\; \Omega.$$

It is possible to prove that the volume fraction $\beta$ is within an interval depending on the temperature. For instance, if $c\partial\beta/\partial t - k\Delta\beta = 0$ and $0 < \beta < 1$, we have

$$\frac{1}{\bar{k}}\left(\frac{L}{T_0}(T - T_0) - K(T,\beta)\right) \le \beta \le \frac{1}{\bar{k}}\left(\frac{L}{T_0}(T - T_0) + K(T,\beta)\right).$$

Thanks to a convenient pseudopotential of dissipation, the predictive theory accounts for the experimental fact: the volume fraction depends on the temperature.

Let us stress the large possibilities offered by the theory: there are so many possible pseudopotentials of dissipation! Mathematical results may be found in [13, 14, 72].

## 4.9 Phase Change with Thermal Memory

The heat flux vector may depend on the past, [40]. We define the new state quantity

$$G(t) = \frac{1}{2}\int_0^{+\infty} h(s)\mathbf{p}(s,t)\cdot\mathbf{p}(s,t)ds,$$

with

$$\mathbf{p}(s,t) = \int_{(t-s)}^{t} \mathrm{grad}\mathrm{T}(\tau)\,\mathrm{d}\tau.$$

This function depends on the history of the temperature gradient. Properties of kernel $h$ may be found in [40]: $h : (0,+\infty) \to \mathbb{R}$ is continuous, positive and decreasing. We assume the free energy depends on $G$. Thus it depends on the past. We may have

$$\Psi(T,\beta,\mathrm{grad}\beta,\mathrm{G}) = -\mathrm{C}\mathrm{T}\ln\mathrm{T} - \beta\frac{\mathrm{L}}{\mathrm{T}_0}(\mathrm{T}-\mathrm{T}_0) + \mathrm{I}(\beta) + \frac{\mathrm{k}}{2}(\mathrm{grad}\beta)^2 + \mathrm{G}.$$

The pseudopotential of dissipation with $\chi = T$ is chosen as

$$\Phi(\mathrm{grad}\mathrm{T},\frac{\mathrm{d}\beta}{\mathrm{dt}},\mathrm{T}) = \frac{\lambda}{2\mathrm{T}}(\mathrm{grad}\mathrm{T})^2 + \frac{\mathrm{c}}{2}\left(\frac{\mathrm{d}\beta}{\mathrm{dt}}\right)^2.$$

The constitutive laws result from upgraded laws of thermodynamics introducing non dissipative entropy flux vector, [40]

$$\sigma = 0\,,\; B \in -\frac{L}{T_0}(T-T_0) + \partial I(\beta) + c\frac{\partial\beta}{\partial t}\,,\; \mathbf{H} = k\mathrm{grad}\beta,$$

$$\mathbf{Q} = -\frac{\lambda}{T}\mathrm{grad}\mathrm{T} - \int_{-\infty}^{\mathrm{t}} \mathrm{k}(\mathrm{t}-\mathrm{s})\mathrm{grad}\mathrm{T}(\mathrm{s})\,\mathrm{ds},$$

$$s = C(1+\ln T) + \frac{L}{T_0}\beta,$$

where function $k$ is defined by $k'(s) = -h(s)$ and $k(+\infty) = 0$. The heat flux vector depends on the past, on the history of the temperature gradient gradT. This model is due to Pierluigi Colli and to Elena Bonetti.

### *4.9.1 The Equations*

We assume that the history term

$$\mathrm{div}\int_{-\infty}^{0} \mathrm{k}(\mathrm{t}-\mathrm{s})\mathrm{grad}\mathrm{T}(\mathrm{s})\,\mathrm{ds},$$

is known and include it in the generalized entropy sources, still denoted by $R$ and $\pi$. Thus the equations of the predictive theory, assuming small perturbations, are

$$C\frac{\partial \ln T}{\partial t} + \frac{L}{T_0}\frac{\partial \beta}{\partial t} - \lambda \Delta \ln T - k * \Delta T = R \, , \; in \; \Omega,$$

$$c\frac{\partial \beta}{\partial t} - k\Delta\beta - \frac{L}{T_0}(T - T_0) + \partial I(\beta) \ni 0 \, , \; in \; \Omega \, ,$$

$$\lambda\frac{\partial \ln T}{\partial N} + \frac{\partial (k * T)}{\partial N} = \pi \, , \; k\frac{\partial \beta}{\partial N} = 0 \, , \; on \; \partial\Omega,$$

$$T(\mathbf{x},0) = T^0(\mathbf{x}) \, , \; \beta(\mathbf{x},0) = \beta^0(\mathbf{x}) \, , \; in \; \Omega,$$

where $*$ denotes the usual time convolution product over the interval $(0,t)$, i.e., for $a$ and $b$

$$(a * b)(t) := \int_0^t a(t-s)b(s)\,ds.$$

Mathematical results may be found in [40, 71]. Other models and memory effects may be found in [91–94].

# Chapter 5
# Shape Memory Alloys

## 5.1 Introduction

Shape memory alloys are mixtures of many martensites and of austenite. The composition of the mixture varies: the martensites and the austenite transform into one another. These phase changes can be produced either by thermal actions or by mechanical actions. The striking properties of shape memory alloys result from interactions between mechanical and thermal actions [29, 136, 183].

We describe a macroscopic predictive theory which can be used for engineering purposes. The phase volume fractions, which are state quantities, are subjected to constraints (for instance their sum is one when there are no voids and no interpenetration in the mixture). It is shown that most of the properties of shape memory alloys result from a careful treatment of those internal constraints, [109, 110, 112, 119].

Other microscopic and macroscopic models, applications and results may be found in [1, 19, 20, 23, 30, 95, 155, 157, 164, 169, 170, 173, 178, 190, 191, 202].

## 5.2 The State Quantities

We deal only with macroscopic phenomena and macroscopic quantities. To describe the deformations of the alloy, both the macroscopic small deformation $\varepsilon$ and the temperature $T$ are chosen as state quantities.

The properties of shape memory alloys result from martensite–austenite phase changes produced either by thermal actions (as it is usual) or by mechanical actions. On the macroscopic level, quantities are needed to take those phase changes into account. For this purpose, the volume fractions $\beta_i$ of the martensites and austenite are chosen as state quantities. We think that this choice is the simplest which can be made. Again to be simple, we assume that only two martensites exist together with austenite. The volume fractions of the martensites are $\beta_1$ and $\beta_2$. The volume

M. Frémond, *Phase Change in Mechanics*, Lecture Notes of the Unione Matematica Italiana 13, DOI 10.1007/978-3-642-24609-8_5,

fraction of austenite is $\beta_3$. We assume there are local microscopic interactions responsible for the twinning and choose the grad$\beta_i$ as state quantities.

The volume fractions are not independent: they satisfy constraints, called internal constraints

$$0 \leq \beta_i \leq 1, \tag{5.1}$$

due to the definition of volume fractions. We assume that no void can appear in the mixture, then $\beta_1 + \beta_2 + \beta_3 \geq 1$, and that no interpenetration of the phases can occur, then $\beta_1 + \beta_2 + \beta_3 \leq 1$. Thus the $\beta_i$'s satisfy an other internal constraint

$$\beta_1 + \beta_2 + \beta_3 = 1\,. \tag{5.2}$$

These internal constraints are physical properties due to their definitions or to their mechanical properties. The macroscopic state quantities are

$$\underline{E} = (\varepsilon, \beta_1, \beta_2, \beta_3, \mathrm{grad}\beta_1, \mathrm{grad}\beta_2, \mathrm{grad}\beta_3, \mathrm{T})\,. \tag{5.3}$$

The case where voids can appear is investigated in Sect. 5.6.

## 5.3 The Equations of Motion

They are for the macroscopic motions

$$\rho\frac{\mathrm{d}\mathbf{U}}{\mathrm{d}t} = \mathrm{div}\sigma + \mathbf{f},\ \mathrm{in}\ \Omega,\ \sigma\mathbf{N} = \mathbf{g},\ \mathrm{on}\ \partial\Omega, \tag{5.4}$$

where $\rho$ is the density, $\mathrm{d}\mathbf{U}/\mathrm{d}t$ the acceleration, $\sigma$ the stresses in the alloy which occupies domain $\Omega$, with boundary $\partial\Omega$ and outward normal vector $\mathbf{N}$. The alloy is loaded by macroscopic body forces $\mathbf{f}$ and by surface tractions $\mathbf{g}$.

### *5.3.1 The Equations for the Microscopic Motions When Voids and Interpenetration Are Possible*

For the microscopic motions, they result from the principle of virtual power

$$\forall \gamma_j, j = 1,2,3,$$

$$-\sum_{j=1}^{3}\int_{\Omega}\{\underline{\mathbf{H}}_j \cdot \mathrm{grad}\gamma_j + \underline{\mathrm{B}}_j\gamma_j\}\,d\Omega + \sum_{j=1}^{3}\int_{\Omega}\underline{A}_j\gamma_j d\Omega + \sum_{j=1}^{3}\int_{\partial\Omega}\underline{a}_j\gamma_j d\Gamma = 0,$$

where the $\gamma_j$ are virtual velocities, the $\underline{\mathbf{H}}_j$ and $\underline{B}_j$ are the internal forces and the $\underline{A}_j$ and $\underline{a}_j$ are the external forces. These equations of motion are

$$\mathrm{div}\underline{\mathbf{H}}_j - \underline{B}_j + \underline{A}_j = 0, \text{ in } \Omega, \ \underline{\mathbf{H}}_j \cdot \mathbf{N} = \underline{a}_j, \text{ on } \partial\Omega, \ j = 1,2,3. \tag{5.5}$$

Let us note that the virtual velocities do not satisfy

$$\gamma_1 + \gamma_2 + \gamma_3 = 0, \tag{5.6}$$

as the actual velocities. The satisfaction of this relationship by the actual velocities will result from the constitutive laws which will introduce reactions to this internal constraint. The set of the virtual velocities is a very large one. This point of view is useful in case voids can appear

$$\beta_1 + \beta_2 + \beta_3 \leq 1,$$

as seen in Sect. 4.5 or in case interpenetration can occur

$$\beta_1 + \beta_2 + \beta_3 \geq 1\,.$$

## 5.4 Shape Memory Alloys When Voids and Interpenetration Are Impossible

An other way to proceed is to choose starting from the beginning that the internal constraint (5.2) is satisfied and to have virtual velocities which satisfy relationship (5.6). The principle gives

$$\begin{aligned}
&\forall \gamma_i, i = 1,2,\\
&-\sum_{i=1}^{2} \int_\Omega \{(\underline{\mathbf{H}}_i - \underline{\mathbf{H}}_3) \cdot \mathrm{grad}\gamma_i + (\underline{B}_i - \underline{B}_3)\gamma_i\} d\Omega\\
&+\sum_{i=1}^{2} \int_\Omega (\underline{A}_i - \underline{A}_3)\gamma_i d\Omega + \sum_{i=1}^{2} \int_{\partial\Omega} (\underline{a}_i - \underline{a}_3)\gamma_j d\Gamma = 0\,.
\end{aligned}$$

By letting

$$\mathbf{H}_i = \underline{\mathbf{H}}_i - \underline{\mathbf{H}}_3,\ B_i = \underline{B}_i - \underline{B}_3,\ A_i = \underline{A}_i - \underline{A}_3,\ a_i = \underline{a}_i - \underline{a}_3,$$

the equations of motion become

$$\mathrm{div}\mathbf{H}_i - B_i + A_i = 0, \text{ in } \Omega, \ \mathbf{H}_i \cdot \mathbf{N} = a_i, \text{ on } \partial\Omega, \ i = 1,2\,. \tag{5.7}$$

In the sequel, we assume no exterior work $A_i$ and $a_i$.

The way relationship (5.2) is used implies that $\beta_3$ is no longer a state quantity. The set of states quantities is

$$E = (\varepsilon, \beta_1, \beta_2, \mathrm{grad}\beta_1, \mathrm{grad}\beta_2, \mathrm{T}) .$$

We denote $\beta$ the vector $(\beta_i,\ i = 1,2)$, $\mathbf{B}$ the vector $(B_i,\ i = 1,2)$ and $H$ the matrix $(\mathbf{H}_i,\ i = 1,2)$.

*Remark 5.1.* The second set of (5.7) may be recovered from the first one (5.5) when voids and no interpenetration are impossible by eliminating the reactions due to internal constraint (5.2).

### 5.4.1 The Entropy Balance

By denoting $s$ the entropy, $\mathbf{Q}$ the entropy flux vector, the entropy balance is

$$\frac{\mathrm{d}s}{\mathrm{d}t} + \mathrm{div}\mathbf{Q} = R + \frac{1}{T}\left\{\sigma^d : \mathsf{D}(\mathbf{U}) + B^d \cdot \frac{\mathrm{d}\beta}{\mathrm{d}t} + H^d : \mathrm{grad}\frac{\mathrm{d}\beta}{\mathrm{dt}}\right\},\ in\ \Omega, \quad (5.8)$$

$$-\mathbf{Q}\cdot\mathbf{N} = \pi,\ on\ \partial\Omega, \quad (5.9)$$

where $R$ is the exterior volume rate of entropy that is supplied to the alloy, $\pi$ is the rate of entropy that is supplied by contact action, $\mathsf{D}(\mathbf{U})$ is the strain rate. The dissipative internal forces are $\sigma^d$, $B^d$, $H^d$. Let us recall that in this setting the second law is

$$T > 0,$$

$$\left\{\sigma^d : \mathsf{D}(\mathbf{U}) + B^d \cdot \frac{\mathrm{d}\beta}{\mathrm{d}t} + H^d : \mathrm{grad}\frac{\mathrm{d}\beta}{\mathrm{dt}}\right\} \geq 0 .$$

### 5.4.2 The Free Energy

As already said, a shape memory alloy is at the macroscopic level a mixture of three martensite and austenite phases with volume fractions $\beta_i$. The volumic free energy of the mixture we choose is

$$\begin{aligned}\Psi &= \sum_{i=1}^{3} \beta_i \Psi_i + Th(E) + I_C(\beta),\\ &= \beta_1\Psi_1(E) + \beta_2\Psi_2(E) + (1-\beta_1-\beta_2)\Psi_3(E) + Th(E) + I_C(\beta)\,, \quad (5.10)\end{aligned}$$

where the $\Psi_i$'s are the volume free energies of the $i$ phases and $Th$ is a free energy describing interactions between the different phases. We have said that internal constraints are physical properties. Being physical properties, we decide to take them into account with the two functions we have to describe the material, i.e., the free energy $\Psi$ and the pseudopotential of dissipation $\Phi$. The pseudopotential describes the kinematic properties, i.e., properties which depend on the velocities. The free energy describes the state properties. Obviously the internal constraints (5.1) and (5.2) are not kinematic properties. Thus we take them into account with the free energy $\Psi$ which involves the quantity $I_C(\beta)$ where $I_C$ is the indicator function (see Appendix A) of the convex set, see Fig. 5.1

$$C = \{(\gamma_1, \gamma_2) \in \mathbb{R}^2; 0 \le \gamma_i \le 1; \gamma_1 + \gamma_2 \le 1\}.$$

The simplest choice we can make for $h(\beta)$ is $h(\beta) = 0$. There is no interaction, besides constraints (5.1) and (5.2), between the different phases in the mixture. For the volume free energies, we choose

$$\Psi_1(E) = \frac{1}{2}\varepsilon : K_1 : \varepsilon + \sigma_1(T) : \varepsilon - C_1 T \ln T,$$

$$\Psi_2(E) = \frac{1}{2}\varepsilon : K_2 : \varepsilon + \sigma_2(T) : \varepsilon - C_2 T \ln T,$$

$$\Psi_3(E) = \frac{1}{2}\varepsilon : K_3 : \varepsilon - \frac{L}{T_0}(T - T_0) - C_3 T \ln T,$$

where $K_i$ are the volume elastic tensors and $C_i$ the volume heat capacities of the phases. The stresses $\sigma_1$ and $\sigma_2$ depend on temperature $T$. The quantity $L$ is the latent heat martensite-austenite volume phase change at temperature $T_0$ (see below).

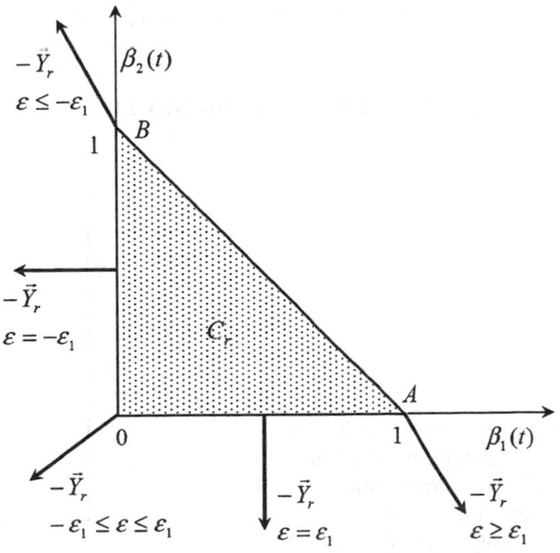


**Fig. 5.1** Vector $-\mathbf{B}^{nd}(E)$ for different deformations $\varepsilon$ at medium temperature

*Remark 5.2.* To make the model more realistic, we can introduce two temperatures to characterize the transformation: $T_0$, the temperature at the beginning of the phase change and $T_f$ the temperature at the end. The interaction free energy is completed by $h(\beta) = (l_2/T_0)(T_0 - T_f)(\beta_3)^2$, [22, 112, 156, 185].

Because we want to describe the main basic properties of the shape memory alloys, we assume that the elastic matrices $K_i$ and the heat capacities $C_i$ are the same for all of the phases

$$\forall i = 1,2,3,\ C_i = C,\ K_i = K\ .$$

Always for the sake of simplicity, we assume that

$$\sigma_1(T) = -\sigma_2(T) = -\tau(T)\ .$$

Concerning the stress $\tau(T)$, it is known that at high temperature the alloy has a classical elastic behaviour. Thus $\tau(T) = 0$ at high temperature, and we choose the schematic simple expression

$$\tau(T) = (T - T_c)\overline{\tau},\ \text{for}\ T \leq T_c,\ \tau(T) = 0,\ \text{for}\ T \geq T_c, \tag{5.11}$$

with $\overline{\tau}_{11} \leq 0$ and assume the temperature $T_c$ is larger than $T_0$. With those assumptions, it results

$$\Psi(E) = \frac{1}{2}\varepsilon : K : \varepsilon - (\beta_1 - \beta_2)\tau(T) : \varepsilon - (1 - \beta_1 - \beta_2)\frac{L}{T_0}(T - T_0) - CT\ln T\ .$$

*Remark 5.3.* In a simple setting, we may choose

$$\tau(T) = \alpha(T)1 = (T - T_c)\overline{\tau} = \begin{cases} 0,\ if\ T \geq T_c, \\ (T - T_c)\overline{\tau}_{11}1,\ if\ T \leq T_c, \end{cases}$$

with $\bar{\tau}_{11} < 0$. Matrix 1 is the identity matrix. Function $\alpha(T)$ is shown in Fig. 5.2.

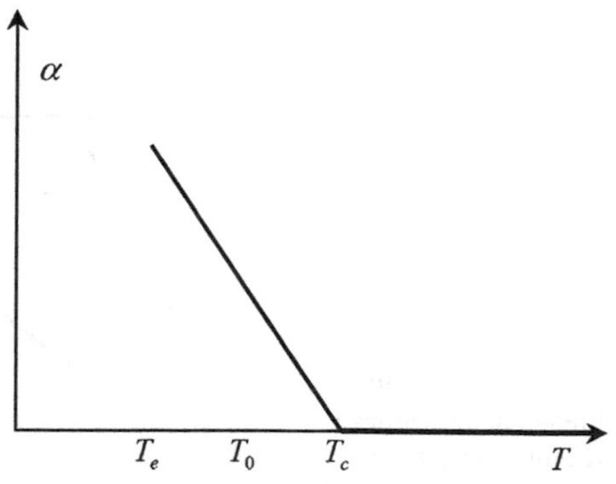


**Fig. 5.2** Function $\alpha(T)$. It is 0 at high temperature. The different characteristic temperatures are in the following order: $T_e < T_0 < T_c$

*Remark 5.4.* It is known that elasticity tensor $K_1$ does depend on temperature $T$, [132]. Thus the entropy of martensite is

$$s_1 = -\frac{\partial \Psi_1}{\partial T} = -\frac{1}{2}\varepsilon : \frac{\partial K_1}{\partial T} : \varepsilon - \frac{\partial_1 \sigma_1}{\partial T} : \varepsilon + C_1(1 + \ln T) .$$

Because free energy has to be a concave function of temperature,

$$-\frac{\partial^2 \Psi_1}{\partial T^2} = \frac{\partial s_1}{\partial T} = -\frac{1}{2}\varepsilon : \frac{\partial^2 K_1}{\partial T^2} : \varepsilon - \frac{\partial_1^2 \sigma_1}{\partial T^2} : \varepsilon + \frac{C_1}{T},$$

has to be non negative. Let us assume, in agreement with experiments, [132], that $-(\partial^2 K_1/\partial T^2)$ is positive definite, i.e., $T \to \varepsilon : K_1(T) : \varepsilon$ is a concave function. Thus we have

$$-\frac{\partial^2 \Psi_1}{\partial T^2} \geq \frac{\partial_1^2 \sigma_1}{\partial T^2} : \frac{\partial^2 K_1}{\partial T^2} : \frac{\partial_1^2 \sigma_1}{\partial T^2} + \frac{C_1}{T} .$$

It results that if

$$C_1 \geq -\frac{T}{2}\left( \frac{\partial_1^2 \sigma_1}{\partial T^2} : \frac{\partial^2 K_1}{\partial T^2} : \frac{\partial_1^2 \sigma_1}{\partial T^2} \right),$$

free energy is a concave function of $T$. This relationship holds in the neighbourhood of $T_c$ in case function $\alpha(T)$ is replaced by a smooth approximation. Previous inequality is easily satisfied if the variation of $\alpha(T)$ from 0 to $\bar{\tau}_{11}$ is spread on a sufficiently large interval of temperature (i.e., if the curvature $\partial_1^2 \sigma_1/\partial T^2$ is not too large in a neighbourhood of $T_c$ and null outside).

In case $s_1$ is discontinuous at temperature $T_c$ as in relationship (5.11)

$$[\Psi] = 0 .$$

We may choose in agreement with constitutive laws (3.36)

$$[T] = 0 .$$

These properties are important to solve the entropy balance. But let us note that within the small perturbation assumption entropy is

$$s_1 = -\frac{\partial \Psi_1}{\partial T} = C_1(1 + \ln T),$$

a concave function of $T$. Thus remaining within this assumption, we assume elasticity tensors do not depend on $T$.

### 5.4.3 The Constitutive Laws

The entropy is

$$s = -\frac{\partial \Psi}{\partial T} = (\beta_1 - \beta_2)\frac{d\tau(T)}{dT} : \varepsilon + (1 - \beta_1 - \beta_2)\frac{L}{T_0} + C(1 + \ln T),$$

or

$$s = -\frac{\partial \Psi}{\partial T} = (1 - \beta_1 - \beta_2)\frac{L}{T_0} + C(1 + \ln T),$$

within the small perturbation assumption.

*Remark 5.5.* With the assumptions of Remark 5.4, we have

$$\begin{aligned} s = -\frac{\partial \Psi}{\partial T} &= -\frac{1}{2}\varepsilon : \frac{\partial K}{\partial T} : \varepsilon + (\beta_1 - \beta_2)\frac{d\tau}{dT} : \varepsilon + C(1 + \ln T) \\ &+ (1 - \beta_1 - \beta_2)\frac{L}{T_0}, \end{aligned}$$

which is an increasing function of $T$ because

$$\begin{aligned} &-\frac{1}{2}\varepsilon : \frac{\partial^2 K_1}{\partial T^2} : \varepsilon + (\beta_1 - \beta_2)\frac{d^2\tau}{dT^2} : \varepsilon + \frac{C}{T} \\ &\geq -\frac{1}{2}\varepsilon : \frac{\partial^2 K_1}{\partial T^2} : \varepsilon - \left|\frac{d^2\tau}{dT^2} : \varepsilon\right| + \frac{C}{T} \geq 0, \end{aligned}$$

where $\sigma_1(T) = -\sigma_2(T) = -\tau(T)$, $K_i(T) = K(T)$ and $C_i = C$, $i = 1,3$.

The non dissipative forces $\sigma^{nd}$, $\mathbf{B}^{nd}$ depending on $E$ and non dissipative reaction $\mathbf{B}^{ndr}$ depending on $(E, \mathbf{x}, t)$ are

$$\sigma^{nd}(E) = \frac{\partial \Psi}{\partial \varepsilon}(E) = K : \varepsilon - (\beta_1 - \beta_2)\tau(T), \tag{5.12}$$

$$\mathbf{B}^{nd}(E) = \frac{\partial \Psi}{\partial \beta}(E) = \{-\tau(T) : \varepsilon + \frac{L}{T_0}(T - T_0), \tau(T) : \varepsilon + \frac{L}{T_0}(T - T_0)\}, \tag{5.13}$$

$$H^{nd}(E) = \frac{\partial \Psi}{\partial(\mathrm{grad}\beta)}(E) = k\,\mathrm{grad}\beta, \tag{5.14}$$

$$\mathbf{B}^{ndr}(E, \mathbf{x}, t) \in \partial I_C(\beta(\mathbf{x}, t))\,. \tag{5.15}$$

Relationships (5.12)–(5.15) are the state laws. From the preceding formulas, the smooth part of the free energy is differentiable and the non-smooth part is subdifferentiable, i.e., the subdifferential set $\partial I_C(\beta)$ is not empty (see Appendix A). The quantity $\mathbf{B}^{ndr}(E,\mathbf{x},t)$ is the thermodynamical reaction to internal constraints (5.1) and (5.2). It is related to $\beta$ by state law (5.15). This implies that the subdifferential set $\partial I_C(\beta)$ is not empty, thus that $\beta \in C$, which means that internal constraints (5.1) and (5.2) are satisfied. State law (5.15) has an other meaning besides implying that the internal constraints are satisfied: it gives the value of the reactions to these internal constraints.

The dissipative forces are defined via a pseudopotential of dissipation $\Phi$. From experiments, it is known that the behaviour of shape memory alloys depends on time, i.e., the behaviour is dissipative. We define pseudopotential of dissipation with $\chi = T$

$$\Phi(\frac{\partial \beta}{\partial t}, \mathrm{gradT}, \mathrm{T}) = \mathrm{c}\left\|\frac{\partial \beta}{\partial \mathrm{t}}\right\| + \frac{\mathrm{k}}{2}\left\|\frac{\partial \beta}{\partial \mathrm{t}}\right\|^2 + \frac{\lambda}{2\mathrm{T}}(\mathrm{gradT})^2,$$

where the euclidean norm is

$$\left\|\frac{\partial \beta}{\partial t}\right\| = \sqrt{\left(\frac{\partial \beta_1}{\partial t}\right)^2 + \left(\frac{\partial \beta_2}{\partial t}\right)^2},$$

with $c \geq 0$ and $k \geq 0$. Of course this expression is induced by experimental results: the first term is related to the permanent deformations exhibited by experiments; the second term is related to the viscous aspect. It has a smoothing effect. The dissipative forces are

$$\partial \Phi(\delta E, \chi) = \left\{\sigma^d, \mathbf{B}^d, \mathbf{Q}^d\right\}, \tag{5.16}$$

where $\partial \Phi$ is the subdifferential of the convex function $\Phi$ with respect to

$$\delta E = \left\{\frac{\partial \varepsilon}{\partial t}, \frac{\partial \beta}{\partial t}, \mathrm{gradT}\right\}, \tag{5.17}$$

see the explicit computation of $\partial \Phi$ below. Relationship (5.16) gives the constitutive laws

$$\sigma = \sigma^d(\delta E, \chi) + \sigma^{nd}(E), \tag{5.18}$$

$$B = \mathbf{B}^d(\delta E, \chi) + \mathbf{B}^{nd}(E) + \mathbf{B}^{ndr}(E, \mathbf{x}, t), \tag{5.19}$$

$$H = H^{nd}(E), \tag{5.20}$$

$$-\mathbf{Q} = \mathbf{Q}^d(\delta E, \chi), \tag{5.21}$$

where $E = (\varepsilon, \beta, \mathrm{grad}\beta, \mathrm{T})$. The functions $\sigma^d$, $\mathbf{B}^d$, and $\mathbf{Q}^d$ are the dissipative or irreversible forces. It can be proved that our choice is such that the internal constraints and the second law of thermodynamics are satisfied [112, 119].

### 5.4.4 An Example of Non Dissipative Evolution

We assume that there is no dissipation and that the material is homogeneous, i.e., $\mathrm{grad}\beta = 0$. This assumption is not very realistic but it is a step toward the complete understanding of the constitutive laws. The results from this non dissipative theory are very schematic and crude, but they give some insight into the important role of the reaction $\mathbf{B}^{ndr}$. Let us consider a unidimensional experiment and assume that $\varepsilon_{11}$ is the only non null deformation. Let us focus on the stress $\sigma_{11}$ as a function of $\varepsilon_{11}$ when the temperature is fixed. From relationship (5.18) with $\sigma^d = 0$ and (5.12) it results

$$\sigma_{11} = K_{1111}\varepsilon_{11} + \tau_{11}(T)(\beta_2 - \beta_1), \tag{5.22}$$

and from relationships (5.19), (5.13) and (5.15)

$$-\frac{\partial \Psi}{\partial \beta}(E) = -\mathbf{B}^{nd}(E) = \mathbf{B}^{ndr} \in \partial I_C(\beta), \tag{5.23}$$

with

$$\mathbf{B}^{nd}(E) = \{-\tau_{11}(T)\varepsilon_{11} + \frac{L}{T_0}(T - T_0), \tau_{11}(T)\varepsilon_{11} + \frac{L}{T_0}(T - T_0)\}\,.$$

The relationship (5.23) means that vector $-\mathbf{B}^{nd}(E)$ is normal to triangle $C$ at the point $\beta_r$ (see Appendix A). Let us consider an experiment at fixed medium temperature ($T_0 < T < T_c$). We denote $\varepsilon = \varepsilon_{11}$ and $\sigma = \sigma_{11}$. The two components of vector $\mathbf{B}^{nd}(E)$ are shown in Fig. 5.3.

When $\varepsilon = 0$, the two components of $-\mathbf{B}^{nd}(E)$ are equal and non positive. Vector $-\mathbf{B}^{nd}(E)$ can be normal to triangle $C$ only at vertex 0 (Fig. 5.1).

Thus $\beta_3 = 1$ and $\beta_1 = \beta_2 = 0$, there is only austenite. Relationship (5.22) gives $\sigma = 0$ for $\varepsilon = 0$. When $\varepsilon \neq 0$, vector $-\mathbf{B}^{nd}(E)$ is normal to $C$ at vertex 0 if its two

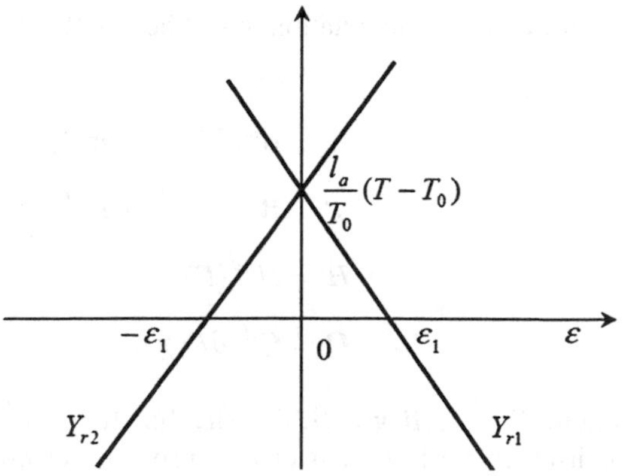


**Fig. 5.3** The components of $\mathbf{B}^{nd}(E)$ at medium temperature

components are non positive, i.e., for

$$-\frac{L(T-T_0)}{T_0\tau_{11}(T)} \leq \varepsilon \leq \frac{L(T-T_0)}{T_0\tau_{11}(T)} = \varepsilon_1 .$$

Thus for $\varepsilon \in ]-\varepsilon_1, \varepsilon_1[$, there is only austenite, $\beta_3 = 1$, and from (5.22), $\sigma = K\varepsilon$ (Fig. 5.4).

For $\varepsilon = \varepsilon_1$, $B_1^{nd}(E) = 0$. Vector $-\mathbf{B}^{nd}(E)$ is normal to triangle $C$ on the side $OA$ (Fig. 5.1) and $\beta_1 + \beta_3 = 1$: there is a mixture of austenite and martensite number one. The stress $\sigma$ can take any value of the segment $[K\varepsilon_1, K\varepsilon_1 - \tau_{11}(T)]$ (Fig. 5.4).

For $\varepsilon > \varepsilon_1$, then $-B_1^{nd}(E) > 0$ and $-B_2^{nd}(E) < 0$. Vector $-\mathbf{B}^{nd}(E)$ is normal to triangle $C$ at vertex $A$ (Fig. 5.1) and $\beta_1 = 1$, there is only the martensite number one and the stress is $\sigma = K\varepsilon - \tau_{11}(T)$.

The increase of deformation produces the martensite–austenite phase change. The same result is obtained when decreasing the deformation: the phase change from austenite to martensite occurs at $\varepsilon = -\varepsilon_1$. When $\varepsilon < -\varepsilon_1$, there is. only martensite number two, $\beta_2 = 1$, and the stress is $\sigma = K\varepsilon + \tau_{11}(T)$, (Fig. 5.4). The resulting constitutive law shown on Fig. 5.4 is typical of the behaviour of shape memory alloys at medium temperature.

### *5.4.5 Latent Heat of Austenite–Martensite Phase Change*

Let us assume the temperature to be fixed and compute

$$Tds = -Td(\frac{\partial \Psi}{\partial T}) = -(\overline{\tau} : \varepsilon + L\frac{T}{T_0})d\beta_2 + (\overline{\tau} : \varepsilon - L\frac{T}{T_0})d\beta_1 - (\beta_2 - \beta_1)\overline{\tau} : d\varepsilon,$$

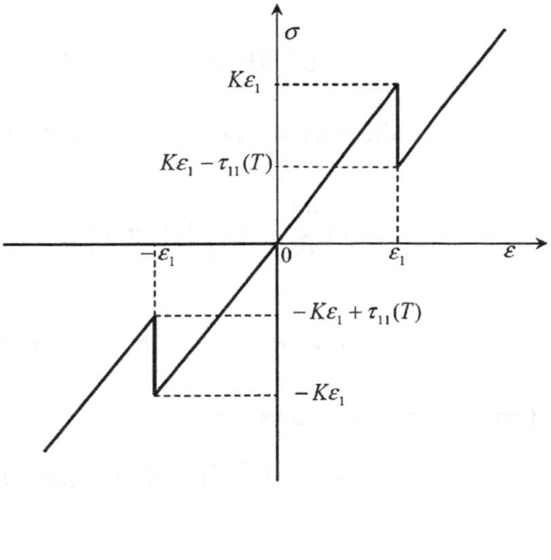


**Fig. 5.4** The constitutive law at medium temperature

where $s$ is the entropy. When a phase change occurs at a fixed medium temperature from austenite to martensite number one at the deformation $\varepsilon = \varepsilon_1$, i.e., when going from $\varepsilon$ slightly lower than $\varepsilon_1$ to $\varepsilon$ slightly greater than $\varepsilon_1$, the reversible heat received by the material is

$$T\Delta s = -L\frac{T}{T_0} + \overline{\tau}_{11}\varepsilon_1,$$

because $\Delta\beta_2 = 0$, $\Delta\beta_1 = 1$. It has been assumed $\overline{\tau}_{11} < 0$, thus $T\Delta s < 0$. The austenite–martensite phase change is exothermic at medium temperature: when the material is deformed heat is produced. This result is in accordance with experiments [136, 183]. Note that the quantity $L$ is the latent heat of austenite–martensite volume phase change at temperature $T_0$ of the undeformed material ($\varepsilon_1 = 0$ and $T = T_0$).

Within the small perturbation assumption, the previous relationships become

$$Tds = -Td(\frac{\partial\Psi}{\partial T}) = -L\frac{T}{T_0}(d\beta_2 + d\beta_1) = L\frac{T}{T_0}d\beta_3,$$

where $\beta_3 = 1 - \beta_1 - \beta_2$ is the austenite volume fraction. Note that the internal energy is

$$e = \Psi + Ts = CT + L\beta_3,$$

within the small perturbation assumption with $L$ the austenite–martensite phase change latent heat.

### 5.4.6 The Dissipative Constitutive Laws

The dissipative forces are

$$\sigma^d(\delta E, \chi) = 0,\ \mathbf{B}^d(\delta E, \chi) \in \partial\Phi(\frac{\partial\beta}{\partial t}),$$

where the subdifferential set of the non-smooth function $(\partial\beta/\partial t) \to \Phi$ is

$$\partial\Phi(\frac{\partial\beta}{\partial t}) = c\frac{\frac{\partial\beta}{\partial t}}{\left\|\frac{\partial\beta}{\partial t}\right\|} + k\frac{\partial\beta}{\partial t},\ \text{if}\ \frac{\partial\beta}{\partial t} \neq 0,$$

$$\partial\Phi(0) = \mathscr{S} = \left\{\mathbf{S} \in R^2; \|\mathbf{S}\| \le c,\right\},\ \text{if}\ \frac{\partial\beta}{\partial t} = 0\,.$$

Then the constitutive laws are

$$\sigma = \sigma^{nd}(E) = K : \varepsilon - (\beta_1 - \beta_2)\tau(T), \tag{5.24}$$

$$\mathbf{B}^{nd}(E) = \{-\tau(T) : \varepsilon + \frac{L}{T_0}(T - T_0), \tau(T) : \varepsilon + \frac{L}{T_0}(T - T_0)\},$$

$$\mathbf{B}^{ndr}(E, \mathbf{x}, t) \in \partial I(\beta(\mathbf{x}, t)),$$

$$\mathbf{B} = \mathbf{B}^{nd}(E) + \mathbf{B}^{ndr}(E, \mathbf{x}, t) + \mathbf{B}^{d}(\frac{\partial \beta}{\partial t}), \tag{5.25}$$

$$H = k \mathrm{grad} \beta,$$

$$-\mathbf{Q} = \mathbf{Q}^{d}(\delta E, T) = \frac{\lambda}{T} \mathrm{gradT}\,. \tag{5.26}$$

### 5.4.7 Evolution of a Shape Memory Alloy

The equations of the predictive theory for the small displacement $\mathbf{u}(x,t)$, the martensites volume fractions $\beta(x,t)$ and the temperature $T(\mathbf{x},t)$ of a structure $\Omega$ made of shape memory alloy are:

- The equation for the macroscopic motion

$$\mathrm{div}(\mathrm{K} : \varepsilon(\mathbf{u}) - (\beta_1 - \beta_2)\tau(\mathrm{T})) + \mathbf{f} = 0, \mathrm{in}\ \Omega,\ \sigma\mathbf{N} = \mathbf{g},\ \mathrm{on}\ \partial\Omega\,. \tag{5.27}$$

- The equations for the microscopic motions

$$\begin{pmatrix} -\tau(T) : \varepsilon(\mathbf{u}) + \frac{L}{T_0}(T - T_0) \\ \tau(T) : \varepsilon(\mathbf{u}) + \frac{L}{T_0}(T - T_0) \end{pmatrix} + \partial\Phi\left(\frac{\partial \beta}{\partial t}\right) + \partial I(\beta) - \begin{pmatrix} k\Delta\beta_1 \\ k\Delta\beta_2 \end{pmatrix} \ni 0,\ in\ \Omega,$$

$$\begin{pmatrix} k\frac{\partial \beta_1}{\partial N} \\ k\frac{\partial \beta_2}{\partial N} \end{pmatrix} = 0,\ on\ \partial\Omega\,. \tag{5.28}$$

- The entropy balance

$$\frac{\partial}{\partial t}\left\{(\beta_1 - \beta_2)\frac{d\tau(T)}{dT} : \varepsilon(\mathbf{u}) + (1 - \beta_1 - \beta_2)\frac{L}{T_0} + C(1 + \ln T)\right\}$$

$$-\lambda\Delta \ln T = \frac{1}{T}\left\{\frac{\partial \beta}{\partial t} \cdot \partial\Phi(\frac{\partial \beta}{\partial t})\right\},\ in\ \Omega,$$

$$\frac{\lambda \partial \ln T}{\partial N} = \pi,\ on\ \partial\Omega, \tag{5.29}$$

where $\pi$ is the surfacic rate of entropy provided to the shape memory structure.

These equations are completed by initial conditions for temperature $T$ and volume fractions $\beta$

$$T(\mathbf{x},0) = T_0(\mathbf{x}),\ \beta(\mathbf{x},0) = \beta_0(\mathbf{x})\ .$$

It is to be emphasized that the three partial differential equations are coupled in agreement with the fact that thermal actions can produce significant deformations. The equations, mainly the entropy balance, may be simplified within the small perturbation assumption (see Remarks 5.4 and 5.5). For instance, we may choose

$$C\frac{\partial T}{\partial t} + L\left(\frac{\partial \beta_1}{\partial t} + \frac{\partial \beta_2}{\partial t}\right) - \lambda \Delta T = 0,$$

as an approximation of the entropy balance or of the energy balance.

To illustrate the behaviour of shape memory alloys, let us consider a isotherm and homogeneous sample, i.e., grad$\beta = 0$. Equations of motion (5.28) result in non-smooth differential equations. The problem to investigate is the evolution of a material subjected to time dependent exterior actions. In this section an unidimensional experiment is investigated. We choose to apply a deformation $\varepsilon(t)$ because it is easier to investigate the structure of the equations and exhibit the hysteretic properties of shape memory alloys. We consider an experiment at medium temperature ($T_0 < T < T_c$): the applied deformation increases from zero then decreases till the stress is zero. The initial mixture is made of the two martensites with equal volume fractions $\beta_1(0) = \beta_2(0) = 1/2$. The point $(\sigma(t), \varepsilon(t))$ in Fig. 5.5 follows the path 1, 2, 3, 4, 5, in agreement with experiments.

In actual experiments, the temperature evolves due to the heat supplied by the phase changes. The results, for instance, in tension experiments, depend on the way the heat is exchanged with the outside of the sample under consideration. Thus the characterization experiments must take into account both the mechanical and the thermal phenomena. The dissipative phenomena are not as important as thought, [64, 65, 159, 185]. To be didactic, the previous example has been assumed, as already said, to be isothermal.

## 5.5 Education of Shape Memory Alloys

Shape memory alloys can exhibit one shape memory effects or two shape memory effects. A shape memory alloy with one shape memory effect remembers only one shape: the material can be permanently deformed at low temperature by an external mechanical action, then submitted to heating it goes back to its initial form but under cooling it does not recover its deformed shape (Fig. 5.6).

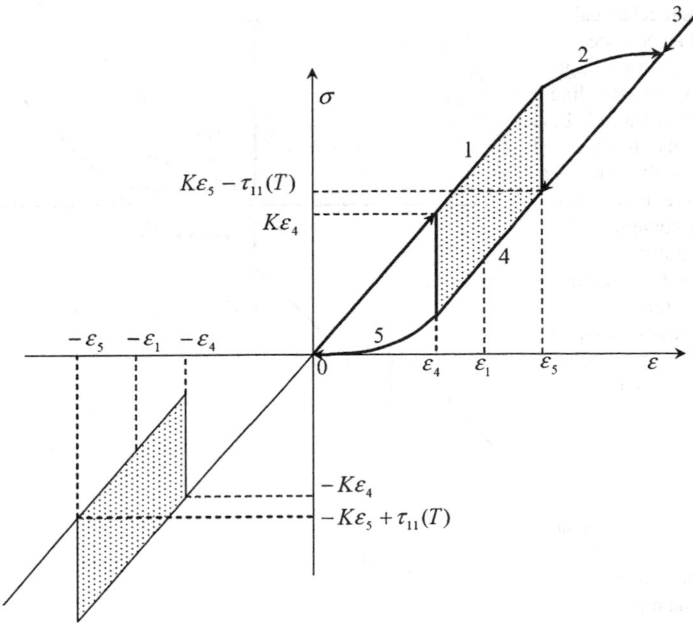


**Fig. 5.5** The stress $\sigma(t)$ versus deformation $\varepsilon(t)$ in an experiment at medium temperature. In the grey domains the alloy can be in equilibrium

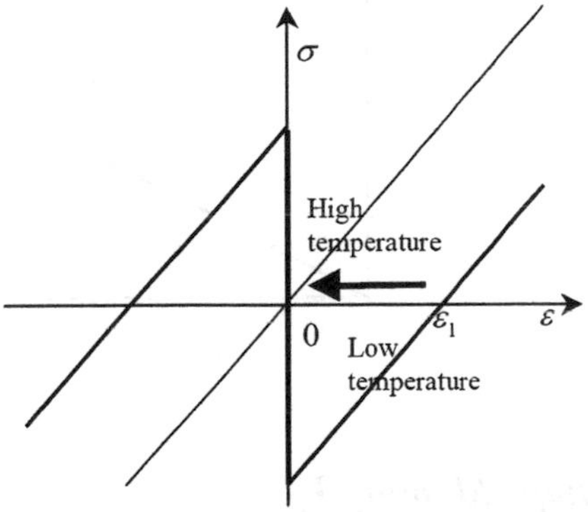


**Fig. 5.6** The mechanical behaviour of a non educated shape memory alloy at low temperature in bold line and at high temperature in standard line. A deformed shape memory alloys heated goes back to the initial configuration but cooling of an hot shape memory alloy does not give it the deformed configuration. It is said the non educated shape memory remembers the undeformed configuration

A shape memory alloy with two shape memory effect remembers two shapes: one at low temperature, an other one at high temperature. Following the temperature evolution, it goes from one shape to the other without external action (Fig. 5.7).

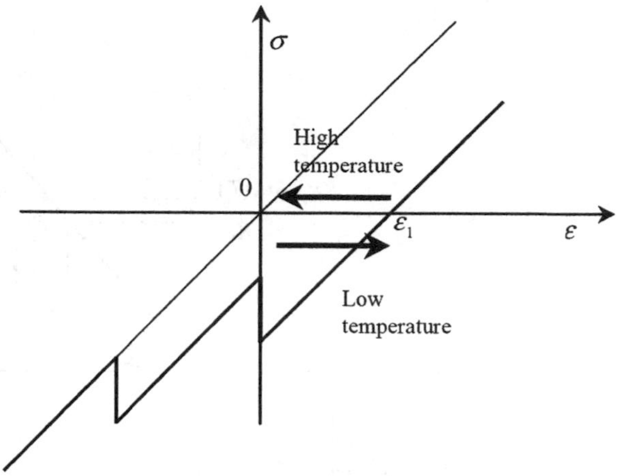


**Fig. 5.7** The mechanical behaviour of an educated shape memory alloy at high temperature in standard line and at low temperature in bold line. The shape of the material changes reversibly following the temperature. It goes from the reference configuration at high temperature to the deformed configuration at low temperature. It is said the educated shape memory alloys remembers the two configurations

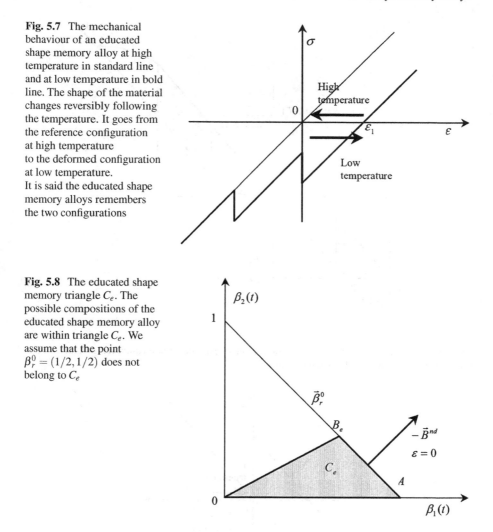

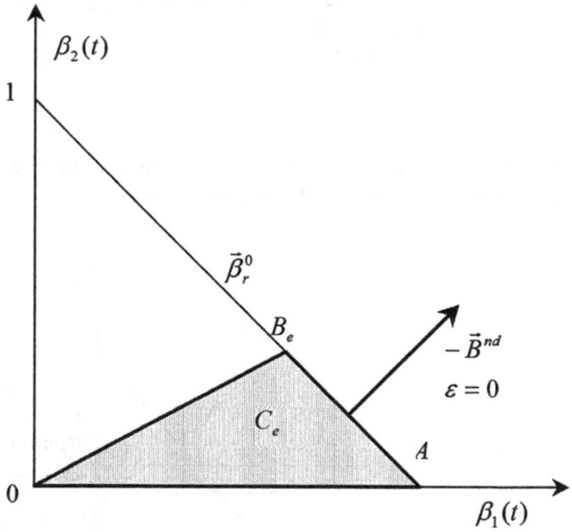


**Fig. 5.8** The educated shape memory triangle $C_e$. The possible compositions of the educated shape memory alloy are within triangle $C_e$. We assume that the point $\beta_r^0 = (1/2, 1/2)$ does not belong to $C_e$

### 5.5.1 The Two-Shape Memory Effect

The two shape memory effect can be obtained by a special thermomechanical treatment of the alloy which then can remember two shapes, one at low temperature, another one at high temperature. Depending on the temperature the shape goes from one to the other. This treatment is called the education [110, 183]. Its effect is to make one martensite dominant. It is much more present than the other one. From our point of view, the effect of education is to replace triangle $C$ by the flattened triangle $C_e$ (Fig. 5.8): the possible compositions of an educated shape memory alloy are within triangle $C_e$ (Fig. 5.8).

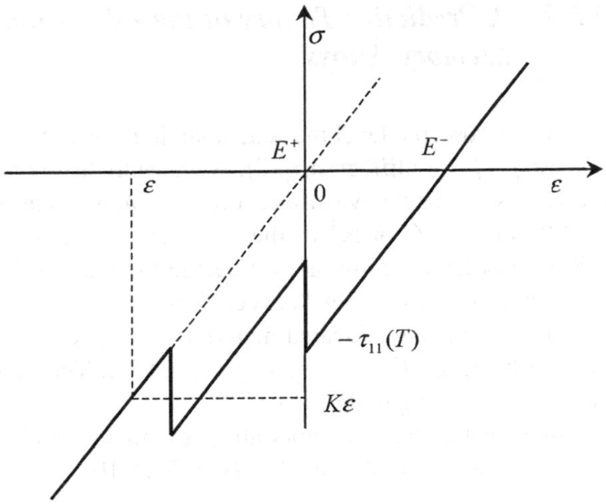


**Fig. 5.9** Constitutive law of a non dissipative, educated, shape memory alloy at low temperature (the bold line) and at high temperature (the dotted line). When temperature goes back and forth between $T^+$ and $T^-$, the alloy goes back and forth between states $E^+$ and $E^-$. It remembers the two states!

It is obvious that martensite number one is present to a greater degree than martensite number two.

Let us consider an unloaded ($\sigma = 0$), educated shape memory alloy at high temperature $T^+$. Its deformation is $\varepsilon = 0$. Let us cool it. The only unloaded ($\sigma = 0$) equilibrium state at low temperature $T^-$ is ($\sigma = 0$, $\varepsilon = \varepsilon_2$). If the temperature is again increased the alloy goes back to the state ($\sigma = 0, \varepsilon = 0$) and so on (Fig. 5.9). The alloy remembers two shapes: the state $E^+(\varepsilon = 0, \sigma = 0, \beta_3 = 1, T^+)$ and the state $E^-(\varepsilon = \varepsilon_2, \sigma = 0, \beta_1 = 1, T^-)$! One at high temperature and an other one at low temperature. The heat which is received from the exterior during the phase change, for instance when going from $E^-$ to $E^+$, is

$$e_r^+ - e_r^- = C(T^+ - T^-) - \frac{1}{2}K\left(\varepsilon_2(T^-)\right)^2 - T_c\overline{\tau}_{11}\varepsilon_2(T^-) + L,$$

with $\varepsilon_2 = \tau_{11}(T)/K = (T - T_c)\overline{\tau}_{11}/K$, ($K = K_{1111}$). Then

$$e_r^+ - e_r^- = C(T^+ - T^-) + L - \{(T^-)^2 - T_c^2\}\frac{\overline{\tau}_{11}^2}{2K}.$$

As we expect, it is positive because $T^- \leq T_c$.

*Remark 5.6.* The one shape memory effect describes what occurs when a non educated alloy is submitted to the same exterior action. The effect is not so striking because it requires mechanical action to go back and forth.

In the sequel it is given a macroscopic predictive theory of this education whose basic features have been given in [110].

## 5.5.2 *A Predictive Theory of the Education of Shape Memory Alloys*

At the macroscopic level the state quantities describing the state of the material are $E = (\varepsilon, \beta_1, \beta_2, \text{grad}\beta_1, \text{grad}\beta_2, \text{T})$. We have to distinguish two materials which have the same state $E$ but have different education. As it has been said the education result in different sets $C$ of $\mathbb{R}^3$ of the physically possible or feasible mixtures $(\beta_1,\beta_2)$, [183]. In order to be simple, we decide to quantify the education by an education index $\delta$ which has values between 0 and 1. The non educated material has index 0 and a completely educated material has index 1. Thus the state of a material is defined by $(\varepsilon, \beta_1, \beta_2, \beta_3, \text{grad}\beta_1, \text{grad}\beta_2, \delta, \text{T})$. The feasible mixtures $(\beta_1, \beta_2)$ with education $\delta$ belong to the set $C(\delta)$.

An educable shape memory alloy is a mixture which can have feasible mixtures $(\beta_1, \beta_2)$ which are in the set $C(0)$ (see Fig. 5.10)

$$C(0) = \{(\gamma_1, \gamma_2) \,|\, 0 \leq \gamma_i \leq 1;\ \gamma_1 + \gamma_2 = 1\},$$

when it is non educated and in $C(1)$ (see Figs. 5.11 or 5.12)

$$C(1) = \{(\gamma_1, \gamma_2) \,|\, 0 \leq \gamma_1 \leq 1;\ 0 \leq \gamma_2 \leq \beta_{max};\ \gamma_1 + \gamma_2 = 1\},$$

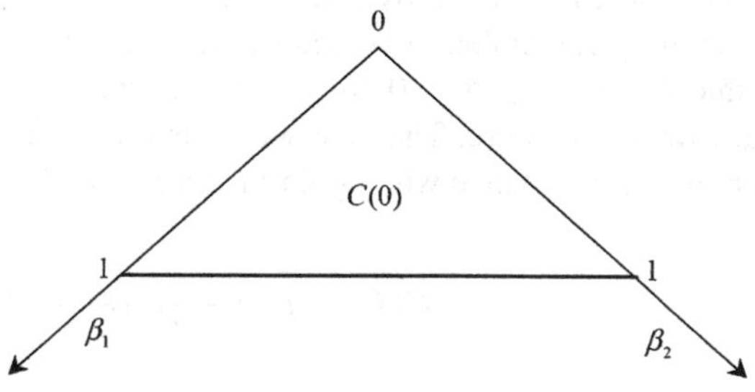


**Fig. 5.10** The triangle $C(0)$ of the feasible mixtures of a non educated material

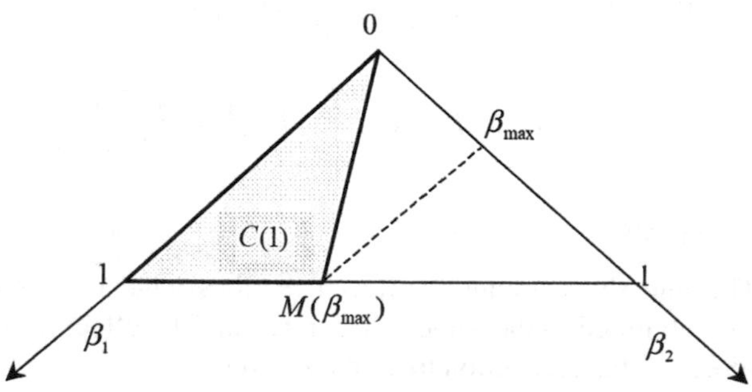


**Fig. 5.11** The possible mixtures of martensites of an educated shape memory alloy are in the triangle $C(1)$. The martensite 1 is much more present than the martensite 2 whose maximum volume fraction is $\beta_{max}$

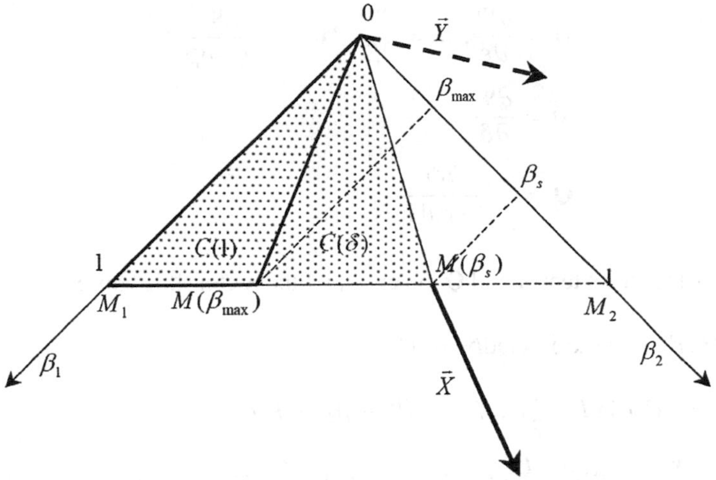


**Fig. 5.12** The triangle $C(1)$ of the feasible mixtures of an educated material. The triangle $C(\delta)$ in grey for the education index $\delta$. The projection **X** of vector **Z** on the plane $(\beta_1,\beta_2)$ in standard line for $\varepsilon$ small and negative and the projection **Y** of the same vector in dotted line for $\varepsilon$ smaller (larger in absolute value) and negative

when it is completely educated and in the set $C(\delta)$ when the education is uncompleted

$$C(\delta) = \{(\gamma_1,\gamma_2)\,|0 \le \gamma_1 \le 1;\ 0 \le \gamma_2 \le \beta_s(\delta);\ \gamma_1+\gamma_2 = 1\},$$

with

$$\delta = \frac{1-\beta_s}{1-\beta_{max}},$$

where $\beta_s$ is the ordinate of the vertex $M(\beta_s)$ of the triangle $C(\delta)$ containing the physically feasible mixtures $(\beta_1,\beta_2)$, where $\beta_{max}$ is the value of $\beta_s$ for a completely educated material ($\delta = 1$), (Figs. 5.11 and 5.12). The volume fraction $\beta_{max}$ is the maximum volume fraction of austenite 2 which can be present in an educated shape memory alloy.

The properties of the material are defined by its free energy $\Psi(E)$ and by its pseudopotential of dissipation $\Phi$, assumed for the sake of simplicity to depend only on $\partial\delta/\partial t = \dot{\delta}$, grad $T$ and $\chi = T$

$$\Phi = \Phi(\frac{\partial\delta}{\partial t}, \mathrm{gradT}, \mathrm{T})\,.$$

The resulting constitutive laws are (see [112] and Sect. 3.3.1)

$$\sigma = \frac{\partial \Psi}{\partial \varepsilon},\ B_i = \frac{\partial \Psi}{\partial \beta_i},\ \mathbf{H}_i = \frac{\partial \Psi}{\partial (\mathrm{grad}\beta_i)},$$
$$0 = \frac{\partial \Psi}{\partial \delta} + \frac{\partial \Phi}{\partial \dot{\delta}},$$
$$\mathbf{Q} = -\frac{\partial \Phi}{\partial (\mathrm{gradT})},$$

where $\sigma$ is the stress tensor and $\mathbf{Q}$ the entropy flux vector. It is chosen

$$\Psi(\varepsilon, \beta_1, \beta_2, \mathrm{grad}\beta_1, \mathrm{grad}\beta_2, \delta, \mathrm{T})$$
$$= -C_c T \ln T + \frac{1}{2}\varepsilon : K : \varepsilon - (\beta_1 - \beta_2)\alpha(T) tr\varepsilon$$
$$+ \frac{k}{2}(\mathrm{grad}\beta)^2 + \frac{L}{T_0}(T - T_0)(\beta_1 + \beta_2) - \frac{L_e}{T_e}(T - T_e)\delta + I_P(\beta_1, \beta_2, \delta),$$

where $C_c$ is the heat capacity, $K$ the elasticity tensor, $L$ the austenite–martensite phase change latent heat at temperature $T_0$, $L_e$ the education latent heat at temperature $T_e$ (it is assumed $T_e < T_0 < T_c$). The thermal expansion function $\alpha(T)$ is shown in Fig. 5.2.

To make the model more realistic, we can introduce two temperatures to characterize the austenite–martensite phase change: $T_0$, the temperature at the beginning of the transformation and $T_f$ the temperature at the end. The free energy is completed by $h(\beta) = (L_2/T_0)(T_0 - T_f)(\beta_3)^2$, [22, 155, 156, 185], (see Remark 4.2 of Chap. 4).

The function $I_P$ is the indicator function of the pyramid $P$ (Fig. 5.13) of all the feasible austenite–martensite mixtures

$$P = \bigcup_{0 \le \delta \le 1} C(\delta)$$
$$= \{(\gamma_1, \gamma_2)\,|0 \le \gamma_i \le 1;\ \gamma_1 + \gamma_2 = 1;\ \gamma_2 \le \frac{1 - \delta(1 - \beta_{\max})}{\delta(1 - \beta_{\max})}\gamma_1\}\,.$$

Even if the pyramid $P$, Fig. 5.13, is not a convex set, generalized derivatives of its indicator function can be defined [181]: they are normal vectors to the pyramid. Their properties are sufficient to ensure the mechanical coherence, for instance to ensure that the second principle of thermodynamics is satisfied.

The chosen pseudopotential of dissipation is

$$\Phi(\frac{\partial \delta}{\partial t}, \mathrm{gradT}, \mathrm{T}) = \frac{\lambda}{2\mathrm{T}}(\mathrm{gradT})^2 + \frac{\mathrm{a}}{2}(\frac{\partial \delta}{\partial \mathrm{t}})^2,$$

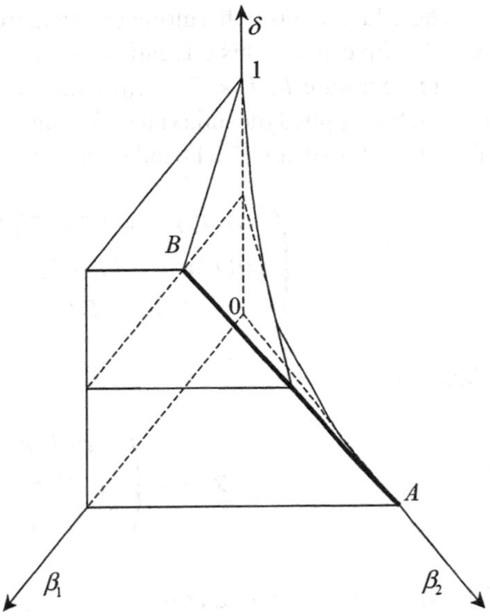


**Fig. 5.13** The pyramid $P$. The education follows the path $AB$

where $\lambda$, the thermal conductivity, and $a$, the parameter characterizing the dissipation of the education, are positive. It results the constitutive laws [112]

$$\sigma = K : \varepsilon + (\beta_2 - \beta_1)\alpha(T)1, \tag{5.30}$$

where 1 is the unit tensor;

$$\begin{pmatrix} B_1 \\ B_2 \\ 0 \end{pmatrix} \in \begin{pmatrix} -\alpha(T)tr\varepsilon + \frac{L}{T_0}(T - T_0) \\ \alpha(T)tr\varepsilon + \frac{L}{T_0}(T - T_0) \\ -\frac{L_e}{T_e}(T - T_e) \end{pmatrix} + \begin{pmatrix} 0 \\ 0 \\ a\frac{\partial \delta}{\partial t} \end{pmatrix} + \partial I_P(\beta_1, \beta_2, \delta), \tag{5.31}$$

and

$$\mathbf{H}_i = k\mathrm{grad}\beta_i,$$

$$\mathbf{Q} = -\frac{\lambda}{T}\mathrm{gradT}.$$

#### 5.5.2.1 The Education in Case There Is No Local Microscopic Interaction

For the sake of exhibiting the basic education features, we assume that there is no local interactions, i.e. $k = 0$. It results the equation of motion for the microscopic motions are

$$\begin{pmatrix} B_1 \\ B_2 \end{pmatrix} = 0.$$

The education is a thermomechanical treatment which makes $\delta$ to evolve from 0 to 1. To investigate this education, let us consider a non educated material at fixed low temperature $T$, $T_e < T < T_0$, with the initial composition $(\beta_1, \beta_2, \delta) = (0,1,0)$ to which is applied the uniaxial deformation $\varepsilon_{11} = \varepsilon < 0$, $(\varepsilon_{ij} = 0 \, for \, (i,j) \neq (1,1))$. The constitutive law (5.31) and equation of motion give with $tr\varepsilon = \varepsilon$

$$\begin{pmatrix} 0 \\ 0 \\ 0 \end{pmatrix} \in \begin{pmatrix} -\alpha(T)\varepsilon + \frac{L}{T_0}(T - T_0) \\ \alpha(T)\varepsilon + \frac{L}{T_0}(T - T_0) \\ -\frac{L_e}{T_e}(T - T_e) \end{pmatrix} + \begin{pmatrix} 0 \\ 0 \\ a\frac{\partial \delta}{\partial t} \end{pmatrix} + \partial I_P(\beta_1, \beta_2, \delta) \,. \quad (5.32)$$

Thus vector

$$\mathbf{Z} = -\begin{pmatrix} -\alpha(T)\varepsilon + \frac{L}{T_0}(T - T_0) \\ \alpha(T)\varepsilon + \frac{L}{T_0}(T - T_0) \\ -\frac{L_e}{T_e}(T - T_e) + a\frac{\partial \delta}{\partial t} \end{pmatrix},$$

is normal to the pyramid at the point $(\beta_1, \beta_2, \delta)$. For the chosen value of $\varepsilon$, vector

$$\mathbf{X} = -\begin{pmatrix} -\alpha(T)\varepsilon + \frac{L}{T_0}(T - T_0) \\ \alpha(T)\varepsilon + \frac{L}{T_0}(T - T_0) \end{pmatrix},$$

projection of vector $\mathbf{Z}$ on the plane $(\beta_1, \beta_2)$ is shown in a standard line in Fig. 5.12.

One can check that vector $\mathbf{Z}$ is normal to the pyramid on the segment $(A,B)$ (vector $\mathbf{X}$ is normal to triangle $C(\delta)$ at point $M(\beta_s)$). It results

$$a\frac{\partial \delta}{\partial t} = \frac{L_e}{T_e}(T - T_e) + 2(1 - \beta_{\max})\alpha(T)\varepsilon > 0,$$

for $\varepsilon$ small enough (in absolute value). The education index increases, the education is achieved when $\delta$ equals 1. The point representing the mixture has followed the path $(A,B)$, (Fig. 5.13).

When $\varepsilon$ is a little bit smaller (larger in absolute value) the point representing the mixture $(\beta_1, \beta_2, \delta)$ moves at the beginning on the segment $AB$. But when the point $C$ is reached (Fig. 5.14), vector $\mathbf{Z}$ whose projection, $\mathbf{Y}$, on the plane $(\beta_1, \beta_2)$ is shown in a dotted line in Fig. 5.12 can no longer be normal to the pyramid on the side $AB$. Vector $\mathbf{Y}$ becomes normal to triangle $C(\delta)$ at point 0. Thus the composition of the mixture changes and becomes austenite, $(\beta_1 = 0, \beta_2 = 0)$. The point representing the mixture goes from point $C$ to point $D$. The education resumes on the side $DE$ following the equation,

$$a\frac{\partial \delta}{\partial t} - \frac{L_e}{T_e}(T - T_e) = 0 \,.$$

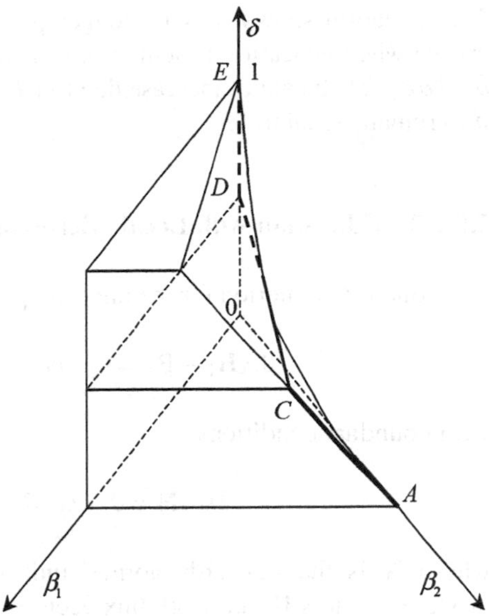


**Fig. 5.14** The education follows the path $(A,C,D,E)$

#### 5.5.2.2 Persistence of the Education

For the sake of simplicity a schematic pseudopotential has been chosen. It allows a loss of education as it can be seen in (5.31). It is not in agreement with experiments which show that education is often strong enough to persist. In order to agree with experiments, the pseudopotential is sophisticated a little bit by assuming the coefficient $a$ to depend on $\partial\delta/dt$,

$$\Phi(\frac{\partial\delta}{\partial t},\mathrm{gradT}) = \frac{\lambda}{2\mathrm{T}}(\mathrm{gradT})^2 + \frac{\mathrm{a}(\partial\delta/\mathrm{dt})}{2}(\frac{\partial\delta}{\partial \mathrm{t}})^2,$$

with $a(\partial\delta/dt) = a$ for $\partial\delta/dt > 0$ and $a(\partial\delta/dt) = a/\eta$ where $\eta$ is small compared to 1, for $\partial\delta/dt < 0$. It even can be forbidden $\partial\delta/dt$ to become negative by choosing,

$$\Phi(\frac{\partial\delta}{\partial t},\mathrm{gradT}) = \frac{\lambda}{2\mathrm{T}}(\mathrm{gradT})^2 + \frac{\mathrm{a}}{2}(\frac{\partial\delta}{\partial \mathrm{t}})^2 + \mathrm{I}_+(\frac{\partial\delta}{\partial \mathrm{t}}),$$

where $I_+$ is the indicator function of $\mathbb{R}^+$. It is equivalent to having $\eta = 0$. In this last situation, (5.31) describing the evolution becomes

$$\begin{pmatrix}0\\0\\0\end{pmatrix} \in \begin{pmatrix}-\alpha(T)tr\varepsilon + \frac{L}{T_0}(T-T_0)\\ \alpha(T)tr\varepsilon + \frac{L}{T_0}(T-T_0)\\ -\frac{L_e}{T_e}(T-T_e)\end{pmatrix} + \begin{pmatrix}0\\0\\ a\frac{\partial\delta}{\partial t} + \partial I_+(\frac{\partial\delta}{\partial t})\end{pmatrix} + \partial I_P(\beta_1,\beta_2,\delta),$$

This equation shows it is no longer possible to have the education index $\delta$ to evolve when education is achieved. In fact when $\delta = 1$, $\delta$ cannot decrease due to $\partial I_+(\partial\delta/dt)$. It cannot increase due to $\partial I_P(\beta_1, \beta_2, \delta)$ which makes compulsory for $\delta$ to remain equal to 1.

#### 5.5.2.3 Education with Local Microscopic Interaction

The equation of motion for the microscopic motions are

$$\mathrm{div}\mathbf{H}_1 - \mathrm{B}_1 = 0, \; and \; \mathrm{div}\mathbf{H}_2 - \mathrm{B}_2 = 0, \; in \; \Omega, \tag{5.33}$$

with boundary conditions

$$\mathbf{H}_1 \cdot \mathbf{N} = 0, \; and \; \mathbf{H}_2 \cdot \mathbf{N} = 0, \; on \; \partial\Omega,$$

where $\mathbf{N}$ is the outwards normal unit vector to domain $\Omega$ occupied by the mixture. Vectors $\mathbf{H}_i$ are work flux vectors describing the work fluxes related to the microscopic motions which occur in the martensites and austenite. Equation (5.33) describe the microscopic motions whereas (5.35) below describe the macroscopic motions. For the sake of completeness it is also assumed that there is dissipation with respect to the $\beta_i$'s, thus the pseudopotential of dissipation becomes

$$\begin{aligned}\Phi\left(\frac{\partial\beta_1}{\partial t}, \frac{\partial\beta_2}{\partial t}, \frac{\partial\delta}{\partial t}, \mathrm{gradT}, \mathrm{T}\right) &= \frac{\lambda}{2T}(\mathrm{gradT})^2 + \frac{\mathrm{a}}{2}\left(\frac{\partial\delta}{\partial \mathrm{t}}\right)^2 \\ &+ \frac{c_1}{2}\left(\frac{\partial\beta_1}{\partial t}\right)^2 + \frac{c_2}{2}\left(\frac{\partial\beta_2}{\partial t}\right)^2 + I_+\left(\frac{\partial\delta}{\partial t}\right),\end{aligned}$$

The equations of motion (5.33) and the constitutive laws give

$$\begin{pmatrix} 0 \\ 0 \\ 0 \end{pmatrix} \in \begin{pmatrix} -\alpha(T)tr\varepsilon + \frac{L}{T_0}(T - T_0) \\ \alpha(T)tr\varepsilon + \frac{L}{T_0}(T - T_0) \\ -\frac{L_e}{T_e}(T - T_e) \end{pmatrix} + \begin{pmatrix} c_1\frac{\partial\beta_1}{\partial t} - k_1\Delta\beta_1 \\ c_2\frac{\partial\beta_2}{\partial t} - k_2\Delta\beta_2 \\ a\frac{\partial\delta}{\partial t} + \partial I_+(\frac{\partial\delta}{\partial t}) \end{pmatrix} + \partial I_P(\beta_1, \beta_2, \delta), \; in \; \Omega, \tag{5.34}$$

with the boundary conditions

$$\frac{\partial\beta_1}{\partial N} = 0, \; and \; \frac{\partial\beta_2}{\partial N} = 0, \; on \; \partial\Omega,$$

where the $\partial\beta_i/\partial N$ are the normal derivatives. This equation of motion is to be coupled with the equation for macroscopic motions

$$\mathrm{div}\sigma + \mathbf{f} = \rho\gamma, \; in \; \Omega, \tag{5.35}$$

where $\rho$ is the density and $\gamma$ the acceleration, and with the entropy balance if the temperature is assumed not to be known

$$\frac{\partial s}{\partial t} - \lambda \Delta \ln T = \frac{1}{T}\left\{a(\frac{\partial \delta}{\partial t})^2 + c_1(\frac{\partial \beta_1}{\partial t})^2 + c_2(\frac{\partial \beta_2}{\partial t})^2\right\}, \ in\ \Omega,$$

where $s$ is the entropy

$$s = -\frac{\partial \Psi}{\partial T} = C_c(1 + \ln T) + \frac{L}{T_0}(\beta_1 + \beta_2) - \frac{L_e}{T_e}\delta + (\beta_1 - \beta_2)\frac{d\alpha(T)}{dt} tr\varepsilon,$$

giving

$$C_c\frac{\partial \ln T}{\partial t} + \frac{L}{T_0}(\frac{\partial \beta_1}{\partial t} + \frac{\partial \beta_2}{\partial t}) - \frac{L_e}{T_e}\frac{\partial \delta}{\partial t} + \frac{d}{dt}\left\{(\beta_1 - \beta_2)\frac{d\alpha(T)}{dt} tr\varepsilon\right\}$$
$$-\lambda \Delta \ln T = \frac{1}{T}\left\{a(\frac{\partial \delta}{\partial t})^2 + c_1(\frac{\partial \beta_1}{\partial t})^2 + c_2(\frac{\partial \beta_2}{\partial t})^2\right\}, \ in\ \Omega\ . \qquad (5.36)$$

This entropy equation can be approximated with the small perturbation assumption

$$C_c\frac{\partial T}{\partial t} + L(\frac{\partial \beta_1}{\partial t} + \frac{\partial \beta_2}{\partial t}) - L_e\frac{T_0}{T_e}\frac{\partial \delta}{\partial t} - \lambda \Delta T = 0, \ in\ \Omega\ .$$

Let us emphasize again how important is this entropy or energy equation because the thermal actions are equivalent to the mechanical actions. In modelling the behaviour of shape memory alloys the energy balance as to be dealt with the same care and importance as the motions equations. The energy balance is actually important since it has been shown that dissipative phenomena (compared to thermal phenomena) are not as important as it was thought, [64, 65]. A good modelling includes the three coupled equations (5.34), (5.35) with (5.30), and (5.36). One can find examples of thermomechanical evolutions in [210]. Let us note that it is possible to upgrade the theory by assuming local interactions in the education phenomenon, i.e., by having grad$\delta$ as a state quantity.

## 5.6 Shape Memory Alloys with the Possibility of Voids

There may be voids in the austenite–austenites mixture, which may appear when the alloy is produced by the aggregation of powders, as it has been recently reported in [195]. Of course, the voids are filled either with gas or air when appearing or when aggregating powders. We do not take into account the gas phase mechanical properties and focus on the mechanical behaviour of the solid mixture, thus we assume the volume fraction of voids is small.

### 5.6.1 The State Quantities

The volume fractions are not independent: they satisfy the following *internal* constraints

$$0 \leq \beta_i \leq 1, \quad i = 1,2,3, \tag{5.37}$$

due to the definition of volume fractions. Since we assume that voids can appear in the martensite–austenite mixture, then the $\beta_i$'s must satisfy the other internal constraint

$$\beta_1 + \beta_2 + \beta_3 \leq 1, \tag{5.38}$$

the quantity $v_o = 1 - (\beta_1 + \beta_2 + \beta_3)$ being the voids volume fraction. This is the case when the alloy is produced by aggregating powders as shown in [195].

We denote by $\beta$ the vector of components $\beta_i$ ($i = 1,2,3$) and the set of the state quantities is

$$E = \{\varepsilon, \beta, \mathrm{grad}\beta, \mathrm{T}\},$$

while the quantities which describe the evolution and the thermal heterogeneity are

$$\delta E = \{\varepsilon(\mathbf{U}), \frac{d\beta}{dt}, \mathrm{grad}\frac{d\beta}{dt}, \mathrm{gradT}\}.$$

The gradient, grad$\beta$, accounts for local interactions of the volume fractions at their neighborhood points.

### 5.6.2 The Mass Balance

Assuming the same constant density $\rho$ (see Sect. 4.6 for phase change with different densities, [125]) and the same velocity $\mathbf{U}$ for each phase, the mass balance reads

$$\rho \frac{d(\beta_1 + \beta_2 + \beta_3)}{dt} + \rho(\beta_1 + \beta_2 + \beta_3)\mathrm{div}\mathbf{U} = 0.$$

Within the small perturbation assumption, this equation gives

$$\rho \frac{\partial(\beta_1 + \beta_2 + \beta_3)}{\partial t} + \rho(\beta_1^0 + \beta_2^0 + \beta_3^0)\mathrm{div}\mathbf{U} = 0,$$

where the $\beta_i^0$'s are the initial values of the $\beta_i$. In agreement with the assumption that the voids volume fraction is small, we assume

$$\beta_1^0 + \beta_2^0 + \beta_3^0 = 1, \tag{5.39}$$

and have

$$\frac{\partial(\beta_1 + \beta_2 + \beta_3)}{\partial t} + \mathrm{div}\mathbf{U} = 0\,. \tag{5.40}$$

Mass balance is a relationship between the quantities of $\delta E$, indeed, its effects will be included in the dissipative properties (see relationship (5.46) below).

### *5.6.3 The Equations of Motion*

They result from the principle of virtual power involving the power of the internal forces,

$$-\int_{\Omega} \{\sigma : \mathsf{D}(\mathbf{V}) + \mathbf{B}\cdot\gamma + \mathsf{H} : \mathrm{grad}\gamma\}\, d\Omega,$$

where $\mathbf{V}$ and $\gamma$ are virtual velocities. The internal forces are the stress $\sigma$, the phase change work vector $\mathbf{B}$, and the phase change work flux tensor $\mathsf{H}$. The equations of motion are

$$\rho\frac{\partial \mathbf{U}}{\partial t} = \mathrm{div}\sigma + \mathbf{f},\ 0 = \mathrm{div}\mathsf{H} - \mathbf{B} + \mathbf{A},\ \text{in } \Omega, \tag{5.41}$$

$$\sigma\mathbf{N} = \mathbf{g},\ \mathsf{H}\mathbf{N} = \mathbf{a},\ on\ \partial\Omega, \tag{5.42}$$

where $\rho$ is the density of the alloy which occupies the domain $\Omega$, with boundary $\partial\Omega$ and outward normal vector $\mathbf{N}$. The alloy is loaded by body forces $\mathbf{f}$ and by surface tractions $\mathbf{g}$, and submitted to body sources of phase change work $\mathbf{A}$ and surfaces sources of phase change work $\mathbf{a}$ (for instance, electric, magnetic or radiative actions producing the evolution of the alloys without macroscopic motion). In the following we will suppose, for simplicity, $\mathbf{A} = \mathbf{a} = 0$.

### *5.6.4 The Free Energy*

As explained above, a shape memory alloy is considered as a mixture of the martensite and austenite phases with volume fractions $\beta_i$. The volumic free energy of the mixture we choose is

$$\Psi = \Psi(E) = \sum_{i=1}^{3} \beta_i \Psi_i(E) + h(E), \tag{5.43}$$

where the $\Psi_i$'s are the volume free energies of the $i$ phases and $h$ is a free energy describing interactions between the different phases. We have assumed that internal constraints are physical properties, hence, we decide to choose properly the two functions describing the material, i.e., the free energy $\Psi$ and the pseudopotential of dissipation $\Phi$, in order to take these constraints into account. Since, the pseudopotential describes the kinematic properties (i.e., properties which depend on velocities) and the free energy describes the state properties, obviously the internal constraints (5.37) and (5.38) are to be taken into account with free energy $\Psi$.

For this purpose, we assume the $\Psi_i$'s are defined over the whole linear space spanned by $\beta_i$ and the free energy is defined by

$$\Psi(E) = \beta_1 \Psi_1(E) + \beta_2 \Psi_2(E) + \beta_3 \Psi_3(E) + h(E)\,.$$

We choose the very simple interaction free energy

$$h(E) = I_C(\beta) + \frac{k}{2} \left|\mathrm{grad}\beta\right|^2,$$

where $I_C$ is the indicator function of the convex set

$$C = \{(\gamma_1, \gamma_2, \gamma_3) \in \mathbb{R}^3;\ 0 \leq \gamma_i \leq 1;\ \gamma_1 + \gamma_2 + \gamma_3 \leq 1\}\,. \tag{5.44}$$

The terms $I_C(\beta) + (k/2)\left|\mathrm{grad}\beta\right|^2$ may be seen as a *mixture or interaction free-energy*.

The only effect of $I_C(\beta)$ is to guarantee that the proportions $\beta_1$, $\beta_2$ and $\beta_3$ take admissible physical values, i.e. they satisfy constraints (5.37) and (5.38) (see also (5.44)). The interaction free energy term $I_C(\beta)$ is equal to zero when the mixture is physically possible ($\beta \in C$) and to $+\infty$ when the mixture is physically impossible ($\beta \notin C$).

Let us note even if the free energy of the voids phase is 0, the voids phase has physical properties due to the interaction free energy term $(k/2)\left|\mathrm{grad}\beta\right|^2$ which depends on the gradient of $\beta$. It is known that this gradient is related to the interfaces properties: $\mathrm{grad}\beta_1$, $\mathrm{grad}\beta_2$ describe properties of the voids–martensites interfaces and $\mathrm{grad}\beta_3$ describes properties of the voids-austenite interface. In this setting, the voids have a role in the phase change and make it different from a phase change without voids. The model is simple and schematic but it may be upgraded by introducing sophisticated interaction free energy depending on $\beta$ and on $\mathrm{grad}\beta$.

For the sake of simplicity, we choose the volume free energies as before

$$\Psi_1(E) = \frac{1}{2}\varepsilon(\mathbf{u}) : K : \varepsilon(\mathbf{u}) - \alpha(T)1 : \varepsilon(\mathbf{u}) - CT\ln T,$$

$$\Psi_2(E) = \frac{1}{2}\varepsilon(\mathbf{u}) : K : \varepsilon(\mathbf{u}) + \alpha(T)1 : \varepsilon(\mathbf{u}) - CT\ln T,$$

$$\Psi_3(E) = \frac{1}{2}\varepsilon(\mathbf{u}) : K : \varepsilon(\mathbf{u}) - \frac{L}{T_0}(T - T_0) - CT\ln T,$$

where $K$ is the volume elastic tensor and $C$ the volume heat capacities of the phases and the quantity $L$ is the latent heat martensite-austenite volume phase change at temperature $T_0$ as already seen (see Sect. 5.4.5).

Concerning the stress $\alpha(T)1$, we keep the schematic simple expression

$$\alpha(T) = (T - T_c)\overline{\tau}_{11}, \text{ for } T \leq T_c, \ \alpha(T) = 0, \text{ for } T \geq T_c,$$

with $\overline{\tau}_{11} \leq 0$ and assume the temperature $T_c$ is greater than $T_0$. With those assumptions, it results

$$\Psi(E) = \frac{(\beta_1 + \beta_2 + \beta_3)}{2}\{\varepsilon(\mathbf{u}) : K : \varepsilon(\mathbf{u})\}$$
$$-(\beta_1 - \beta_2)\alpha(T)1 : \varepsilon(\mathbf{u}) - \beta_3\frac{L}{T_0}(T - T_0)$$
$$-(\beta_1 + \beta_2 + \beta_3)CT\ln T + \frac{k}{2}|\mathrm{grad}\beta|^2 + I_C(\beta) \,. \quad (5.45)$$

## 5.6.5 *The Pseudopotential of Dissipation*

The dissipative forces are defined via a pseudopotential of dissipation $\Phi$. As already remarked, the mass balance (5.40) is a relationship between velocities of $\delta E$. Thus we take it into account in order to define the pseudopotential and introduce the indicator function $I_0$ of the origin of $\mathbb{R}$ as follows

$$I_0(\frac{\partial(\beta_1 + \beta_2 + \beta_3)}{\partial t} + \mathrm{div}\mathbf{U}) \,.$$

From experiments, it is known that the behaviour of shape memory alloys depends on time, i.e., the behaviour is dissipative. We define a pseudopotential of dissipation with $\chi = T$

$$\Phi\left(\frac{\partial\beta}{\partial t}, \mathrm{grad}\frac{\partial\beta}{\partial \mathrm{t}}, \mathrm{gradT}, \mathrm{T}\right) = \frac{\mathrm{c}}{2}\left|\frac{\partial\beta}{\partial \mathrm{t}}\right|^2 + \frac{\upsilon}{2}|\mathrm{grad}\frac{\partial\beta}{\partial \mathrm{t}}|^2$$
$$+\frac{\lambda}{2T}|\mathrm{gradT}|^2 + I_0(\frac{\partial(\beta_1 + \beta_2 + \beta_3)}{\partial t} + \mathrm{div}\mathbf{U}), \quad (5.46)$$

where $\lambda \geq 0$ represents the thermal conductivity and $c \geq 0$, $\upsilon \geq 0$ stand for phase change viscosities. We have assumed dissipation with respect to $\mathrm{grad}(\partial\beta/\partial \mathrm{t})$ to show how it intervene.

### 5.6.6 The Constitutive Laws

The internal forces are split between non-dissipative forces $\sigma^{nd}$, $\mathbf{B}^{nd}$ and $\mathsf{H}^{nd}$ depending on $(E,\mathbf{x},t)$ and dissipative forces by

$$\left\{\sigma^d,\mathbf{B}^d,\mathsf{H}^d,-\mathbf{Q}\right\},$$

depending on

$$\delta E = \{\frac{\partial\varepsilon}{\partial t},\frac{\partial\beta}{\partial t},\mathrm{grad}\frac{\partial\beta}{\partial \mathrm{t}},\mathrm{gradT}\}, \tag{5.47}$$

and on $(E,\mathbf{x},t)$

$$\sigma = \sigma^{nd} + \sigma^d,\ \mathbf{B} = \mathbf{B}^{nd} + \mathbf{B}^d,\ \mathsf{H} = \mathsf{H}^{nd} + \mathsf{H}^d\ .$$

The non dissipative forces are defined with the free energy

$$\sigma^{nd}(E) = \frac{\partial\Psi}{\partial\varepsilon(\mathbf{u})}(E) = (\beta_1+\beta_2+\beta_3)K:\varepsilon(\mathbf{u}) - (\beta_1-\beta_2)\alpha(T)1, \tag{5.48}$$

$$\mathbf{B}^{nd}(E,\mathbf{x},t) = \frac{\partial\Psi}{\partial\beta}(E)$$

$$= \frac{1}{2}\begin{pmatrix} \varepsilon(\mathbf{u}):K:\varepsilon(\mathbf{u}) - 2\alpha(T)1:\varepsilon(\mathbf{u}) \\ \varepsilon(\mathbf{u}):K:\varepsilon(\mathbf{u}) + 2\alpha(T)1:\varepsilon(\mathbf{u}) \\ \varepsilon(\mathbf{u}):K:\varepsilon(\mathbf{u}) - 2\frac{L}{T_0}(T-T_0) \end{pmatrix} + \mathbf{B}^{ndr}(E,\mathbf{x},t), \tag{5.49}$$

$$\mathbf{B}^{ndr}(E,\mathbf{x},t) \in \partial I_C(\beta), \tag{5.50}$$

$$\mathsf{H}^{nd} = \frac{\partial\Psi}{\partial(\mathrm{grad}\beta)}(E) = k\,\mathrm{grad}\beta\ . \tag{5.51}$$

The dissipative forces are defined with the pseudopotential of dissipation

$$\left\{\sigma^d,\mathbf{B}^d,\mathsf{H}^d,-\mathbf{Q}^d\right\} \in \partial\Phi(\delta E,\chi), \tag{5.52}$$

where the subdifferential of $\Phi$ is with respect to $\delta E$. Relationship (5.52) gives

$$\sigma^d = -p1, \tag{5.53}$$

$$\mathbf{B}^d = -p \begin{pmatrix} 1 \\ 1 \\ 1 \end{pmatrix} + c\frac{\partial \beta}{\partial t}, \tag{5.54}$$

$$\mathsf{H}^d = \upsilon \text{grad} \frac{\partial \beta}{\partial \mathrm{t}}, \tag{5.55}$$

$$-\mathbf{Q}^d = \frac{\lambda}{T} \text{gradT}, \tag{5.56}$$

where $p$ is the pressure in the mixture and defined by

$$-p \in \partial I_0 \left( \frac{\partial(\beta_1 + \beta_2 + \beta_3)}{\partial t} + \text{div}\mathbf{U} \right). \tag{5.57}$$

The state laws (5.48)–(5.51), besides implying that the internal constraints are satisfied, give also the value of the reactions, during the evolution, to these internal constraints.

Relationships (5.48)–(5.51) and (5.53)–(5.57) give the constitutive laws

$$\sigma = (\beta_1 + \beta_2 + \beta_3) K : \varepsilon(\mathbf{u}) - ((\beta_1 - \beta_2)\alpha(T) + p)1, \tag{5.58}$$

$$\mathbf{B}(E, \delta E, \mathbf{x}, t) = \frac{1}{2} \begin{pmatrix} \varepsilon(\mathbf{u}) : K : \varepsilon(\mathbf{u}) - 2\alpha(T)1 : \varepsilon(\mathbf{u}) - p \\ \varepsilon(\mathbf{u}) : K : \varepsilon(\mathbf{u}) + 2\alpha(T)1 : \varepsilon(\mathbf{u}) - p \\ \varepsilon(\mathbf{u}) : K : \varepsilon(\mathbf{u}) - 2\frac{L}{T_0}(T - T_0) - p \end{pmatrix}$$

$$+ \mathbf{B}^{ndr}(E, \mathbf{x}, t) + c\frac{\partial \beta}{\partial t}, \tag{5.59}$$

$$\mathbf{B}^{ndr}(E, \mathbf{x}, t) \in \partial I_C(\beta), \tag{5.60}$$

$$-p \in \partial I_0 \left( \frac{\partial(\beta_1 + \beta_2 + \beta_3)}{\partial t} + \text{div}\mathbf{U} \right), \tag{5.61}$$

$$\mathsf{H} = k \text{grad} \beta + \upsilon \text{grad} \frac{\partial \beta}{\partial \mathrm{t}}, \tag{5.62}$$

$$-\mathbf{Q}(E, \delta E) = -\mathbf{Q}^d(E, \delta E) = -\frac{\lambda}{T} \text{gradT}. \tag{5.63}$$

It can be proved that our choice is such that the internal constraints and the second law of thermodynamics are satisfied, (see [112, 119] and the next Sect. 5.6.7).

### 5.6.7 The Entropy Balance

By denoting

$$s = -\frac{\partial \Psi}{\partial T} = (\beta_1 + \beta_2 + \beta_3) C (1 + \ln T) + \beta_3 \frac{L}{T_0}, \tag{5.64}$$

the entropy balance is

$$\begin{aligned} \frac{\partial s}{\partial t} + \operatorname{div} \mathbf{Q} &= R + \frac{1}{T}\left\{\sigma^d : \frac{\partial \varepsilon}{\partial t} + \mathbf{B}^d \cdot \frac{\partial \beta}{\partial t} + \mathsf{H}^d : \mathrm{grad}\frac{\partial \beta}{\partial \mathrm{t}} - \mathbf{Q}\cdot \mathrm{gradT}\right\} \\ &= R + \frac{1}{T}\left\{c|\frac{\partial \beta}{\partial t}|^2 + \upsilon|\mathrm{grad}\frac{\partial \beta}{\partial \mathrm{t}}|^2 + \frac{\lambda}{\mathrm{T}}|\mathrm{gradT}|^2\right\}, \ in\ \Omega, \end{aligned} \tag{5.65}$$

$$-\mathbf{QN} = \pi, \ on\ \partial\Omega, \tag{5.66}$$

because

$$p\left(\frac{\partial(\beta_1 + \beta_2 + \beta_3)}{\partial t} + \mathrm{div}\mathbf{U}\right) = 0,$$

due to (5.57), the pressure is workless. Vector $T\mathbf{Q}$ is the heat flux vector, $RT$ is the exterior volume rate of heat that is supplied to the alloy, $T\pi$ is the rate of heat that is supplied by contact action, $\partial\varepsilon/\partial t = \varepsilon(\mathbf{U})$ is the strain rate. The constitutive laws, within the small perturbation assumption and (5.39), become

$$\sigma = K : \varepsilon(\mathbf{u}) - ((\beta_1 - \beta_2)\alpha(T) + p)1, \tag{5.67}$$

$$-p \in \partial I_0\left(\frac{\partial(\beta_1 + \beta_2 + \beta_3)}{\partial t} + \mathrm{div}\mathbf{U}\right),$$

$$\mathbf{B} = \begin{pmatrix} -\alpha(T)1 : \varepsilon(\mathbf{u}) - p \\ \alpha(T)1 : \varepsilon(\mathbf{u}) - p \\ -\frac{L}{T_0}(T - T_0) - p \end{pmatrix} + \mathbf{B}^{ndr} + c\frac{\partial \beta}{\partial t},$$

$$\mathbf{B}^{ndr} \in \partial I_C(\beta),$$

$$\mathsf{H} = k\,\mathrm{grad}\beta + \upsilon\,\mathrm{grad}\frac{\partial \beta}{\partial \mathrm{t}},$$

$$\mathbf{Q}(E, \delta E) = \mathbf{Q}^d(E, \delta E) = -\frac{\lambda}{T}\mathrm{gradT}\,. \tag{5.68}$$

The entropy becomes

$$s = -\frac{\partial \Psi}{\partial T} = C(1 + \ln T) + \beta_3 \frac{L}{T_0}, \tag{5.69}$$

### 5.6.8 *The Set of Partial Differential Equations*

We assume also quasi-static evolution and, using again the small perturbation assumption (i.e. neglecting the higher order contributions in the velocities in (5.65) which are smaller than the other quantities), we get the following set of partial differential equations coupling the equations of motion (5.41), the entropy balance (5.65) with entropy (5.69) and constitutive laws (5.67)–(5.68)

$$\mathrm{div}\,(K:\varepsilon(\mathbf{u})-((\beta_1-\beta_2)\alpha(T)+p)1)+\mathbf{f}=0, \tag{5.70}$$

$$-p\in\partial I_0(\frac{\partial(\beta_1+\beta_2+\beta_3)}{\partial t}+\mathrm{div}\mathbf{U}),$$

$$c\frac{\partial\beta}{\partial t}-\upsilon\Delta\frac{\partial\beta}{\partial t}-k\Delta\beta+\begin{pmatrix}-\alpha(T)1:\varepsilon(\mathbf{u})-p\\ \alpha(T)1:\varepsilon(\mathbf{u})-p\\ -\frac{L}{T_0}(T-T_0)-p\end{pmatrix}+\mathbf{B}^{ndr}=0,$$

$$\mathbf{B}^{ndr}\in\partial I_C(\beta),$$

$$C\frac{\partial\ln T}{\partial t}+\frac{L}{T_0}\frac{\partial\beta_3}{\partial t}-\lambda\Delta\ln T=R\,.$$

This set is completed by suitable initial conditions and the following boundary conditions:

$$\sigma\mathbf{N}=\mathbf{g},\ on\ \Gamma_1, \tag{5.71}$$

$$\mathbf{u}=\mathbf{U}=0,\ on\ \Gamma_0,$$

$$\upsilon\frac{\partial}{\partial N}(\frac{\partial\beta}{\partial t})+k\frac{\partial\beta}{\partial N}=0,\ on\ \partial\Omega,$$

$$\lambda\frac{\partial}{\partial N}(\ln T)=\pi,\ on\ \partial\Omega,$$

where $\mathbf{g}$ is the exterior contact force applied to $\Gamma_1$, where $(\Gamma_0,\Gamma_1)$ is a partition of $\partial\Omega$. It may be proved that this set of partial differential equations has solutions in a convenient mathematical setting, insuring the model is coherent from both the mathematical and the mechanical point of view allowing numerical approximations, [126].

## 5.7 Conclusion

The evolution of a structure made of shape memory alloys, i.e., the computation of

$$E(\mathbf{x},t)=(\varepsilon(\mathbf{x},t),\beta_1(\mathbf{x},t),\beta_2(\mathbf{x},t),\beta_3(\mathbf{x},t),$$

$$\mathrm{grad}\beta_1(\mathbf{x},\mathrm{t}),\mathrm{grad}\beta_2(\mathbf{x},\mathrm{t}),\mathrm{grad}\beta_3(\mathbf{x},\mathrm{t}),\mathrm{T}(\mathbf{x},\mathrm{t})),$$

together with the small displacement $\mathbf{u}(\mathbf{x},t)$ depending on the point $\mathbf{x}$ of the domain occupied by the structure and on time $t$, can be performed by solving numerically the set of partial differential equations, (5.27)–(5.29), resulting from the equations of motion, the entropy balance and the constitutive laws (5.24)–(5.26), completed by convenient initial and boundary conditions (or equations of motion (5.41), (5.42), entropy balance (5.65), (5.66) and constitutive laws (5.58)–(5.63)), [36, 37, 44, 63, 69, 74, 112, 179, 210]. The models we have described are able to account for the different features of the shape memory alloy macroscopic, mechanical and thermal properties. We have used schematic free energies and schematic pseudopotentials of dissipation.

There are many possibilities to upgrade the basic choices we have made to take into account the practical properties of shape memory alloys. Let us, for instance, mention that the pseudopotential of dissipation can be modified to describe more precisely the hysteretic properties. There is no difficulty in having more than two martensites, for instance, to take care of 24 possible martensites! In the same way, it is possible to introduce different forms one martensite may exhibit, as in [156].

Note that the physical quantities for characterizing an educated shape memory alloys are $K$, $C$, $L$, $T_0$, $T_c$, $\tau$, the two martensite volume fractions, coordinates of point $B_e$ defining triangle $C_e$, for the free energy and $c$, $k$, $\lambda$ for the pseudopotential of dissipation. Those are not so many to have a complete multidimensional model which can be used for engineering purposes.

Let us note the very important role of internal constraints and of the reaction $\mathbf{B}^{ndr}$ to those internal constraints which are responsible for many properties. Let us also note that when there are voids, the pressure is the reaction to the kinematic constraint resulting from the mass balance. From this point of view, pressure $p$ is involved in the equations in a logic and clear way.

# Chapter 6
# Damage

It is known that damage results from microscopic motions in the structures, as it is caused by microfractures and microcavities resulting in the decreasing of the material stiffness. There is wide literature on this topic and it is difficult to be exhaustive. Let us mention [145, 151, 152] modelling damage within the framework of continuum mechanics. The damage quantities are internal quantities which are involved in the free energy of materials [104, 129, 138, 151, 152]. The many possible expressions for the free energy yield numerous, versatile constitutive laws [4,7,9,26,80,138,151,196]. The damage quantities may also appear in the equations of motions [50,75,104,107,112,120,137,148,165,175]. They may also approximate fractures [6, 52, 53, 77, 81, 100–102, 176]. Predictive theories of damage of living materials are also developed, [163].

At the macroscopic level, damage is represented by the state quantity $\beta$ with values between 0 and 1. When $\beta = 0$, the material is completely damaged, when $\beta = 1$, the material is sound or undamaged. When $0 < \beta < 1$, the material is partly damaged. State quantity $\beta$ may be thought as the volume fraction of microfractures and microcavities. In many practical situations, the evolution is irreversible. But there are cases where it is not. For instance, some polymers may recover there strength if left at rest after a damaging loading. This is also the case for some bituminous materials which damage during hot days and mend by themselves during cool nights. The state quantities we choose are

$$E = (T, \varepsilon, \beta, \mathrm{grad}\beta),$$

where $\varepsilon = \varepsilon(\mathbf{u})$ is the small deformation and $\mathbf{u}$ the small displacement. The gradient of the volume fraction is introduced to take into account local interactions or the influence of a material point on its neighbourhood. What are the velocities we need to describe the motion? We have the macroscopic velocities $\mathbf{U}$. But we have said that damage results from microscopic motions, thus we have to account for these motions. Let us remark that when there are microscopic motions, the damage quantity $\beta$ evolves and when there are not microscopic motions, $\beta$ remains constant.

M. Frémond, *Phase Change in Mechanics*, Lecture Notes of the Unione Matematica Italiana 13, DOI 10.1007/978-3-642-24609-8_6,

Thus we choose to account at the macroscopic level for the microscopic motions by velocity

$$\frac{d\beta}{dt}.$$

The actual or virtual velocities have two components $(\mathbf{U}(\mathbf{x},t),(d\beta/dt)(\mathbf{x},t))$ and $(\mathbf{V}(\mathbf{x}),\gamma(\mathbf{x}))$ where $\mathbf{V},\gamma$ are macroscopic and microscopic virtual velocities. By stretch of language we may say that $d\beta/dt$ is the microscopic velocity whereas we should say that it is the damage velocity accounting for the microscopic velocity. The theory has been established in Chaps. 2 and 3 with a different physical meaning for $\beta$.

## 6.1 The Equations of Motion

The equations of motion established in Chaps. 2 and 3 with a different physical meaning for $\beta$ are

$$\rho\frac{d\mathbf{U}}{dt} - \mathrm{div}\sigma = \mathbf{f},\ in\ \Omega,\ \sigma\mathbf{N} = \mathbf{g},\ on\ \partial\Omega,$$

$$\rho_0\frac{d^2\beta}{dt^2} - \mathrm{div}\mathbf{H} = A - B,\ in\ \Omega, \mathbf{H}\cdot\mathbf{N} = a,\ on\ \partial\Omega. \tag{6.1}$$

The second equation of motion (6.1) is new. It accounts for the microscopic motions. Both the partial differential equation and the boundary condition have a precise physical meaning: they describe how work is provided to the structure without macroscopic motion.

In the sequel, we consider only quasi-static evolutions. Results involving the accelerations are reported in [49, 123].

## 6.2 Free Energy and Pseudopotential of Dissipation

The free energy and pseudopotential of dissipation with $\chi = T$, we choose, are assuming small perturbations

$$\Psi(\varepsilon,\beta,\mathrm{grad}\beta,\mathrm{T}) = -CT\ln T + \frac{\beta}{2}\{\lambda_e(tr\varepsilon)^2 + 2\mu_e\varepsilon:\varepsilon\} + w(1-\beta)$$
$$+I(\beta) + \frac{k}{2}(\mathrm{grad}\beta)^2,$$

$$\Phi(\mathrm{gradT},\frac{\partial\beta}{\partial \mathrm{t}},\mathrm{T}) = \frac{\lambda}{2T}(\mathrm{gradT})^2 + \frac{\mathrm{c}}{2}\left(\frac{\partial\beta}{\partial \mathrm{t}}\right)^2 + \mathrm{I}_-(\frac{\partial\beta}{\mathrm{dt}}),$$

where $\lambda_e$ and $\mu_e$ are the elasticity Lamé parameters and, $c$ is the viscosity of damage, i.e., the viscosity of the microscopic motions, $k$ is the microscopic interaction coefficient and $w$ is the cohesion energy. The constitutive are

$$\sigma = \beta\{\lambda_e(tr\varepsilon)1 + 2\mu_e\varepsilon\},$$

$$B \in \frac{1}{2}\{\lambda_e(tr\varepsilon)^2 + 2\mu_e\varepsilon : \varepsilon\} - w + c\frac{\partial\beta}{\partial t} + \partial I(\beta) + \partial I_-(\frac{\partial\beta}{\partial t}),\ \mathbf{H} = k\mathrm{grad}\beta,$$

$$\mathbf{Q} = -\frac{\lambda}{T}\mathrm{gradT},$$

$$s = C(1 + \ln T).$$

Let us note that the actual Lamé parameters are proportional to $\beta$: $\beta\lambda_e$ and $\beta\mu_e$. Thus the Young modulus is proportional to $\beta$. The linear function $w(1-\beta)$ in free energy, is responsible for a threshold in the constitutive law for the damage. Depending of the value of the elastic energy $(1/2)\{\lambda_e(tr\varepsilon)^2 + 2\mu_e\varepsilon : \varepsilon\}$ with respect to $w$, damage occurs or does not occur, as it results from the equation of microscopic motion. Information on the measurements of the physical parameters may be found in [175].

## 6.3 The Equations

The equation of the predictive theory are

$$C\frac{\partial \ln T}{\partial t} - \lambda\Delta \ln T = R,\ in\ \Omega,$$

$$c\frac{\partial\beta}{\partial t} - k\Delta\beta + \partial I(\beta) + \partial I_-(\frac{\partial\beta}{\partial t})$$

$$\ni w - \frac{1}{2}\{\lambda_e(tr\varepsilon)^2 + 2\mu_e\varepsilon : \varepsilon\} + A,\ in\ \Omega, \tag{6.2}$$

$$\mathrm{div}\left(\beta\left(\lambda_e(\mathrm{tr}\varepsilon(\mathbf{u})1 + 2\mu_e\varepsilon(\mathbf{u})\right)\right) + \mathbf{f} = 0,\ \mathrm{in}\ \Omega, \tag{6.3}$$

$$\lambda\frac{\partial \ln T}{\partial N} = \pi,\ k\frac{\partial\beta}{\partial N} = a,\ \sigma\mathbf{N} = \mathbf{g},\ on\ \partial\Omega,$$

$$T(\mathbf{x},0) = T^0(\mathbf{x}),\ \beta(\mathbf{x},0) = \beta^0(\mathbf{x}),\ in\ \Omega.$$

The damage sources $A$ and $a$ take into account chemical, electrical, radiative actions, [21, 24], which can break the links insuring the cohesion of the material. Note that in this setting the thermal problem is decoupled from the mechanical problem. The thermal effects of damage due to the dissipation are investigated in [142]. Mechanical results may be found in [112, 113, 120–122, 168, 175, 180] and mathematical results in dimension one may be found in [117, 118], and in dimension two and three in [46, 116].

## 6.4 The Macroscopic Motions Become Microscopic

In this predictive theory, there are relationships between macroscopic and microscopic motions during the damaging process. Indeed, as the model is based on the separation between the description of macroscopic and microscopic motions, it is natural to investigate what occurs when macroscopic motions vanish progressively and become microscopic. In order to answer this question let apply exterior actions to a structure in such a way that the amplitude of the resulting macroscopic motions become smaller and smaller

$$\mathbf{u}_n \to 0,$$

but the strain rates are highly oscillating and to not tend to 0

$$\frac{1}{2}\{\lambda_e(tr\varepsilon(\mathbf{u}_n))^2 + 2\mu_e\varepsilon(\mathbf{u}_n) : \varepsilon(\mathbf{u}_n)\} \to d \geq 0,$$

in some weak sense. The macroscopic motions vanish and it is no longer possible to consider that they are macroscopic: one has to consider them as microscopic. One may wonder if their damaging effects also vanish. If they do not, how are they taken into account by the theory? Is there a transfer from (6.3) which describes the macroscopic motions towards (6.2) which describes the microscopic motions?

The theory shows that when macroscopic motions become microscopic, their damaging effects remain as a source of damage $d$, which is clearly related to microscopic motions, [42, 113, 168]. The limit equations for $\beta = \lim \beta_n$ and $\mathbf{u} = \lim \mathbf{u}_n = 0$, are

$$C\frac{\partial \ln T}{\partial t} - \lambda\Delta \ln T = R, \; in \; \Omega,$$

$$c\frac{\partial \beta}{\partial t} - k\Delta\beta + \partial I(\beta) + \partial I_-(\frac{\partial \beta}{\partial t}) \ni w - d, \; in \; \Omega,$$

$$\mathbf{u} = 0,$$

$$T(\mathbf{x},0) = T^0(\mathbf{x}), \; \beta(\mathbf{x},0) = \beta^0(\mathbf{x}), \; in \; \Omega.$$

The work which is provided to the structure by the macroscopic forces becomes the work provided by the damage source $d$.

In the limit, the rapid oscillations with vanishing amplitude are no longer seen because $\mathbf{u} = 0$, but their effect is still present via the external damage source $d$. We conclude that the damaging effect of a vanishing macroscopic motion does not

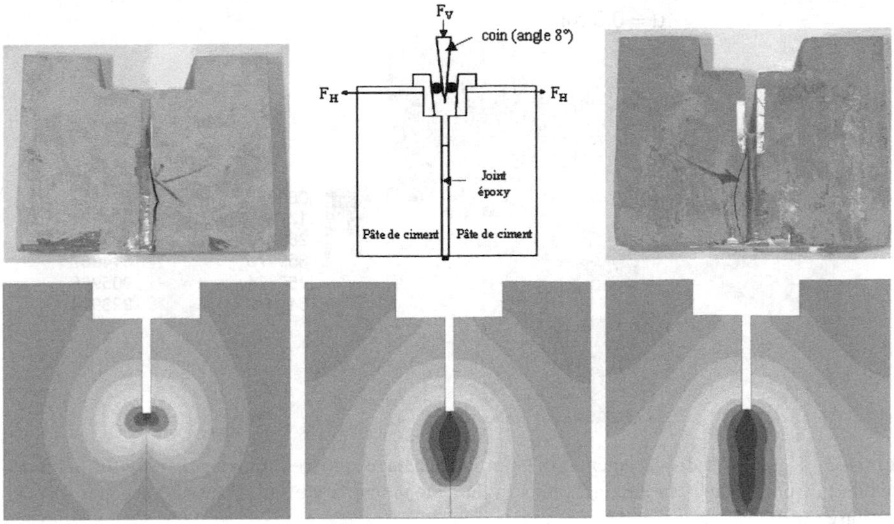


**Fig. 6.1** The wedge spitting test. The progressive concrete damage. The color scales are slightly different in the pictures. In the red domains, damage is at its maximum which is different in each picture. The green domains are not damaged

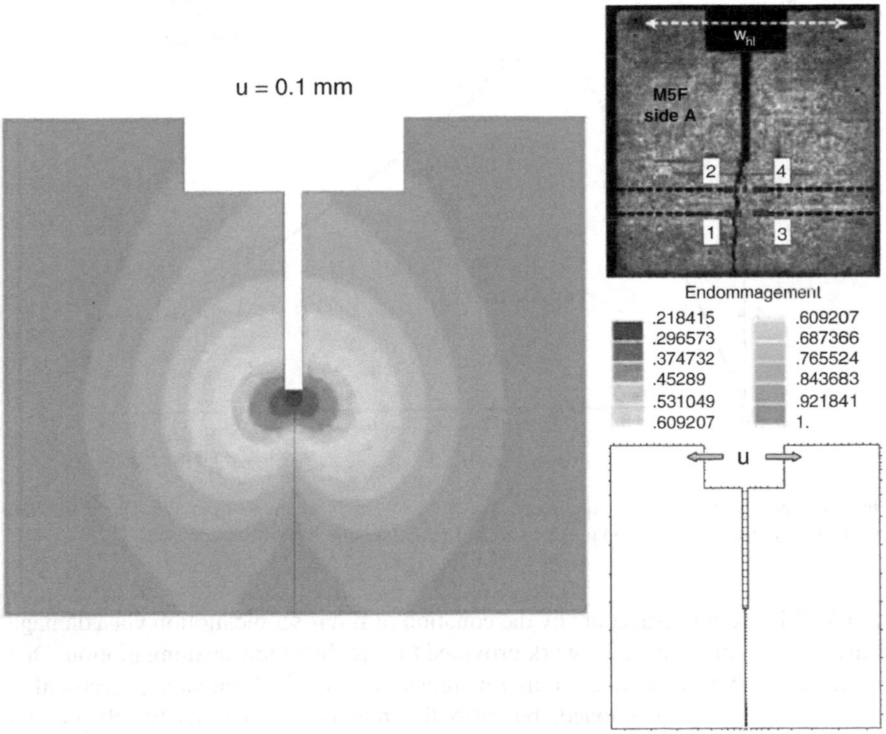


**Fig. 6.2** The wedge splitting test experiment. A displacement is applied as shown on the figure. A splitted sample and the value of damage in the sample when displacement is 0.1 mm

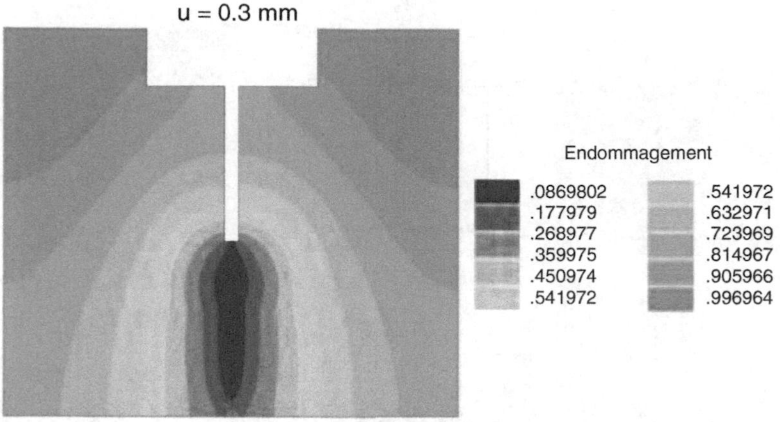


**Fig. 6.3** The applied displacement is 0.3 mm. The damage is more important than for the 0.1 mm. applied displacement. The displacements in the damage zone are important and account for the fracture

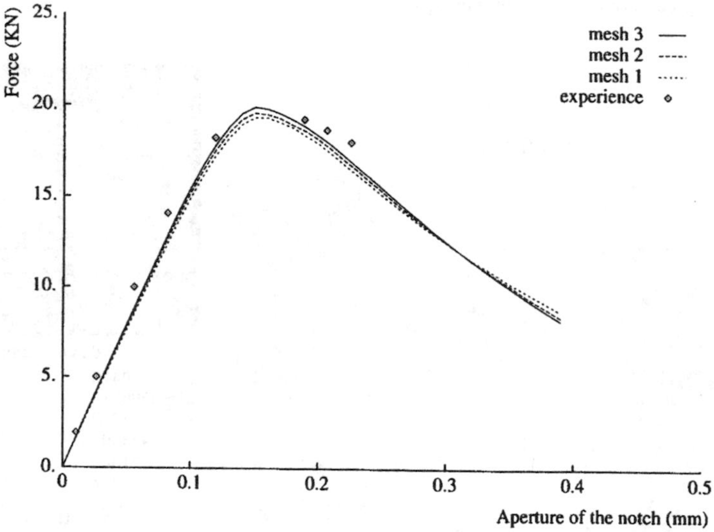


**Fig. 6.4** The wedge splitting experiment. Force versus displacement is plotted. The points are experimental results. The continuous line are numerical results for different meshes, [175]

vanish. It is taken into account by the equation of microscopic motion via a damage source whose intensity is the work provided to establish the vanishing motion. One can also say that the damage source includes the chemical, radiative, electrical,... actions, as already mentioned, but also the mechanical actions which are not accounted for by the macroscopic displacement **u**, for instance rapid microscopic oscillations. In this case the external action $A$ includes the work $d$ provided to the structure by the rapid oscillations.

## 6.5 An Example: The Wedge Splitting Test

The wedge splitting experiment is shown in Figs. 6.1 and 6.2, together with the picture of samples after splitting into two pieces, [134]. Some numerical results due to Francesco Freddi are shown in Figs. 6.1, 6.2 and 6.3.

Experimental and numerical results due Boumediene Nedjar are shown in Fig. 6.4, [175]. The softening of the curve force versus-displacement result from damage, i.e., from the decrease of $\beta$.

# Chapter 7
# Contact with Adhesion

Let us consider two solids $\Omega_1$ and $\Omega_2$ glued on one another on their contact surface $\Gamma$. In order to take into account the adhesive properties of the glue which result from fibers connecting the contact surfaces and breaking progressively, Fig. 7.1, [57, 79], we choose as state quantity the surface fraction of active glue fibers, $\beta(\mathbf{x},t)$, [31, 58, 67, 76, 82, 104, 106–108, 168, 180, 187, 204, 206]. When $\beta(\mathbf{x},t) = 0$ all the glue fibers are broken and Signorini contact properties are valid: the interactions of the two solids result only from the impenetrability condition, [11, 98, 161, 182, 199]. When $\beta(\mathbf{x},t) = 1$, the glue is active: it prevents the separation when tension is applied. The interactions between the two solids result both from the impenetrability condition and the adhesion properties. When $\beta(\mathbf{x},t)$ is between 0 and 1, part of the glue fibers are active. The glue is partly damaged: the interaction between the two solids is in between the two extreme interactions. The evolution of the surface fraction of active fibers is produced by microscopic motions which break or mend the glue fibers. We think that the power of these microscopic motions are to be taken into account in a predictive theory. The previous damage example and theory of Sect. 3.1 have shown how to build a macroscopic theory: we choose $\mathrm{d}\beta/\mathrm{d}t$ on the contact surface to account at the macroscopic level for the microscopic motions.

The state quantities of the system made of the two solids are besides $\beta$, the two small deformations $\varepsilon_1$, $\varepsilon_2$ and the gap $\mathbf{u}_2 - \mathbf{u}_1$ on the contact surface $\Gamma$, where the $\mathbf{u}_i$ are the small displacements of the two solids. For the sake of simplicity, we neglect the thermal phenomena and do not choose the temperature as a state quantity.

As well as for the volume damage, there is an abundant bibliography on contact with adhesion, involving peeling [25, 166, 167, 184, 186]. Let us also mention results on damage of dimension two solids, [205].

M. Frémond, *Phase Change in Mechanics*, Lecture Notes of the Unione Matematica Italiana 13, DOI 10.1007/978-3-642-24609-8_7,

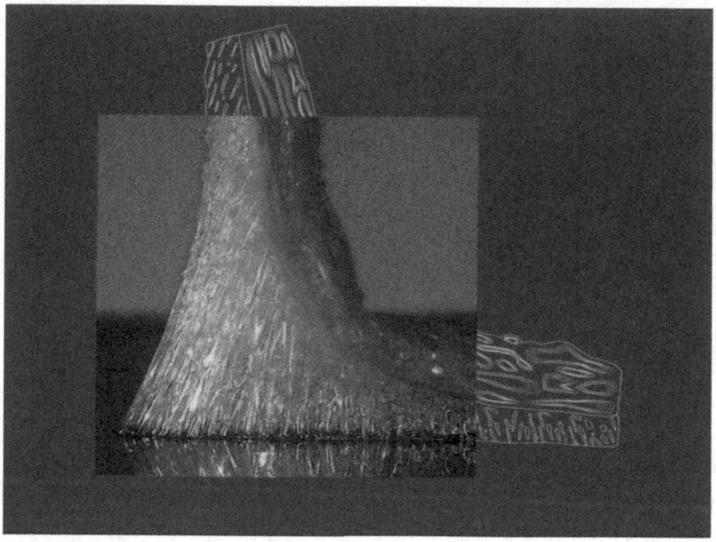

**Fig. 7.1** Peeling experiment: long fibers connect two solids: a flat one and a thin one which is slightly grey. The fibers are shiny. Experiment of M. Horgnies, [79]

## 7.1 The Equations of Motion

They result from the principle of virtual power which involves the virtual power of the internal forces

$$\begin{aligned}\mathscr{P}_{int}(\mathbf{V}_1,\mathbf{V}_2,\gamma) = &-\int_{\Omega_1}\sigma_1 : \mathsf{D}(\mathbf{V}_1)d\Omega_1 - \int_{\Omega_2}\sigma_2 : \mathsf{D}(\mathbf{V}_2)d\Omega_2\\ &-\int_{\Gamma}(B\gamma+\mathbf{H}\cdot \mathrm{grad}_s\gamma)\mathrm{d}\Gamma - \int_{\Gamma}\mathbf{R}\cdot(\mathbf{V}_2-\mathbf{V}_1)\mathrm{d}\Gamma,\end{aligned}$$

where $\mathbf{V}_1, \mathbf{V}_2$ are macroscopic virtual velocities of the two solids, $\gamma$ is a microscopic virtual velocity and $\mathrm{grad}_s$ is the surface gradient. The force $\mathbf{R}$ is the action of solid 1 on solid 2 on contact surface. This vector describes for instance the friction forces. The virtual power of the exterior forces is

$$\begin{aligned}\mathscr{P}_{ext}(\mathbf{V}_1,\mathbf{V}_2,\gamma) = &\int_{\Omega_1}\mathbf{f}_1\cdot\mathbf{V}_1 d\Omega_1 + \int_{\Omega_2}\mathbf{f}_2\cdot\mathbf{V}_2 d\Omega_2\\ &+\int_{\partial\Omega_1\backslash\Gamma}\mathbf{g}_1\cdot\mathbf{V}_1 d\Gamma + \int_{\partial\Omega_2\backslash\Gamma}\mathbf{g}_2\cdot\mathbf{V}_2 d\Gamma\\ &+\int_{\Gamma}A\gamma d\Gamma + \int_{\partial\Gamma}a\gamma d\Gamma_s,\end{aligned}$$

where the $\mathbf{f}_i$ and $\mathbf{g}_i$ are the body and surface forces applied to the solids. The $A$ and $a$ are respectively the surface and line exterior sources of microscopic work which can damage the glue (the line $\partial\Gamma$ is the boundary of the surface $\Gamma$). It is known that chemical action and optical actions damage glue (light damages some type of glue). It is assumed these works are known. The virtual work of acceleration forces is

$$\mathscr{P}_{acc}(\mathbf{V}_1,\mathbf{V}_2,\gamma) = \int_{\Omega_1} \rho_1 \frac{\mathrm{d}\mathbf{U}_1}{\mathrm{d}t} \cdot \mathbf{V}_1 d\Omega_1 + \int_{\Omega_2} \rho_2 \frac{\mathrm{d}\mathbf{U}_2}{\mathrm{d}t} \cdot \mathbf{V}_2 d\Omega_2 + \int_\Gamma \rho_0 \frac{\mathrm{d}^2\beta}{\mathrm{d}t^2} \gamma d\Gamma,$$

where $\rho$ is the density of the solid and $\rho_0$ is proportional to the density of the glue.

The principle of virtual power

$$\forall \mathbf{V}_1,\mathbf{V}_2,\gamma,\ \mathscr{P}_{acc}(\mathbf{V}_1,\mathbf{V}_2,\gamma) = \mathscr{P}_{int}(\mathbf{V}_1,\mathbf{V}_2,\gamma) + \mathscr{P}_{ext}(\mathbf{V}_1,\mathbf{V}_2,\gamma),$$

gives easily the equations of motion

$$\begin{aligned}
\rho_1 \frac{\mathrm{d}\mathbf{U}_1}{\mathrm{d}t} - \mathrm{div}\sigma_1 &= \mathbf{f}_1,\ in\ \Omega_1,\ \sigma_1\mathbf{N}_1 = \mathbf{g}_1,\ on\ \partial\Omega_1,\\
\rho_2 \frac{\mathrm{d}\mathbf{U}_2}{\mathrm{d}t} - \mathrm{div}\sigma_2 &= \mathbf{f}_2,\ in\ \Omega_2,\ \sigma_2\mathbf{N}_2 = \mathbf{g}_2,\ on\ \partial\Omega_2,\\
\sigma_1\mathbf{N}_1 - \mathbf{R} &= 0,\ \sigma_2\mathbf{N}_2 + \mathbf{R} = 0,\\
\rho_0 \frac{\mathrm{d}^2\beta}{\mathrm{d}t^2} + B - \mathrm{div}_s\mathbf{H} &= A,\ on\ \Gamma,\ \mathbf{H}\cdot\mathbf{n}_s = a,\ on\ \partial\Gamma,
\end{aligned} \tag{7.1}$$

where the vector $\mathbf{N}_\mathrm{i}$ is the outside normal vector to $\Omega_i$, $\mathbf{n}_s$ is the surface outside vector to $\Gamma$ and $\mathrm{div}_\mathrm{s}$ is the surface divergence.

## 7.2 The Constitutive Laws

They result from volume and surface free energies and pseudopotential of dissipation. We choose

$$\Psi_i(\varepsilon_i) = \frac{1}{2}\{\lambda_i(tr\varepsilon_i)^2 + 2\mu_i\varepsilon_i : \varepsilon_i\},$$

$$\Phi_i(\frac{\partial\varepsilon_i}{\partial t}) = 0,$$

for the volume functions. The small deformations $\varepsilon_i$ due to small displacement $\mathbf{u}_i$ are

$$\varepsilon_i = \varepsilon(\mathbf{u}_i) = \frac{1}{2}(\mathrm{grad}\mathbf{u}_\mathrm{i} + (\mathrm{grad}\mathbf{u}_\mathrm{i})^\mathrm{T}),$$

where $(\mathrm{grad}\mathbf{u}_\mathrm{i})^\mathrm{T}$ is the transposed matrix of $\mathrm{grad}\mathbf{u}_\mathrm{i}$. Thus each solid is elastic. We choose

$$\Psi(\mathbf{u}_2-\mathbf{u}_1,\beta,\mathrm{grad}_s\beta) = \frac{k}{2}\beta(\mathbf{u}_2-\mathbf{u}_1)^2 + w_s(1-\beta) + \frac{k_s}{2}(\mathrm{grad}_s\beta)^2$$
$$+I(\beta) + I_-((\mathbf{u}_2-\mathbf{u}_1)\cdot\mathbf{N}_2),$$
$$\Phi(\frac{\partial\beta}{\partial t}) = \frac{c_s}{2}\left(\frac{\partial\beta}{\partial t}\right)^2 + I_-(\frac{\partial\beta}{\partial t}),$$

for the contact surface functions. Function $I_-$ is the indicator function of $\mathbb{R}^-$, taking into account the impenetrability of the two solids

$$(\mathbf{u}_2-\mathbf{u}_1)\cdot\mathbf{N}_2 \leq 0.$$

This contact surface has an elastic reaction whose rigidity is proportional to $\beta$. The evolution of the glue is dissipative with coefficient of viscosity $c_s$ and, irreversible: once the glue fibers are broken they do not mend by themselves. This irreversible behaviour characterizes solid glue. In case we consider fresh and liquid glue, the glue fibers once broken are rebuilt if the solids get close to one another. In this situation, the evolution is reversible and the indicator function $I_-(\partial\beta/dt)$ is removed from the surface pseudopotential. The cohesion of the glue, called the Dupré's energy [166], is $w_s$. The non dissipative local interactions of the glue fibers is characterized by $k_s$. The constitutive laws are

$$\sigma_i = \lambda_i(tr\varepsilon_i)1 + 2\mu_i\varepsilon_i,$$

in the solids. On the contact surface the constitutive laws are

$$\mathbf{R} \in k\beta(\mathbf{u}_2-\mathbf{u}_1) + \partial I_-((\mathbf{u}_2-\mathbf{u}_1)\cdot\mathbf{N}_2)\mathbf{N}_2,$$
$$B \in \frac{k}{2}(\mathbf{u}_2-\mathbf{u}_1)^2 - w_s + c_s\frac{\partial\beta}{\partial t} + \partial I(\beta) + \partial I_-(\frac{\partial\beta}{\partial t}),$$
$$\mathbf{H} = k_s\mathrm{grad}_s\beta.$$

## 7.3 The Equations

We assume that no work involving microscopic motions is provided to the system, i.e., there are no chemical, radiative, optical or electrical actions: $A = 0$ and $a = 0$. We assume solid 2 is fixed to a rigid support on part $\Gamma_2$. The equations of the predictive theory for a quasi-static evolution are

$$c_s \frac{\partial \beta}{\partial t} - k_s \Delta_s \beta + \partial I(\beta) + \partial I_-(\frac{\partial \beta}{\partial t}) \ni w_s - \frac{k}{2}(\mathbf{u}_2 - \mathbf{u}_1)^2, \ on\ \Gamma, \tag{7.2}$$

$$\text{div}\,\{\lambda_i \text{tr}\varepsilon(\mathbf{u}_i)1 + 2\mu_i \varepsilon_i(\mathbf{u}_i)\} + \mathbf{f}_i = 0, \text{ in } \Omega_i, \tag{7.3}$$

$$\sigma_1 \mathbf{N}_1 = \mathbf{g}_1, \ on\, \partial\Omega_1 \backslash \Gamma,$$

$$\sigma_2 \mathbf{N}_2 = \mathbf{g}_2, \ on\, \partial\Omega_2 \backslash (\Gamma \cup \Gamma_2), \ \mathbf{u_2} = 0, \ on\ \Gamma_2,$$

$$\sigma_1 \mathbf{N}_1 - k\beta(\mathbf{u}_2 - \mathbf{u}_1) - \partial I_-((\mathbf{u}_2 - \mathbf{u}_1)\cdot \mathbf{N}_2)\mathbf{N}_2 \ni 0, \ on\ \Gamma,$$

$$\sigma_2 \mathbf{N}_2 + k\beta(\mathbf{u}_2 - \mathbf{u}_1) + \partial I_-((\mathbf{u}_2 - \mathbf{u}_1)\cdot \mathbf{N}_2)\mathbf{N}_2 \ni 0, \ on\ \Gamma,$$

$$k\frac{\partial \beta}{\partial n_s} = 0, \ on\ \partial\Gamma,$$

$$\beta(\mathbf{x}, 0) = \beta^0(\mathbf{x}), \ on\ \Gamma,$$

where $\beta^0$ is the initial surface fraction of unbroken glue fibers. For instance, $\beta^0 = 1$ when the glue is sound. The elements of the subdifferential set $\partial I_-((\mathbf{u}_2 - \mathbf{u}_1)\cdot \mathbf{N}_2)$ are the impenetrability reactions which are active only when normal gap $(\mathbf{u}_2 - \mathbf{u}_1)\cdot \mathbf{N}_2$ is null.

## 7.4 Examples

They are given in the next chapter where volume damage is coupled to surface damage.

# Chapter 8
# Damage of Solids Glued on One Another: Coupling of Volume and Surface Damages

Let us consider two pieces of concrete glued on one another. Both the concrete and the glue responsible for the adhesion, can damage due to external actions. The volume and surface damages are coupled. The interesting part of the predictive theory is the set of equations on the contact surface, [99, 100]. Mathematical results are reported in [43].

## 8.1 State Quantities and Quantities Describing the Evolution

We neglect the thermal effects and do not take the temperature into account. The state quantities and quantities describing the evolution are:

in $\Omega_1$ and in $\Omega_2$

$$E_1 = \{\varepsilon_1, \beta_1, \text{grad}\,\beta_1\}, \delta E_1 = \left\{\frac{\partial \varepsilon_1}{\partial t}, \frac{\partial \beta_1}{\partial t}, \text{grad}\frac{\partial \beta_1}{\partial t}\right\},$$

$$E_2 = \{\varepsilon_2, \beta_2, \text{grad}\,\beta_2\},\ \delta E_2 = \left\{\frac{\partial \varepsilon_2}{\partial t}, \frac{\partial \beta_2}{\partial t}, \text{grad}\frac{\partial \beta_2}{\partial t}\right\},$$

where the $\varepsilon$'s are the small deformations and the $\beta$'s the volume damages;

in $\partial\Omega_1 \cap \partial\Omega_2$

$$E_s = \{\mathbf{u}_2 - \mathbf{u}_1, \beta_s, \text{grad}_s\,\beta_s, \beta_1, \beta_2\},$$

$$\delta E_s = \left\{\mathbf{U}_2 - \mathbf{U}_1, \frac{\partial \beta_s}{\partial t}, \text{grad}_s\frac{\partial \beta_s}{\partial t}, \frac{\partial \beta_1}{\partial t}, \frac{\partial \beta_2}{\partial t}\right\},$$

with

$$\mathbf{U}_1 = \frac{\partial \mathbf{u}_1}{\partial t},\ \mathbf{U}_2 = \frac{\partial \mathbf{u}_2}{\partial t},$$

M. Frémond, *Phase Change in Mechanics*, Lecture Notes of the Unione Matematica Italiana 13, DOI 10.1007/978-3-642-24609-8_8,

where the $\mathbf{u}$'s are the small displacements with velocities $\mathbf{U} = \partial \mathbf{u}/dt$, $\beta_s$ is the surface or glue damage. On the contact surface, we use the gap $\mathbf{u}_2 - \mathbf{u}_1$ and its velocity as it is usual in contact mechanics. But we will have to deal with damage resulting from the elongation or the stretching of the contact surface. In an elongation where

$$\forall \mathbf{x},\ \mathbf{u}_2(\mathbf{x}) - \mathbf{u}_1(\mathbf{x}) = 0,$$

the gap is 0 because the two displacements are equal. Thus we choose to introduce a non local deformation quantity

$$g(\mathbf{y},\mathbf{x}) = 2(\mathbf{y}-\mathbf{x}) \cdot (\mathbf{u}_2(\mathbf{y}) - \mathbf{u}_1(\mathbf{x})),$$

which does not vanish in an elongation.

$$E_{s,1,2} = \{g(\mathbf{y},\mathbf{x}) = 2(\mathbf{y}-\mathbf{x}) \cdot (\mathbf{u}_2(\mathbf{y}) - \mathbf{u}_1(\mathbf{x})), \beta_s(\mathbf{x}), \beta_s(\mathbf{y})\},$$

$$\delta E_{s,1,2} = \left\{ D_{1,2}(\mathbf{U}_1,\mathbf{U}_2), \frac{\partial \beta_s}{\partial t}(\mathbf{x}), \frac{\partial \beta_s}{\partial t}(\mathbf{y}) \right\},$$

with

$$D_{1,2}(\mathbf{U}_1,\mathbf{U}_2)(\mathbf{x},\mathbf{y}) = 2(\mathbf{y}-\mathbf{x}) \cdot (\mathbf{U}_2(\mathbf{y}) - \mathbf{U}_1(\mathbf{x})).$$

The last quantity is the velocity of the non local deformation $g(\mathbf{y},\mathbf{x})$. Let us note that this velocity is 0 in any rigid body velocity, $\mathbf{V} + \boldsymbol{\varpi} \times \mathbf{x}$. This property is obvious because $D_{1,2}$ is the time derivative of the square of the distance of two material points

$$D_{1,2}(\mathbf{U}_1,\mathbf{U}_2)(\mathbf{x},\mathbf{y}) = \frac{\mathrm{d}}{\mathrm{d}t}(\mathbf{y}(t) - \mathbf{x}(t))^2.$$

## 8.2 Equations of Motion

They result from the principle of virtual power which involves the powers of the interior forces, of the exterior forces and of the acceleration forces, [112].

### 8.2.1 *Virtual Power of the Interior Forces*

Both volume damage and surface damage result from microscopic motions whose power is taken into account in the power of the interior forces. We choose the velocities $\partial \beta/dt$ to account for the microscopic velocities at the macroscopic level. In order to take into account local interactions, we introduce the gradients $\mathrm{grad}(\partial \beta/dt)$. Assuming $\mathbf{V} = (\mathbf{V}_1, \mathbf{V}_2)$ and $\gamma = (\gamma_1, \gamma_2, \gamma_s)$ to be macroscopic and microscopic virtual velocities, the virtual power of the interior forces, which is linear function of the virtual velocities, is chosen as

$$
\begin{aligned}
\mathscr{P}_{int}(\mathbf{V},\gamma) = &-\int_{\Omega_1} \sigma_1 : \mathsf{D}(\mathbf{V}_1) d\Omega - \int_{\Omega_1} \{B_1\gamma_1 + \mathbf{H}_1 \cdot \operatorname{grad}\gamma_1\} d\Omega \\
&-\int_{\Omega_2} \sigma_2 : \mathsf{D}(\mathbf{V}_2) d\Omega - \int_{\Omega_2} \{B_2\gamma_2 + \mathbf{H}_2 \cdot \operatorname{grad}\gamma_2\} d\Omega \\
&-\int_{\partial\Omega_1 \cap \partial\Omega_2} \mathbf{R} \cdot (\mathbf{V}_2 - \mathbf{V}_1) d\Gamma \\
&-\int_{\partial\Omega_1 \cap \partial\Omega_2} \{B_s\gamma_s + \mathbf{H}_s \cdot \operatorname{grad}_s \gamma_s + B_{1,s}(\gamma_1 - \gamma_s) + B_{2,s}(\gamma_2 - \gamma_s)\} d\Gamma \\
&+\int_{\partial\Omega_1 \cap \partial\Omega_2} \int_{\partial\Omega_1 \cap \partial\Omega_2} M(\mathbf{x},\mathbf{y}) D_{1,2}(\mathbf{V}_1,\mathbf{V}_2)(\mathbf{x},\mathbf{y}) d\mathbf{x} d\mathbf{y} \\
&+\int_{\partial\Omega_1 \cap \partial\Omega_2} \int_{\partial\Omega_1 \cap \partial\Omega_2} \{B_{s,1}(\mathbf{x},\mathbf{y})\gamma_s(\mathbf{x}) + B_{s,2}(\mathbf{x},\mathbf{y})\gamma_s(\mathbf{y})\} d\mathbf{x} d\mathbf{y}.
\end{aligned}
$$

The different quantities which contribute to the power of the interior forces are products of kinematic quantities by interior forces. The kinematic quantities are the quantities that intervene in the motion we intend to describe. Their choice is of paramount importance since they determine the whole predictive theories. They are chosen by experimenting phenomena which are volume and surface deformations together with volume and surface damage, i.e., micro-voiding and micro-cracking. Thus there are quantities with surface and volume densities depending on the quantities we have chosen to describe the evolutions or the deformations of the system. Some of them are classical others are new. Most of them are local but a few are non local because there is a kinematic quantity which is non local. Let us comment on the different power densities:

- The usual strain rate $\mathsf{D}$ introduces the stress $\sigma$.
- The damage velocity, $\partial\beta/dt$ is a scalar, thus the associated interior force is also a scalar, $B$. It is a work, the internal damage work which is responsible for the evolution of the damage in the volume and in the surface.
- The gradient of the damage velocity, $\operatorname{grad}(\partial\beta/dt)$ is a vector, thus the interior force is a vector, $\mathbf{H}$. It is a work flux vector which is responsible for the interaction of the damage at a point on the damage of its neighbourhood. Its physical meaning is to be given by the boundary condition of the equation of motion as the physical meaning of the stress is given by the boundary condition of the equation of motion.
- The gap velocity $\mathbf{U}_2 - \mathbf{U}_1$ on the contact surface introduces the classical macroscopic interaction force $\mathbf{R}$.
- In the same way the difference between the damage velocities $\partial\beta_i/dt - \partial\beta_s/dt$ introduces a damage work flux on the surface, $B_{i,s}$ which describes the interaction of volume damage and surface damage.
- The elongation velocity, $D_{1,2}(\mathbf{U}_1,\mathbf{U}_2)(\mathbf{x},\mathbf{y})$ which is non local, introduces a non local scalar $M(\mathbf{x},\mathbf{y})$ interior force. It describes the effects of the elongation. It results in the equations of motion as a classical force. The interaction

macroscopic mechanical force has a non local part and a classical local part, the force **R** (see formula (8.2) below). Since we are going to assume the interior force $M(\mathbf{x},\mathbf{y})$ depending on the surface damage $\beta_s$, it is wise to add an extra non local power depending on damage velocity $\partial\beta_s/dt$. It describes the effect of damage at point **x** on damage at point **y**. The interior forces $B_{s,i}(\mathbf{x},\mathbf{y})$ have the same effect than $M$: they introduce a non local internal source of damage work. The microscopic mechanical force has a non local part and three local parts, $B_s$ due to the glue and the two $B_{i,s}$ due to the interactions with the volumes (see formula (8.1) below).

Let us note that even if the interior forces are numerous and some of them are unusual, all of them are simple and precisely committed to take into account a particular aspect of the coupling of volume and surface, and of microscopic and macroscopic evolution of the system.

### 8.2.2 Virtual Power of the Exterior Forces

We assume no exterior microscopic, either surface or volume, source of damage such as radiative, electrical or chemical damaging actions. Thus we have

$$\begin{aligned}\mathscr{P}_{ext}(\mathbf{V},\gamma) &= \int_{\Omega_1} \mathbf{f}_1\cdot\mathbf{V}_1 d\Omega + \int_{\partial\Omega_1\setminus(\partial\Omega_1\cap\partial\Omega_2)} \mathbf{g}_1\cdot\mathbf{V}_1 d\Gamma \\ &+ \int_{\Omega_2} \mathbf{f}_2\cdot\mathbf{V}_2 d\Omega + \int_{\partial\Omega_2\setminus(\partial\Omega_1\cap\partial\Omega_2)} \mathbf{g}_2\cdot\mathbf{V}_2 d\Gamma,\end{aligned}$$

where the **f** and **g** are the body and surface exterior forces.

### 8.2.3 Virtual Power of the Acceleration Forces

For the sake of simplicity, we assume a quasi-static problem. Thus

$$\mathscr{P}_{acc}(\mathbf{V},\gamma) = 0.$$

### 8.2.4 The Principle of Virtual Power

It is

$$\begin{aligned}&\forall\mathbf{V} = (\mathbf{V}_1,\mathbf{V}_2),\ \gamma = (\gamma_1,\gamma_2,\gamma_s),\\ &\mathscr{P}_{acc}(\mathbf{V},\gamma) = \mathscr{P}_{int}(\mathbf{V},\gamma) + \mathscr{P}_{ext}(\mathbf{V},\gamma).\end{aligned}$$

### 8.2.5 The Equations of Motion

They result from the principle. By choosing convenient virtual velocities, we get easily for domain $\Omega_1$

$$\operatorname{div}\sigma_1 + \mathbf{f}_1 = 0,\ -B_1 + \operatorname{div}\mathbf{H}_1 = 0,\ in\ \Omega_1,$$
$$\sigma_1\mathbf{N}_1 = \mathbf{g}_1,\ \mathbf{H}_1 \cdot \mathbf{N}_1 = 0,\ in\ \partial\Omega_1 \backslash (\partial\Omega_1 \cap \partial\Omega_2).$$

They are the volume equations of motion accounting for macroscopic and microscopic effects. The equations of motion for domain $\Omega_2$ are the same. For the sake of simplicity, we consider the domains $\Omega_1$ and $\Omega_2$ have the same mechanical properties and give only the mechanical relationships for domain $\Omega_1$. The related boundary conditions and the equation of motion on $\partial\Omega_1 \cap \partial\Omega_2$ involve non local forces.

$$\begin{aligned}
\sigma_1\mathbf{N}_1(\mathbf{x}) &= \mathbf{R}(\mathbf{x}) + \int_{\partial\Omega_1\cap\partial\Omega_2} 2(\mathbf{x}-\mathbf{y})M(\mathbf{x},\mathbf{y})d\mathbf{y},\ \mathbf{x} \in \partial\Omega_1 \cap \partial\Omega_2,\\
\sigma_2\mathbf{N}_2(\mathbf{y}) &= -\mathbf{R}(\mathbf{y}) + \int_{\partial\Omega_1\cap\partial\Omega_2} 2(\mathbf{y}-\mathbf{x})M(\mathbf{x},\mathbf{y})d\mathbf{x},\ \mathbf{y} \in \partial\Omega_1 \cap \partial\Omega_2,\\
\mathbf{H}_1 \cdot \mathbf{N}_1 &= -B_{1,s},\ \mathbf{H}_2 \cdot \mathbf{N}_2 = -B_{2,s},\\
&\quad -B_s(\mathbf{x}) + \operatorname{div}_s \mathbf{H}_s(\mathbf{x}) + B_{1,s}(\mathbf{x}) + B_{2,s}(\mathbf{x})\\
&\quad - \int_{\partial\Omega_1\cap\partial\Omega_2} B_{s,1}(\mathbf{x},\mathbf{y}) + B_{s,2}(\mathbf{y},\mathbf{x})d\mathbf{y} = 0,\ \mathbf{x} \in \partial\Omega_1 \cap \partial\Omega_2, \qquad (8.1)\\
\mathbf{H}_s \cdot \mathbf{n}_s &= 0,\ on\ \partial(\partial\Omega_1 \cap \partial\Omega_2),
\end{aligned}$$

where $\mathbf{n}_s$ is the normal vector to the boundary $\partial(\partial\Omega_1 \cap \partial\Omega_2)$ of $\partial\Omega_1 \cap \partial\Omega_2$. Constitutive laws are needed for the numerous interior forces. As usual, we choose to define them with free energies depending on the state quantities $E$ and pseudopotential of dissipation depending on the velocities $\delta E$.

*Remark 8.1.* Function $M(\mathbf{x},\mathbf{y})$ is not symmetric. Because there are at a distance interactions, there is no clear difference between the contact surface considered as a structure and contact surface considered as a material. In case a subdomain of $\partial\Omega_1 \cap \partial\Omega_2$ is considered, the non local actions result in interior non local actions and exterior non local actions. This point of view is developed in [112].

## 8.3 The Constitutive Laws

Because the thermal phenomenon are not taken into account, the second law of thermodynamics is

$$\frac{d\Psi_1}{\partial t}(\varepsilon_1,\beta_1,\operatorname{grad}\beta_1) \le \sigma_1 : \mathsf{D}(\mathbf{U}_1) + B_1\frac{\partial\beta_1}{\partial t} + \mathbf{H}_1\cdot\operatorname{grad}\frac{\partial\beta_1}{\partial t},\ in\ \Omega_1,$$

$$\frac{d\Psi_s}{\partial t}(\mathbf{u}_2-\mathbf{u}_1,\beta_s,\operatorname{grad}_s\beta_s,\beta_1-\beta_s,\beta_2-\beta_s) \le \mathbf{R}\cdot(\mathbf{U}_2-\mathbf{U}_1) + B_s\frac{\partial\beta_s}{\partial t}$$

$$+\mathbf{H}_s\cdot\operatorname{grad}_s\frac{\partial\beta_s}{\partial t} + B_{1,s}(\frac{\partial\beta_1}{\partial t}-\frac{\partial\beta_s}{\partial t}) + B_{2,s}(\frac{\partial\beta_2}{\partial t}-\frac{\partial\beta_s}{\partial t}),\ on\ \partial\Omega_1\cap\partial\Omega_2,$$

$$\frac{d\Psi_{s,1,2}}{\partial t}(g(\mathbf{y},\mathbf{x}),\beta_s(\mathbf{x}),\beta_s(\mathbf{y})) \le -M(\mathbf{x},\mathbf{y})D_{1,2}(\mathbf{U}_1,\mathbf{U}_2)(\mathbf{x},\mathbf{y})$$

$$-B_{s,1}(\mathbf{x},\mathbf{y})\frac{\partial\beta_s}{\partial t}(\mathbf{x}) - B_{s,2}(\mathbf{x},\mathbf{y})\frac{\partial\beta_s}{\partial t}(\mathbf{y}),\ in\ (\partial\Omega_1\cap\partial\Omega_2)\times(\partial\Omega_1\cap\partial\Omega_2).$$

These relationships are used to define the constitutive laws with pseudopotential of dissipation. The $+$ or $-$ signs appearing in the constitutive laws result from the $+$ or $-$ signs which are in the right hand sides of the inequalities. The right hand sides are the opposite of the densities of the actual powers of the interior forces.

The free energy and pseudopotential of dissipation are the same for each domain. For domain $\Omega_1$, they are

$$\Psi_1(E_1) = \Psi_1(\varepsilon_1,\beta_1,\operatorname{grad}\beta_1) = w_1(1-\beta_1) + \frac{k_1}{2}(\operatorname{grad}\beta_1)^2 + I(\beta_1)$$

$$+\frac{\beta_1}{2}\left\{\lambda_1(tr\varepsilon_1)^2 + 2\mu_1\varepsilon_1:\varepsilon_1\right\},$$

$$\Phi_1(\delta E_1) = \Phi_1\left(\frac{\partial\beta_1}{\partial t}\right) = \frac{c_1}{2}\left(\frac{\partial\beta_1}{\partial t}\right)^2 + I_-\left(\frac{\partial\beta_1}{\partial t}\right),$$

where $w_1$ is the damage threshold, $k_1$ the damage interaction parameter which quantifies the influence of volume damage on its neighbourhood, $\lambda_1$ and $\mu_1$ the Lamé parameters. They are the more simple energies coupling elasticity and volume damage. They give the constitutive laws

$$\sigma_1 = \frac{\partial\Psi_1}{\partial\varepsilon_1} = \beta_1\{\lambda_1 tr\varepsilon_1 1 + 2\mu_1\varepsilon_1\},$$

$$B_1 = \frac{\partial\Psi_1}{\partial\beta_1} + \frac{\partial\Phi_1}{\partial(\partial\beta_1/dt)}$$

$$= -w_1 + \frac{1}{2}\left\{\lambda_1(tr\varepsilon_1)^2 + 2\mu_1\varepsilon_1:\varepsilon_1\right\} + \partial I(\beta_1) + c\left(\frac{\partial\beta_1}{\partial t}\right) + \partial I_-\left(\frac{\partial\beta_1}{\partial t}\right),$$

$$\mathbf{H}_1 = \frac{\partial\Psi_1}{\partial(\operatorname{grad}\beta_1)} = k_1\operatorname{grad}\beta_1,$$

where 1 is the identity matrix.

The free energy and pseudopotential of the glued contact surface are

$$\Psi_s(E_s) = \Psi_s(\mathbf{u}_2 - \mathbf{u}_1, \beta_s, \mathrm{grad}_s\,\beta_s, \beta_1 - \beta_s, \beta_2 - \beta_s)$$
$$= w_s(1-\beta_s) + \frac{k_s}{2}(\mathrm{grad}_s\,\beta_s)^2 + I(\beta_s) + I_-((\mathbf{u}_2 - \mathbf{u}_1)\cdot\mathbf{N}_2)$$
$$+\frac{\beta_s\hat{k}_s}{2}(\mathbf{u}_2 - \mathbf{u}_1)^2 + \frac{k_{s,1}}{2}(\beta_1 - \beta_s)^2 + \frac{k_{s,2}}{2}(\beta_2 - \beta_s)^2,$$
$$\Phi_s(\delta E_s) = \Phi_s\left(\frac{\partial\beta_s}{\partial t}\right) = \frac{c_s}{2}\left(\frac{\partial\beta_s}{\partial t}\right)^2 + I_-\left(\frac{\partial\beta_s}{\partial t}\right),$$

where $w_s$ is the surface damage threshold, $k_s$ the surface damage interaction parameter, $\hat{k}_s$ is the surface rigidity, $k_{s,1}$ and $k_{s,2}$ are the damage surface-volume interaction parameter, $c_s$ is the damage viscosity. The function $I_-((\mathbf{u}_2 - \mathbf{u}_1)\cdot\mathbf{N}_2)$ takes into account the impenetrability of the two pieces of concrete on their contact surface. The surface free energies and pseudopotential of dissipation are also the more simple we may choose. They account for elastic, viscous and damage properties. They give the constitutive laws

$$\mathbf{R} = \frac{\partial\Psi_s}{\partial(\mathbf{u}_2 - \mathbf{u}_1)} \in \beta_s\hat{k}_s(\mathbf{u}_2 - \mathbf{u}_1) + \partial I_-((\mathbf{u}_2 - \mathbf{u}_1)\cdot\mathbf{N}_2)\mathbf{N}_2, \tag{8.2}$$

$$B_s = \frac{\partial\Psi_s}{\partial\beta_s} \in -w_s + \frac{\hat{k}_s}{2}(\mathbf{u}_2 - \mathbf{u}_1)^2 + \partial I(\beta_s) + c_s\frac{\partial\beta_s}{\partial t} + \partial I_-\left(\frac{\partial\beta_s}{\partial t}\right),$$

$$\mathbf{H}_s = \frac{\partial\Psi_s}{\partial(\mathrm{grad}_s\,\beta_s)} = k_s\,\mathrm{grad}_s\,\beta_s,$$

$$B_{1,s} = \frac{\partial\Psi_s}{\partial(\beta_1 - \beta_s)} = k_{s,1}(\beta_1 - \beta_s),$$

$$B_{2,s} = \frac{\partial\Psi_s}{\partial(\beta_2 - \beta_s)} = k_{s,2}(\beta_2 - \beta_s).$$

The force in $\partial I_-((\mathbf{u}_2 - \mathbf{u}_1)\cdot\mathbf{N}_2)\mathbf{N}_2$ is the impenetrability reaction. The non-local free energy on the glued contact surface is

$$\Psi_{s,1,2}(E_{s,1,2}(\mathbf{x},\mathbf{y})) = \frac{k_{s,1,2}}{2}g^2(\mathbf{y},\mathbf{x})(\beta_s(\mathbf{x})\beta_s(\mathbf{y}))\exp\left(-\frac{|\mathbf{x}-\mathbf{y}|^2}{d^2}\right),$$

with

$$g(\mathbf{y},\mathbf{x}) = 2(\mathbf{y}-\mathbf{x})\cdot(\mathbf{u}_2(\mathbf{y}) - \mathbf{u}_1(\mathbf{x})).$$

The exponential function with distance $d$, describes the attenuation of non-local actions with distance $|\mathbf{x}-\mathbf{y}|$ between points $\mathbf{x}$ and $\mathbf{y}$. We assume no dissipation with respect to $\delta E_{s,1,2}(\mathbf{x},\mathbf{y})$ and have the constitutive law

$$-B_{s,1}(\mathbf{x},\mathbf{y}) = \frac{\partial \Psi_{s,1,2}}{\partial \beta_s(\mathbf{x})}(E_{s,1,2}(\mathbf{x},\mathbf{y})) = \frac{k_{s,1,2}}{2} g^2(\mathbf{x},\mathbf{y})\beta_s(\mathbf{y})\exp(-\frac{|\mathbf{x}-\mathbf{y}|^2}{d^2}),$$

$$-B_{s,2}(\mathbf{x},\mathbf{y}) = \frac{\partial \Psi_{s,1,2}}{\partial \beta_s(\mathbf{y})}(E_{s,1,2}(\mathbf{x},\mathbf{y})) = \frac{k_{s,1,2}}{2} g^2(\mathbf{x},\mathbf{y})\beta_s(\mathbf{x})\exp(-\frac{|\mathbf{x}-\mathbf{y}|^2}{d^2}),$$

$$-M(\mathbf{x},\mathbf{y}) = \frac{\partial \Psi_{s,1,2}}{\partial g(\mathbf{y},\mathbf{x})}(E_{s,1,2}(\mathbf{x},\mathbf{y})) = k_{s,1,2} g(\mathbf{x},\mathbf{y})\left(\beta_s(\mathbf{x})\beta_s(\mathbf{y})\right)\exp(-\frac{|\mathbf{x}-\mathbf{y}|^2}{d^2}).$$

Let us note that all the constitutive laws involve the reactions to the internal constraints when needed, which are clearly non linear relationships, and linear relationships between the forces and the state quantities and velocities. Thus they are simple and we think that they have to account for the main physical phenomena: non linear constitutive laws are to be chosen only to make the results more precise and adapted to deal with a particular situation. But the linear relationships have to be sufficient to capture the main properties.

## 8.4 The Equations

They result from the equations of motion and the constitutive laws. They are:

### *8.4.1 On the Contact Surface*

$$c_s \frac{\partial \beta_s}{\partial t} - k_s \Delta_s \beta_s + \partial I(\beta_s) + \partial I_-\left(\frac{\partial \beta_s}{\partial t}\right) \ni$$

$$w_s - \frac{\hat{k}_s}{2}(\mathbf{u}_2 - \mathbf{u}_1)^2 + k_{s,1}(\beta_1 - \beta_s) + k_{s,2}(\beta_2 - \beta_s)$$

$$-\int_{\partial\Omega_1\cap\partial\Omega_2} \frac{k_{s,1,2}}{2}(g^2(\mathbf{y},\mathbf{x}) + g^2(\mathbf{x},\mathbf{y})\beta_s(\mathbf{y})\exp(-\frac{|\mathbf{x}-\mathbf{y}|^2}{d^2})d\mathbf{y},\ in\ \partial\Omega_1\cap\partial\Omega_2,$$

$$k_s \frac{\partial \beta_s}{\partial n_s} = 0,\ on\ \partial(\partial\Omega_1\cap\partial\Omega_2).$$

The last but one term is not 0 when

$$\mathbf{u}_2 - \mathbf{u}_1 = 0.$$

It is responsible for the damage resulting from elongation. The glue damage source in the right hand side results from the gap between the two solids, from the elongation (the non-local effect) and from the flux of damaging work coming from the concrete. It is proportional to the difference of damage between the concrete and the glue. Thus it is more difficult to damage the glue when the concrete is not damaged. In this case the glue cohesion is $w_s + k_{s,1} + k_{s,2}$ whereas it is $w_s$ when the concrete is completely damaged. The contact boundary conditions on the glued contact surface $\partial\Omega_1 \cap \partial\Omega_2$ are

$$\forall \mathbf{x} \in \partial\Omega_1 \cap \partial\Omega_2,\ \sigma_1 \mathbf{N}_1(\mathbf{x}) = \beta_s \hat{k}_s (\mathbf{u}_2 - \mathbf{u}_1) + \partial I_-((\mathbf{u}_2 - \mathbf{u}_1)\cdot \mathbf{N}_2)\mathbf{N}_2$$

$$-\int_{\partial\Omega_1 \cap \partial\Omega_2} 2(\mathbf{x}-\mathbf{y}) k_{s,1,2}\, g(\mathbf{x},\mathbf{y})\, (\beta_s(\mathbf{x})\beta_s(\mathbf{y})) \exp\left(-\frac{|\mathbf{x}-\mathbf{y}|^2}{d^2}\right) d\Gamma(\mathbf{y}),$$

$$\forall \mathbf{y} \in \partial\Omega_1 \cap \partial\Omega_2,\ \sigma_2 \mathbf{N}_2(\mathbf{y}) = -\beta_s \hat{k}_s (\mathbf{u}_2 - \mathbf{u}_1) - \partial I_-((\mathbf{u}_2 - \mathbf{u}_1)\cdot \mathbf{N}_2)\mathbf{N}_2,$$

$$-\int_{\partial\Omega_1 \cap \partial\Omega_2} 2(\mathbf{y}-\mathbf{x}) k_{s,1,2}\, g(\mathbf{x},\mathbf{y})\, (\beta_s(\mathbf{x})\beta_s(\mathbf{y})) \exp\left(-\frac{|\mathbf{x}-\mathbf{y}|^2}{d^2}\right) d\Gamma(\mathbf{x}),$$

$$k_1 \frac{\partial \beta_1}{\partial N_1} = k_{s,1}(\beta_s - \beta_1),\ k_1 \frac{\partial \beta_2}{\partial N_2} = k_{s,2}(\beta_s - \beta_2). \tag{8.3}$$

In the following numerical applications, we have neglected the non local mechanical effect on the contact surface stresses because it has not an important overall mechanical effect. The values of parameters $\hat{k}_s \gg k_{s,1,2}$ of the constitutive laws we choose in the sequel agree with this assumption. Let us recall that this non-local effect has been introduced to take into account damage which is produced by the elongation, i.e. by displacements such that

$$\mathbf{u}_2(\mathbf{x}) - \mathbf{u}_1(\mathbf{x}) = 0,$$

with

$$\mathbf{u}_2(\mathbf{x}) - \mathbf{u}_1(\mathbf{y}) \neq 0,\ for\, \mathbf{x} \neq \mathbf{y}.$$

The boundary condition (8.3) means that the damaging work flux in the concrete is proportional to the difference of damage between the glue and the concrete.

### 8.4.2 In the Domains

As already said, they are identical for the two domain. For domain $\Omega_1$, they are

$$\operatorname{div}(\beta_1 \{\lambda_1 tr\varepsilon_1(\mathbf{u}_1)1 + 2\mu_1 \varepsilon_1(\mathbf{u}_1)\}) = 0,$$

$$c_1 \frac{\partial \beta_1}{\partial t} - k_1 \Delta \beta_1 + \partial I(\beta_1) + \partial I_-\left(\frac{\partial \beta_1}{\partial t}\right) \ni w_1 - \frac{1}{2}\left\{\lambda_1 (tr\varepsilon_1)^2 + 2\mu_1 \varepsilon_1 : \varepsilon_1\right\},$$

with initial conditions

$$\beta_1(\mathbf{x},0) = \beta_1^0(\mathbf{x}),\ in\ \Omega_1,$$
$$\beta_s(\mathbf{x},0) = \beta_1^0(\mathbf{x}),\ on\ \partial\Omega_1 \cap \partial\Omega_2,$$

and boundary conditions

$$\sigma_1 \mathbf{N}_1 = \mathbf{g}_1,\ on\ \partial\Omega_1 \backslash (\partial\Omega_1 \cap \partial\Omega_2),$$
$$k_1 \frac{\partial \beta_1}{\partial N_1} = 0,\ on\ \partial\Omega_1 \backslash (\partial\Omega_1 \cap \partial\Omega_2).$$

## 8.5 Examples

The following examples show how important are the interaction parameters $k_{s,1}$ and $k_{s,2}$ which couple the damages of solids 1 and 2: when solid 1 damages in the neighbourhood of solid 2, solid 2 damages also. The examples confirm also that it is more difficult to damage the glue when the concrete is not damaged than when the concrete is damaged: the glue cohesion or threshold is $w_s + k_{s,1} + k_{s,2}$ when the concrete is not damaged whereas it is $w_s$ when concrete is completely damaged in the two solids. The examples show also the important effect of the stretching described by the non local interactions. All the computations are due to Francesco Freddi, [10, 27, 28, 99, 100].

### 8.5.1 *Four Points Bending*

Some experimental results due to Marie Paule Thaveau, [203], are reported on Fig. 8.1. One piece of concrete or two pieces of concrete glued on one another are tested. Their length is 0.4 m, their width is 0.3 m, their height is either 0.1 m or $2 \times 0.05$ m. The Young modulus is 38 $G$Pa, the Poisson modulus is 0.2. The maximum load is 14.3 kN for the one piece specimen and 18.2 kN for the two pieces specimen. Numerical results are shown on Figs. 8.2–8.5. They concern three models:

1. There is no damaging interaction between solids 1, 2 and the glue: $k_{s,1} = k_{s,2} = 0$. Stretching is not taken into account: $k_{s,1,2} = 0$.
2. There is damaging interaction between solids 1, 2 and the glue: $k_{s,1} \neq 0$, $k_{s,2} \neq 0$. Stretching is not taken into account: $k_{s,1,2} = 0$.
3. There is damaging interaction between solids 1, 2 and the glue: $k_{s,1} \neq 0$, $k_{s,2} \neq 0$. Stretching is taken into account: $k_{s,1,2} \neq 0$.

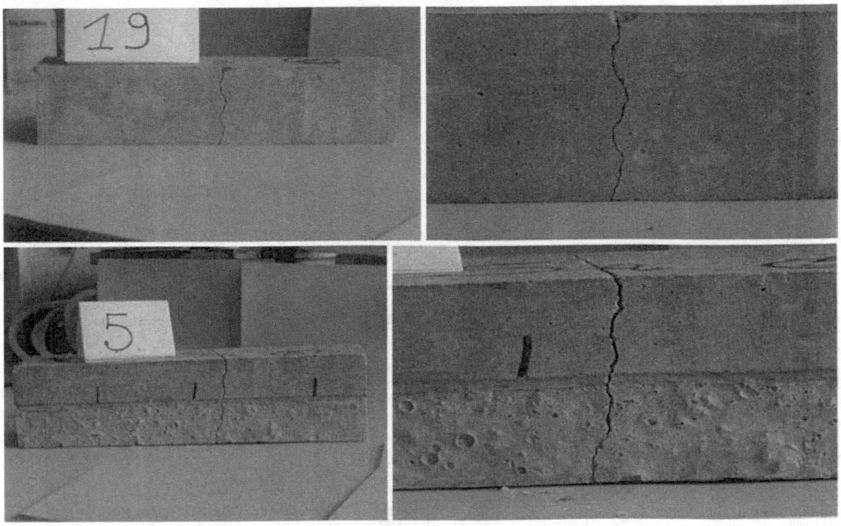


**Fig. 8.1** Four points bending, experimental results due to Marie Paule Thaveau, [203]. The experiments are either with two pieces of concrete glued on one another (number 5) or with one piece of concrete (number 19)

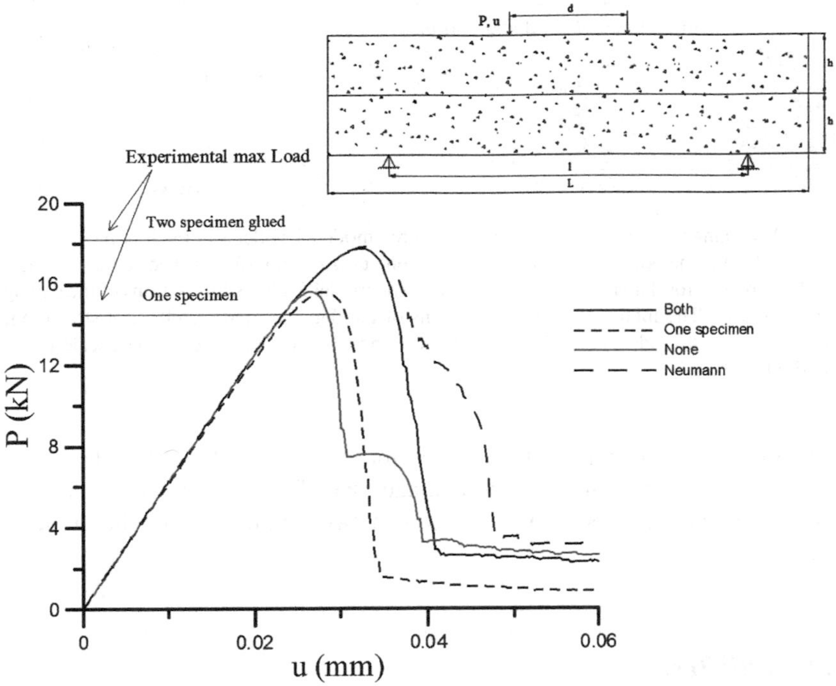


**Fig. 8.2** Load versus displacement curves for one piece and two pieces concrete specimens in four point bending. The short dashed line is for the one piece specimen. The red line, long dashed line and continuous line are for the two pieces concrete specimen and models 1, 2 and 3 respectively

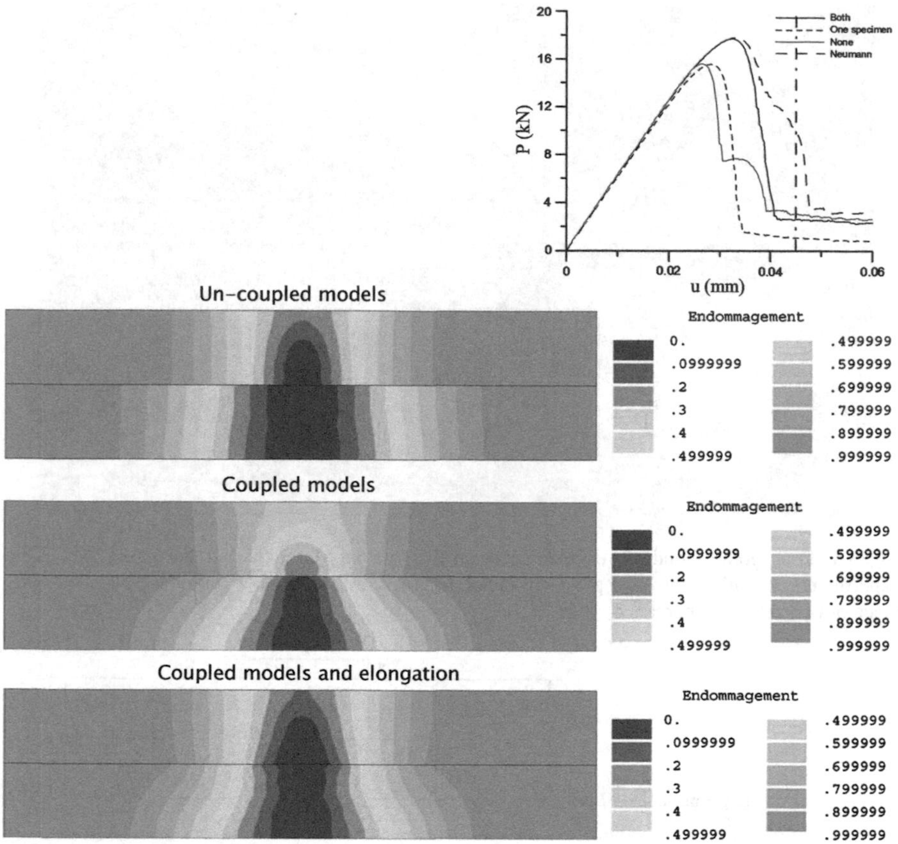


**Fig. 8.3** The damage for the two pieces and the three models (model 1 at the top, model 2 in the middle, model 3 at the bottom). The applied displacement is $u = 0.045$ mm. The contact surface of model 1 is a barrier for damage. It is not very good. The model 2 allows interactions of the damages of the two pieces. The model 3 allows interactions of the damages and the effect of stretching. It seems to be the best model. The color scales are different in the pictures (the color scale is given on the right of them)

The first model is unable to account for experimental results: the contact surface is a barrier which stops the damage (see Fig. 8.4). The two other models are good but the best is the one which takes into account the stretching effect.

### 8.5.2 *Pull Test*

A vertical force is applied to two glued pieces of concrete (Fig. 8.6). The relative stiffness of the concrete and of the glue governs the behaviour of the structure.

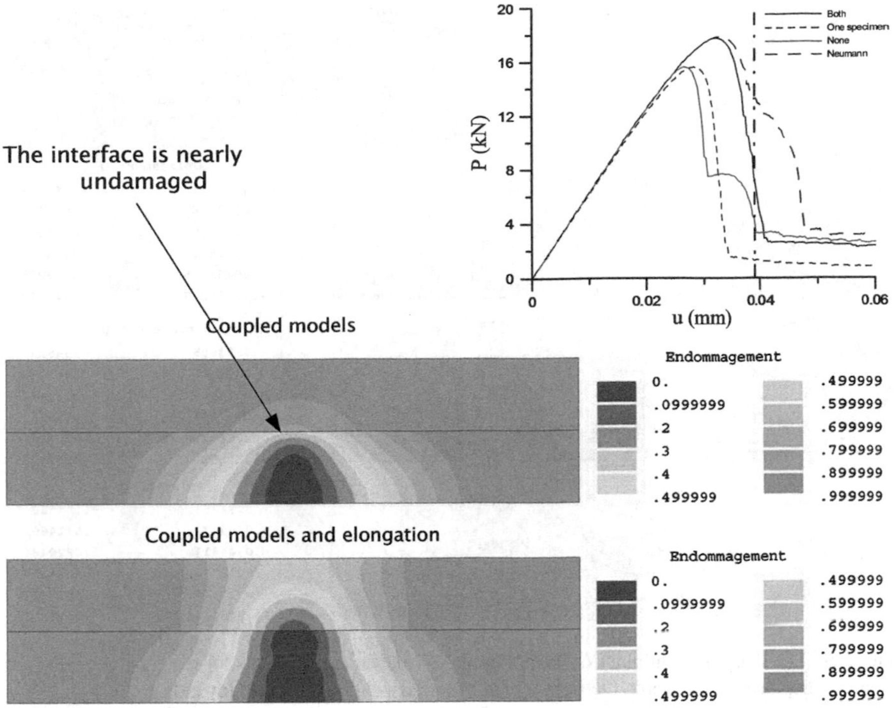


**Fig. 8.4** The contact surface is a barrier in model 1 whereas the effect of damage interaction and the effect of stretching are important in the best model 3. The applied displacement is $u = 0.039$ mm

When glue is solid and the concrete is weak, damage occurs in concrete just under the contact surface (Fig. 8.7). If the glue is weak and the concrete is solid, separation of the two pieces occurs on the contact surface and the concrete is not damaged, (Fig. 8.8).

### 8.5.3 Fibre-Reinforced Polymers-Concrete Delamination

The experiment is described in Fig. 8.9. Experimental and numerical results are shown in Fig. 8.10. It appears a thin damaged zone in the concrete as well as large displacements: they correspond to a layer of concrete which remains glued on the Fibre-reinforced polymers (FRP, [15–17, 59, 103]) in the experiments. The determination of the parameters of the predictive theory from practical engineering knowledge is described in [28] together with the practical computation of the maximal load which a structure can bear.

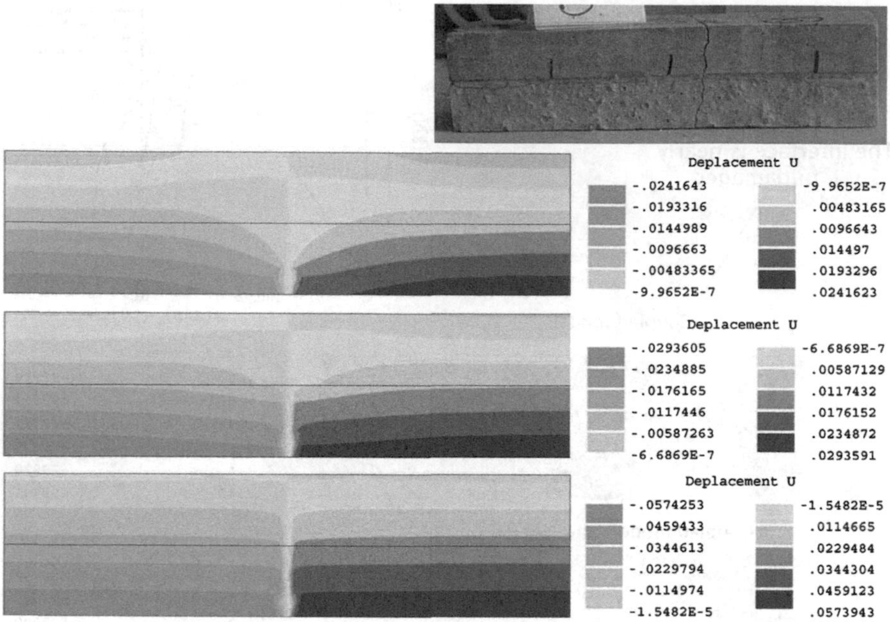


**Fig. 8.5** The sharp discontinuity of the horizontal displacement accounts for the fracture. Red is a positive displacement towards the right, green is a negative displacement towards the left

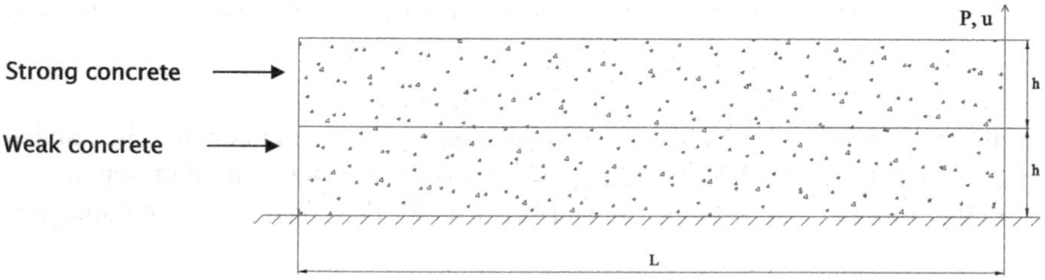


2 CASES:

1. The interface is stronger than the lower (weaker) concrete
2. The interface is weaker than than the lower (weaker) concrete

| L (mm) | h (mm) | t (mm) |
|---|---|---|
| 400 | 50 | 200 |

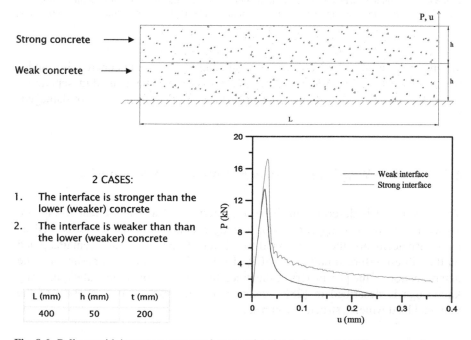

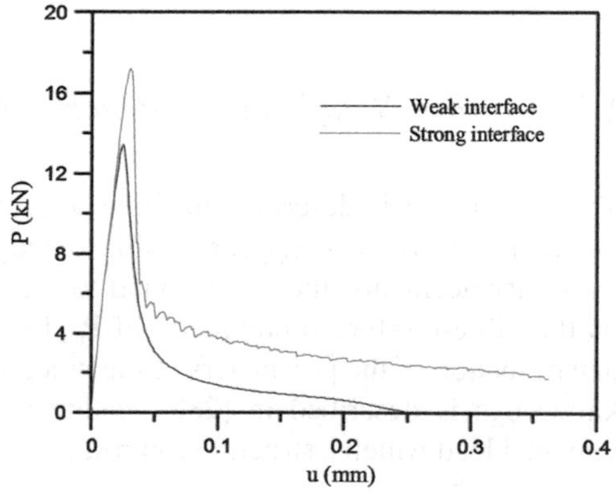


**Fig. 8.6** Pull test with important contrasts between the glue and concrete stiffness properties

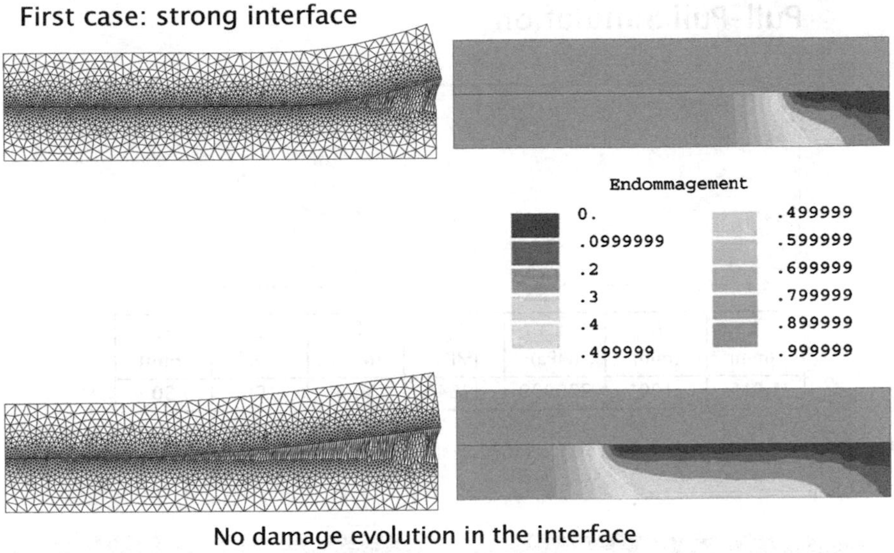


**Fig. 8.7** The glue is much more solid than the concrete; The damage occurs within the concrete just under the contact surface whereas the glue does not break

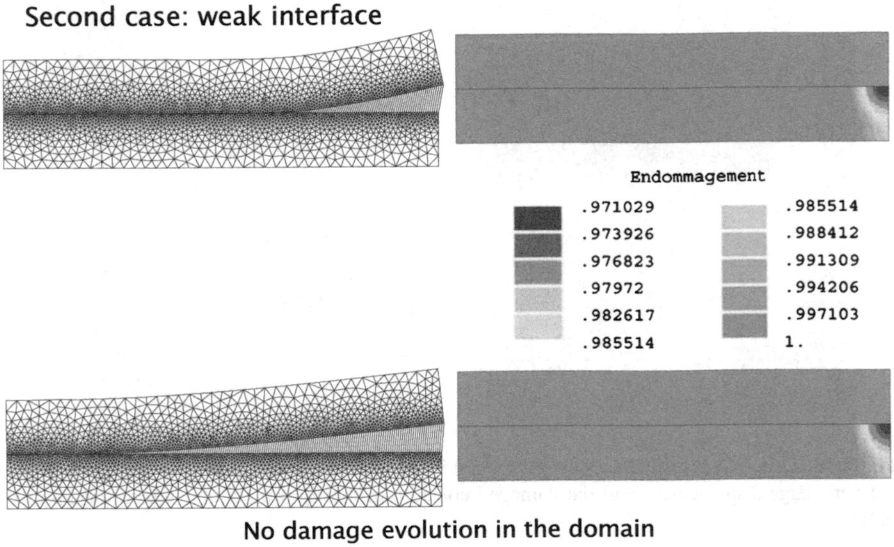


**Fig. 8.8** The glue is weaker than the concrete. Damage occurs only in the glue. The concrete even if it is somewhere red is not damaged or almost not damaged. Let recall that the color scales are unique for each picture

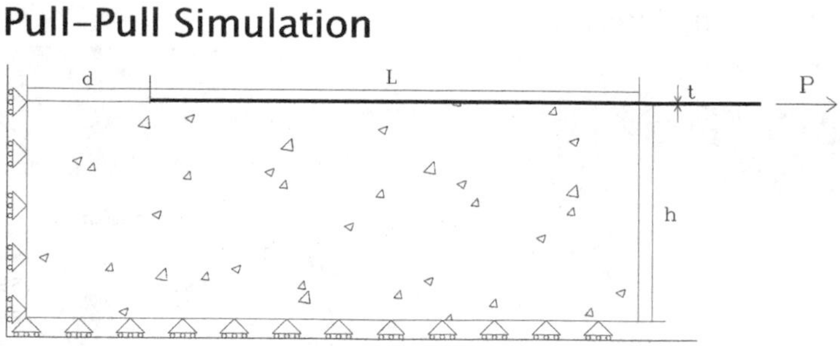


| t (mm) | h (mm) | $E_p$ (MPa) | $E_c$ (MPa) | $f_c$ (MPa) | L (mm) | d (mm) |
|---|---|---|---|---|---|---|
| 1.016 | 100 | 230000 | 33640 | 36.4 | 100 | 50 |

**Fig. 8.9** A pull experiment for a FRP reinforced structure

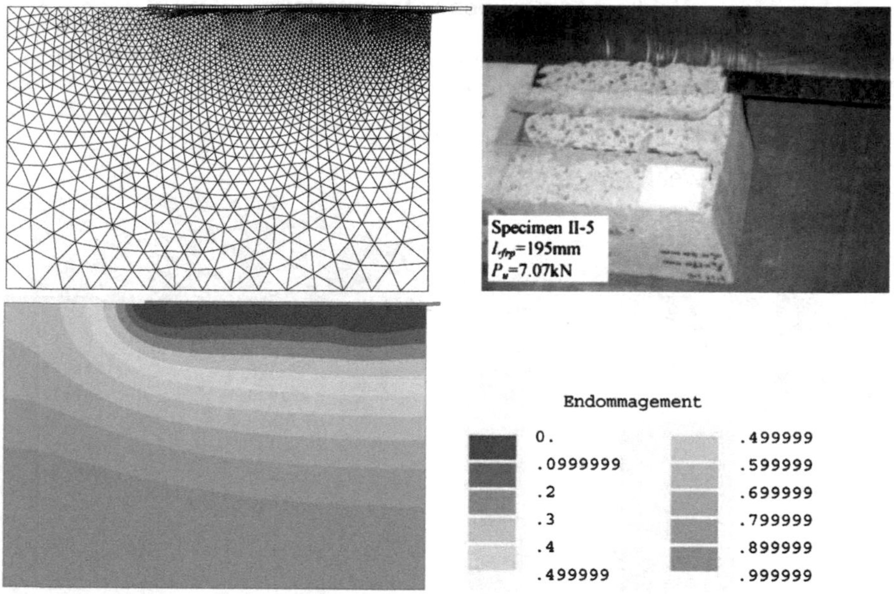


**Fig. 8.10** Damage occurs in the concrete and not in the FRP in agreement with experiments. Note the very large displacement and the damaged concrete in the right up corner where the FRP is pulled

*Remark 8.2.* In collisions which are investigated in following chapters, the damage $\beta$ may be discontinuous with respect to space jumping from $\beta = 1$, sound material, to $\beta = 0$, completely damaged material. When $\beta$ becomes 0 on a line whereas it is 1 elsewhere, a fracture is created. Fractures resulting from collisions may be investigated with these tools. This point of view is developed in [100].

# Chapter 9
# Phase Change with Discontinuity of Temperature: Warm Water in Contact with Cold Ice

Let us consider warm water on cold ice, for instance water flowing on a frozen lake. On the surface of the lake, there is a discontinuity of temperature between the temperature $T_2$ of the liquid water which is larger than the water freezing temperature $T_0$ and the temperature $T_1$ of the ice of the lake which is lower than $T_0$. It is to be known if the water freezes or if the ice thaws, [111, 153].

In order to answer this question, let consider water with temperature $T_2 > T_0$ in contact with ice with temperature $T_1 < T_0$ on a surface $\Gamma(t)$ which is a free boundary. We intend to describe the evolution of the temperatures $T_1$ and $T_2$ and the motion of the free surface which separates water and ice. For the sake of simplicity, we assume the water and ice do not move and that their densities are the same, $\rho$.

## 9.1 Warm Water in Contact with Cold Ice

Outside the free boundary, we assume the free energy and pseudopotential of dissipation (4.1) chosen for the Stefan problem. The basic equations of mechanics on the free boundary between water and ice have been given in Sect. 3.5. Let us choose the pseudopotential, [111, 153] as

$$\Phi(m,[T]) = I_M(m) + \frac{k}{2}[T]^2 , \tag{9.1}$$

where $k$ is a positive constant and $I_M$ is the indicator function of the convex set $M$

$$M = \left\{ m \mid m_{\text{inf}} \le m \le m_{\text{sup}} \right\} ,$$

with $m_{\text{inf}} \le 0 \le m_{\text{sup}}$. We have

$$\Psi(T,\beta) = -CT \ln T - \beta \frac{L}{T_0}(T - T_0) ,$$

M. Frémond, *Phase Change in Mechanics*, Lecture Notes of the Unione Matematica Italiana 13, DOI 10.1007/978-3-642-24609-8_9,

$$[\Psi] = -[CT \ln T] - \frac{L}{T_0}(T_2 - T_0)\,,$$

$$\underline{s} = C(1 + \frac{\ln T_1 + \ln T_2}{2}) + \frac{L}{2T_0}\,,$$

and

$$-[\Psi] - \underline{s}[T] = [CT \ln T] - C(1 + \frac{\ln T_1 + \ln T_2}{2})[T] + \frac{L}{T_0}(\underline{T} - T_0)$$

$$= \mathscr{O}([T]^2) + \frac{L}{T_0}(\underline{T} - T_0)\,,$$

because $\beta_1 = 0$ and $\beta_2 = 1$, where $\mathscr{O}([T]^2)$ is of order $[T]^2$. Within the small perturbation assumption, we have

$$-[\Psi] - \underline{s}[T] = \frac{L}{T_0}(\underline{T} - T_0)\,.$$

The constitutive laws (3.36) are

$$-[\Psi] - \underline{s}[T] \in \partial I_M(m)\,,$$

which is within the small perturbation assumption

$$\frac{L}{T_0}(\underline{T} - T_0) \in \partial I_M(m)\,, \tag{9.2}$$

and

$$\underline{Q} = k[T]\,. \tag{9.3}$$

The law (9.3) is a Fourier law: the entropy flux through the free boundary is proportional to the difference of temperature. Parameter $k$ is the thermal conductivity between the cold and warm domains. The other law (9.2) describes the movement of the free boundary. The mass flow through the free boundary is bounded: it cannot be too large. The constitutive laws gives the difference between the average temperature and the phase change temperature versus the mass flow or versus the velocity of the free boundary:

1. If $\underline{T} - T_0 < 0$ then $m = m_{\text{inf}} = -\rho W \le 0$, the water freezes if the average temperature is low.
2. If $\underline{T} - T_0 > 0$ then $m = m_{\text{sup}} = -\rho W \ge 0$, the ice melts if the average temperature is high.
3. If $\underline{T} - T_0 = 0$ then $m \in [m_{\text{inf}}, m_{\text{sup}}]$, depending on the sign of $m$, either the water freezes or the ice melts.

This situation occurs when solid and liquid in contact have different temperatures. The partial differential equations describing the evolution result from the chosen free energy and pseudopotential of dissipation (4.1). They are

$$C\frac{\partial \ln T_1}{\partial t} - \lambda_1 \Delta \ln T_1 = 0\,,$$

the entropy balance in the ice or the solid phase;

$$C\frac{\partial \ln T_2}{\partial t} - \lambda_2 \Delta \ln T_2 = 0\,,$$

the entropy balance in the water or the liquid phase;

$$\frac{T_1+T_2}{2} - T_0 \in \partial I_M(-\rho W)\,,$$

$$\frac{1}{2}(\lambda_1\frac{\partial \ln T_1}{\partial N} + \lambda_2\frac{\partial \ln T_2}{\partial N}) = k(T_2 - T_1)\,,$$

$$-\rho LW = \lambda_2\frac{\partial T_2}{\partial N} - \lambda_1\frac{\partial T_1}{\partial N}\,,$$

the two constitutive laws (3.36) and the energy balance (3.29) on the free boundary $\Gamma(t)$, the contact surface between solid and liquid. The equations are completed by boundary conditions and the initial position of the free boundary, $\Gamma(0)$.

*Remark 9.1.* The predictive theory given above investigates the contact of two materials at different temperatures (in this case $m = 0$). The problem of warm rain falling on frozen ground or on ice has to be investigated by using both thermal equations and mechanical equations. The effect of the collision of the rain droplets with the ground has to be taken into account. Both the temperatures and the velocities are discontinuous. The above thermal equations have to be coupled with the equations of motion resulting from collision theory [112, 114]. This problem is investigated in a simplified version in the following chapter where solids colliding are points and in a complete but more sophisticated model in Chap. 11.

## 9.2 Mixture of Ice and Water in Contact with Cold Ice

Let us consider a mixture of water and ice at temperature, $T_2 = T_0$ with $0 < \beta_2 < 1$, in contact with cold ice. Constitutive laws (9.2) becomes

$$\frac{\beta_2 L}{T_0}(T_1 - T_0) \in \partial I_M(m)\,.$$

The equations of the predictive theory become

$$C\frac{\partial \ln T_1}{\partial t} - \lambda_1 \Delta \ln T_1 = 0\,,$$

the entropy balance in the ice or the solid phase;

$$L\frac{\partial \beta_2}{\partial t} = 0\,,$$

the entropy balance in the ice water mixture. In the mixture the temperature is the phase change temperature $T_0$;

$$\begin{aligned} T_1 - T_0 &\in \partial I_M(-\rho W)\,, \\ \frac{\lambda_1}{2}\frac{\partial \ln T_1}{\partial N} &= k(T_0 - T_1)\,, \\ -\rho L \beta_2 W &= -\lambda_1 \frac{\partial T_1}{\partial N}\,, \end{aligned} \tag{9.4}$$

the two constitutive laws (3.36) and the energy balance (3.29) on the free boundary $\Gamma(t)$, the contact surface between ice solid phase and mixture.

The equations are completed by boundary conditions and the initial position of the free boundary, $\Gamma(0)$. Because $T_1 - T_0 < 0$, we have

$$-\rho W = m_{\text{inf}}\,.$$

It results that $W > 0$ with $W = \mathbf{W} \cdot \mathbf{N}$ where normal vector $\mathbf{N}$ is directed from the ice domain with index $_1$ toward mixture domain with index $_2$ (see Sect. 3.5). Thus the mixture of water and ice freezes and the ice warms up as shown by relationship (9.4).

## 9.3 An Example: Rain Falling on a Frozen Ground

Let consider rain falling on a frozen ground or on ice. We investigate the evolution of the water: does it freeze? Does it remain liquid producing a liquid layer on top of ice or frozen ground. The rain is considered as an homogeneous mixture of air and water with water mixture density $\rho_2$, temperature $T_2$ and vertical rain velocity $U_2 \leq 0$ (the vertical direction is directed upward). The rain falls on the ice or on the possible unfrozen water layer due to the rain water already fallen. We assume that the ice and possible water layer have null velocity. The mechanical effects are neglected and only the thermal effects are taken into account. The rain mass water flow is

$$m_{rain} = \frac{-\rho_1 \rho_2 U_2}{\rho_2 - \rho_1}.$$

Note that it is negative (mass flow from side 1 toward side 2). The maximal flow of water which can be frozen when reaching the ground is $-m_{rain}$. Thus we let $m_{rain} = m_{\text{inf}}$. From experiments, we assume that the heat exchanges on the freezing line are proportional to the differences of temperature between the ice and the water. For the sake of simplicity, we also assume that the mass flow is bounded from above: $m \leq m_{\text{sup}}$, with $m_{\text{sup}} \geq 0$. From those assumptions, it results the pseudopotential of dissipation (9.1) is valid.

The parameter $k$ is the thermal conductivity of the ice-liquid water interface. A more sophisticated and realistic pseudopotential of dissipation can be chosen [112, 153].

The problem which is considered is to find the evolution of the free boundary: either it remains at the surface in contact with the air (the rain freezes completely) or it is covered progressively by a layer of water (the rain does not freeze). The constitutive law (9.2) is the criterion for the rain to freeze or not to freeze, see Fig. 9.1:

1. If $\underline{T} - T_0 < 0$, the rain freezes immediately when touching the ice or the frozen ground and $m = m_{rain}$, giving the velocity of the freezing line

$$W = -\frac{m_{rain}}{\rho_1}.$$

2. If $\underline{T} - T_0 > 0$, the rain does not freeze and the ice or the frozen ground thaws. The velocity of the thawing line is $m = m_{\text{sup}}$,

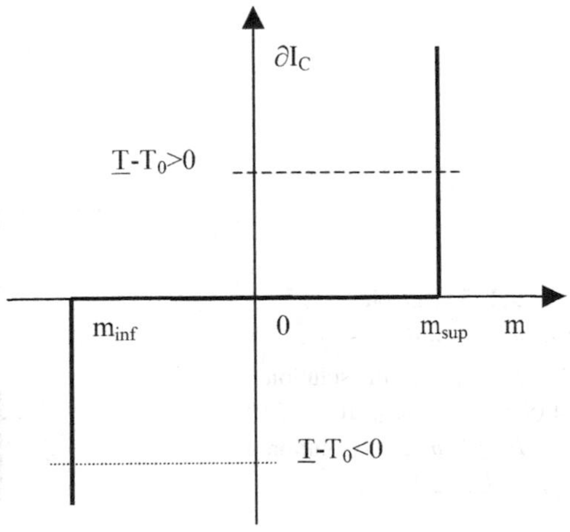


**Fig. 9.1** Relationship (9.2) means that $m = m_{\text{sup}}$ if $\underline{T} - T_0 > 0$, $m = m_{\text{inf}}$ if $\underline{T} - T_0 < 0$ and, $m \in [m_{\text{inf}}, m_{\text{sup}}]$, if $\underline{T} - T_0 = 0$

$$W = -\frac{m_{\text{sup}}}{\rho_1}.$$

3. If $\underline{T} - T_0 = 0$, the velocity of the freezing or thawing line is such that

$$W \in \left[-(m_{\text{sup}}/\rho_1), -(m_{rain}/\rho_1)\right],$$

because $m \in \left[m_{rain}, m_{\text{sup}}\right]$. A part of the water freezes if $W > 0$ and the ice or the frozen ground thaws if $W < 0$.

### 9.3.1 A Dissipative Behaviour

Let us choose the pseudopotential of dissipation, [111, 112, 153]

$$\Phi(m, [T]) = \frac{\widehat{k}}{2}m^2 + I_{\widehat{C}}(m) + \frac{k}{2}[T]^2,$$

where $\widehat{k}$ is a positive constant and $\widehat{C}$ is the convex set,

$$\widehat{C} = \{m \mid m_{\text{inf}} \leq m\},$$

with $m_{\text{inf}} < 0$. The constitutive law (9.3) is still valid. The new constitutive law is, see Fig. 9.2

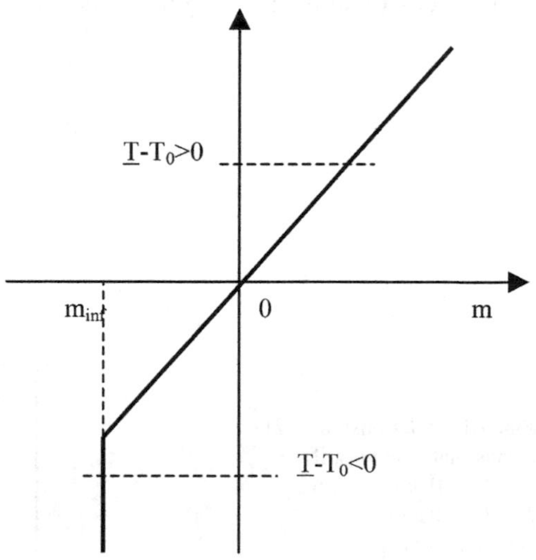


**Fig. 9.2** The graph $\partial I_{\widehat{C}}(m) + \widehat{k}m$. If $\underline{T} - T_0 \leq \frac{\widehat{k}L}{T_0} m_{\text{inf}}$, the solution of (9.5) is $m = m_{\text{inf}}$. If $\underline{T} - T_0 \geq \frac{\widehat{k}L}{T_0} m_{\text{inf}}$, the solution is $m = \frac{T_0(\underline{T} - T_0)}{\widehat{k}L}$

$$\frac{L}{T_0}(\underline{T} - T_0) \in \widehat{k}m + I_{\widehat{C}}(m)\,. \tag{9.5}$$

Let consider rain with temperature 1 °C, density $\rho_2 = 0.1\ \mathrm{kg}/\mathrm{m}^3$, mass flow $m_{rain} = -2\times 10^{-4}\ \mathrm{kg}/(\mathrm{m}^2\,\mathrm{s})$ which falls on a 5 mm ice layer at −3 °C on a frozen ground with thickness 1m. The values of the dissipative parameters are $\widehat{k} = 183\,\mathrm{m}/\mathrm{s}$, $m_{\inf} = m_{rain} = -2\times 10^{-4}\ \mathrm{kg}/(\mathrm{m}^2\,\mathrm{s})$, $k = 3.66\ \mathrm{W}/(\mathrm{m}^2\,\mathrm{K}^2)$. At the beginning the water freezes when falling on the ice. The ice-layer thickens and the heat provided by the rain warms up the ice.

As long as the temperature of the ice is such that

$$\frac{L}{T_0}(\underline{T} - T_0) \le \widehat{k}m_{rain},$$

the rain freezes, see Fig. 9.3. When the ice is warmed enough, i.e., when its temperature is such that

$$\frac{L}{T_0}(\underline{T} - T_0) \ge \widehat{k}m_{rain},$$

it thaws. The rain no longer freezes and a water layer forms on top of the ice layer which is thawing, see Fig. 9.3. This phenomenon begins at time $t = 954$ s. At that time the temperature of the ice is −1.016 °C and the thickness of the ice layer is

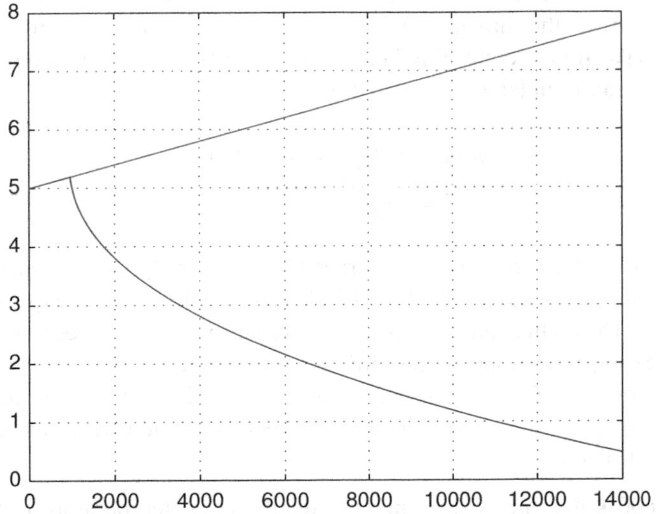


**Fig. 9.3** The position (mm) of the top of the structure versus time (s), the ice at the beginning (in blue), the water at the end (in red), versus time (s). The position of the free surface (in blue) which separates the ice which is thawing from the water layer made of rain and thawed ice

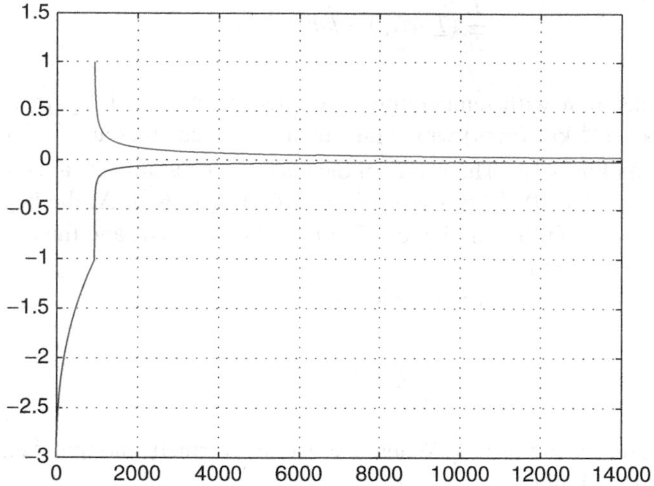


**Fig. 9.4** The temperature of the top of the ice layer in blue and the temperature of the bottom of the water layer in red versus time. The difference of temperature becomes rapidly almost 0 °C if thermal conductivity $k$ is large. In this example, it has been chosen small to emphasis the phenomena

5.12 mm. Then the temperatures of the ice and water in contact tend rapidly to 0 °C, Fig. 9.4.

*Remark 9.2.* When the rain freezes, the ice receives heat from the rain but it also receives heat from the surrounding air. Thus the equations have to be slightly modified to take into account this heat intake. The basic modification are on the energy balance and fundamental inequality which become,

$$m[e] = [TQ] + T_1 B_1 + T_2 B_2,$$
$$H = m[s] - [Q] - \underline{B} \geq 0,$$

where $T_1 B_1$ is the heat received by the rain from the air, $T_2 B_2$ is the heat received by the ice from the air and, $\underline{B} = (B_1 + B_2)/2$. It is easy to take these quantities into account, [112, 153]. One can, for instance, assume that the heat received from the air by the ice is proportional to the difference of the temperatures of the air and ice $T_2 B_2 = -\alpha_2(T_2 - T_{air})$ and, assume that the rain is in thermal equilibrium with the air $T_1 B_1 = T_1 Q_2$. The results shown have been computed with this assumption ($\alpha_2 = 10$ W/(m$^2$ K)).

At time 600 s the rain is freezing. The depth versus temperature in the whole ice-ground structure is shown in Fig. 9.5. One can see the effect of the rain which warms up the top of the structure.

Figure 9.6 shows depth versus temperature just after the beginning of the ice thaw ($t = 1029$ s). It is possible to see in red the very thin layer of water made of thawed

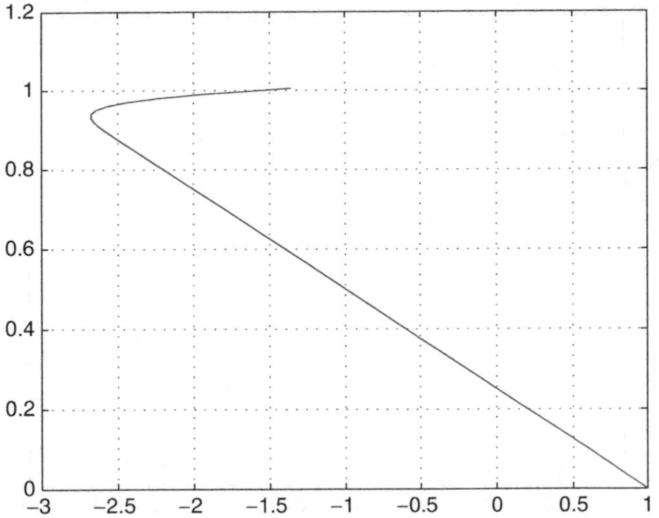


**Fig. 9.5** At time 600 s, the depth in the ground (m) versus temperature. The rain warms up the ground

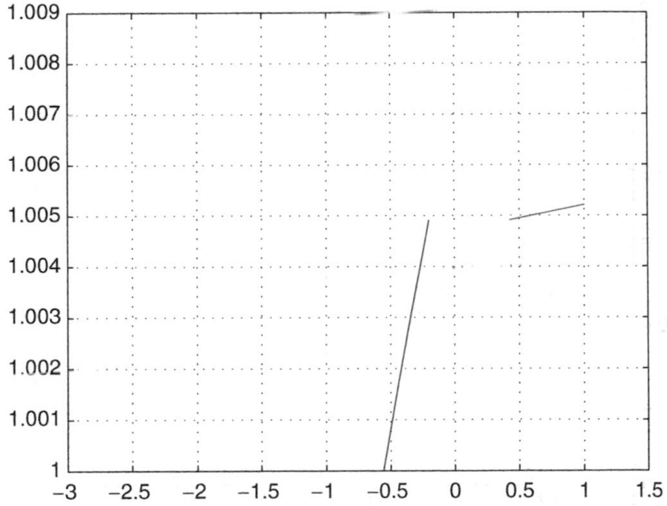


**Fig. 9.6** The depth versus temperature in the ice layer and the thin water layer just after the rain has stopped to freeze. The discontinuity of temperature between the water and ice is decreasing with time

ice and rain. The discontinuity of temperature is still visible. Figure 9.7 shows the temperature inside the whole structure.

After four hours of rain, the layer of ice is almost thawed, see Fig. 9.8. And the discontinuity of temperature is no longer noticeable.

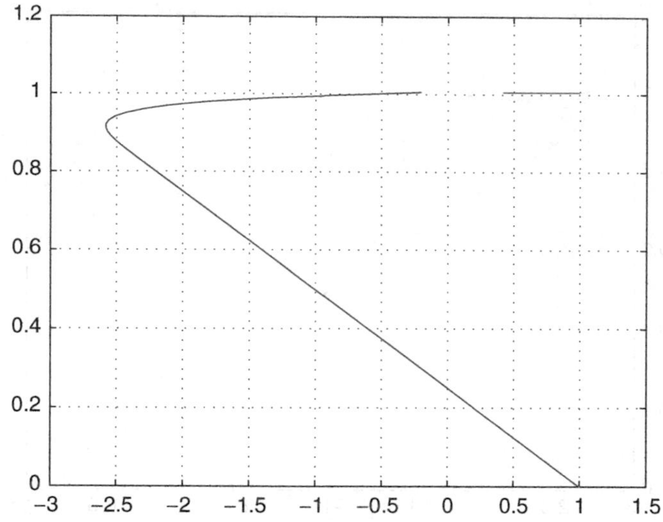


**Fig. 9.7** The depth in the ground (m) versus temperature when the rain no longer freezes. Note a thin layer of water on top of the ice layer

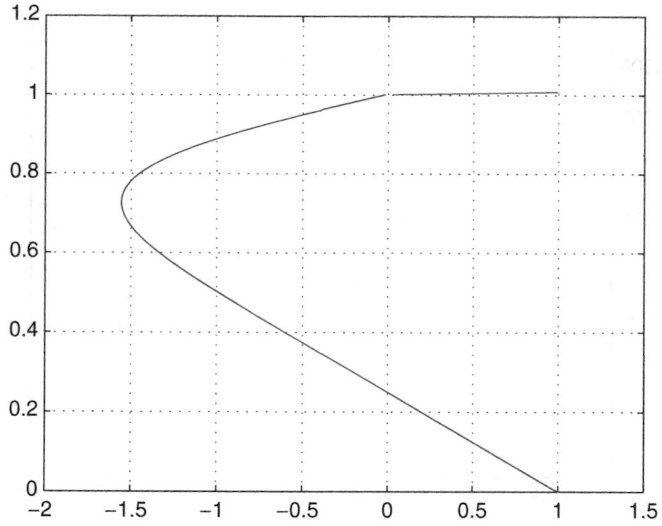


**Fig. 9.8** The depth versus temperature in the structure after four hours. The ice layer is almost thawed. There is no longer a temperature discontinuity between ice and water

*Remark 9.3.* The value of the thermal conductivity $k$ is crucial for the decrease of the temperature discontinuity, [153]. For $k = 11\ \mathrm{W}/(\mathrm{m}^2\,\mathrm{K}^2)$, the temperatures of the water and of the ice become almost immediately equal to $0\,^\circ\mathrm{C}$ as soon as the rain no longer freezes. This value is more realistic than the value chosen in the present example which has the advantage to emphasis what occurs.

# Chapter 10
# Phase Change and Collisions

Let us consider warm rain falling on a frozen ground. On the surface of the ground, there is a discontinuity of temperature between the temperature $T_2$ of the rain which is larger than the water freezing temperature $T_0$ and the temperature $T_1$ of the frozen ground which is lower than $T_0$. It is to be known if the rain freezes or if the frozen ground thaws. In this problem, the temperatures are discontinuous both with respect to space and with respect to time. Let us also note that the velocities are also discontinuous. In order to investigate this problem, we recall the theory of collisions without thermal effects and the theory of collision with thermal effects. For the sake of simplicity, we consider first collisions of balls schematized by points then in next chapter, collisions of continuous media either solid or liquid. This theory is developed in [114].

## 10.1 Collisions of Two Balls: The Mechanical Theory

Let us consider two balls moving on a line, Fig. 10.1. For the sake of simplicity, we assume they are points with mass $m_i$. They have position $x_i(t)$ and velocity $U_i(t) = (\mathrm{d}x_i/\mathrm{d}t)(t)$. Because the points can collide, we suppose that the velocities can be discontinuous, assuming the collisions to be instantaneous.

The velocity before a collision at time $t$ is $U_i^-(t)$ and the velocity after is $U_i^+(t)$. In terms of mathematics, it seems interesting to think that the velocities are bounded variation functions of time. In the sequel we will use virtual velocities: a set of virtual velocities $V = (V_1, V_2)$ is such that the two functions $t \to V_i(t)$ are bounded variation functions.

We denote $S(U,t_1,t_2)$ the set of discontinuity times of velocity $U = (U_1,U_2)$ between times $t_1$ and $t_2$ ($t_1 < t_2$)

$$S(U,t_1,t_2) = \left\{t_k \in ]t_1,t_2[ \,\middle|\, U^+(t_k) \neq U^-(t_k)\right\}.$$

M. Frémond, *Phase Change in Mechanics*, Lecture Notes of the Unione Matematica Italiana 13, DOI 10.1007/978-3-642-24609-8_10,

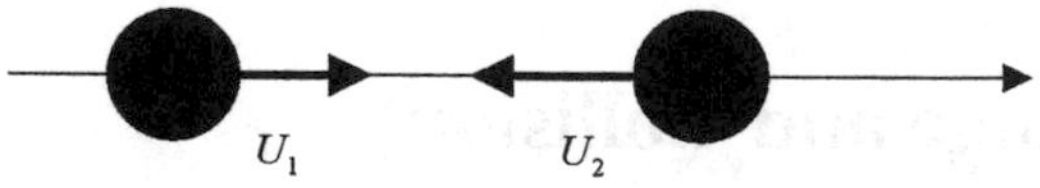


**Fig. 10.1** Two balls schematized by points move on an axis (their mass moments of inertia are 0). The mass of the balls are $m_1$ and $m_2$, they have velocities $U_1$ and $U_2$

This set is numerable because the two functions $U_i$ are functions of bounded variation. We denote

$$[U_i(t_k)] = U_i^+(t_k) - U_i^-(t_k),$$

the velocity discontinuity and in general $[A(t_k)] = A^+(t_k) - A^-(t_k)$ the discontinuity of function $t \to A(t)$.

More precisely, we assume that velocities $U_i$ are *special bounded variation functions*, [5, 18, 54, 172]: their differential

$$\mathrm{d}U_i = \left\{\frac{\mathrm{d}U_i}{\mathrm{d}t}\right\}\mathrm{d}t + \sum_{t_k \in S(U,t_1,t_2)} [U_i(t_k)]\,\delta(t - t_k),$$

is the sum of a Lebesgue measure, the smooth part of the differential, whose density is $\{\mathrm{d}U_i/\mathrm{d}t\}$ and of a Dirac measure, the non-smooth part of the differential, whose density at point $t_k$ is discontinuity $[U_i(t_k)]$.

*Remark 10.1.* Velocity $U^+(t)$ depends on the future. It intervenes in the theory, particularly in the constitutive laws. This seems to contradict the causality principle. This is not the case, as already mentioned in Sect. 3.5. Velocity $U^+(t)$ sums up the sophisticated evolution taking place during the physical phenomenon responsible for the velocity discontinuity. Thus it depends on the past.

### *10.1.1 The Velocities of Deformation*

It is easy to write the equations of motion for the two points. But in order to be more precise and have results which can be easily adapted to more general settings, we derive carefully these equations by using the principle of virtual work. Before we introduce it, let us remark that there is no reason to speak of deformation when dealing with an isolated point. But if we consider the *system* made of the two points, its *form changes* because the distance of the two points may change. Thus it is wise to consider that the *system is deformable*. A way to characterize the system velocity of deformation is to consider the relative velocity of the two points

$$D(U) = U_1 - U_2,$$

where $U = (U_1, U_2)$ is the set of the two actual velocities of the points. A *rigid system set of velocities* is such that the form of the system does not change: it is easy to see that they are characterized by

$$D(U) = U_1 - U_2 = 0,$$

because when the form of the system does not change, the velocities of the two points are equal.

### 10.1.2 The Principle of Virtual Work

The principle is

*The virtual work of the acceleration forces is equal to the sum of the virtual work of the interior forces and the virtual work of the exterior forces. More precisely, for any time $t_1$, any time $t_2 > t_1$ and any virtual velocity $V$*

$$\mathcal{T}_{acc}(t_1,t_2,U,V) = \mathcal{T}_{int}(t_1,t_2,U,V) + \mathcal{T}_{ext}(t_1,t_2,U,V),$$

*where $\mathcal{T}_{acc}(t_1,t_2,U,V)$ is the virtual work of the acceleration forces between times $t_1$ and $t_2$, $\mathcal{T}_{int}(t_1,t_2,U,V)$ and $\mathcal{T}_{ext}(t_1,t_2,U,V)$ are the virtual works of the interior and exterior forces between the same times. These works satisfy:*

1. *The virtual work of the acceleration forces is a linear function of the virtual velocity $V$ , when $S(V,t_1,t_2)$ is fixed. The actual work of the acceleration forces is equal to the variation of the kinetic energy between times $t_1$ and $t_2$*

   $$\mathcal{T}_{acc}(t_1,t_2,U,U) = \mathcal{K}(U^-(t_2)) - \mathcal{K}(U^+(t_1)),$$

   *where $\mathcal{K}$ is the system kinetic energy. This property and the principle of virtual work with the actual velocities, i.e., $V = U$, are sometimes called the theorem of expanded energy (teorema dell'energia cinetica in Italian and le théorème de l'énergie cinétique in French).*
2. *The virtual work of the interior forces is a linear function of the virtual velocity $V$ , when $S(V,t_1,t_2)$ is fixed. It satisfies*

   $$\mathcal{T}_{int}(t_1,t_2,U,V) = 0,$$

   *for any rigid system velocity. This relationship is often called the axiom of the principle of virtual work, [132, 197].*
3. *The virtual work of the exterior forces is a linear function of the virtual velocity $V$, when $S(V,t_1,t_2)$ is fixed.*

*Remark 10.2.* The principle of virtual work we use here is not to be confused with the principle of virtual power where the velocities are understood as small displacements (This relationship is often also called misleadingly the principle of virtual work).

### 10.1.3 The Virtual Works

Now we may introduce *interior forces to the deformable system.* A productive way is to *define these generalized forces by their work* (or by duality in terms of mathematics). The work of the interior forces of the system between times $t_1$ and $t_2$ ($t_1 < t_2$) is a linear function of the velocity of deformation. We choose as virtual work of the interior forces

$$\mathcal{T}_{int}(t_1,t_2,U,V) = -\int_{t_1}^{t_2} R^{int}(\tau)D(V)(\tau)d\tau - \sum_{t_k \in S(U,t_1,t_2)\cup S(V,t_1,t_2)} \left\{ P^{int}(t_k) D\left(\frac{V^+(t_k)+V^-(t_k)}{2}\right)\right\},$$

which is a linear function of $D(V) = (V_1 - V_2)$ when $S(V,t_1,t_2)$ is fixed. Sign minus has no importance, it is chosen in accordance with habits of continuum mechanics. Its advantage will appear down below where a classical inequality has its classical sign. The virtual work defines an interior force $R^{int}(\tau)$ which intervene in the smooth evolution when the two points evolve without colliding ($R^{int}(\tau) = 0$ when the two points do not interact at a distance). It defines also an interior percussion $P^{int}(t)$ which intervene when collisions occur.

*Remark 10.3.* The linear function of $D(V)$ we have chosen is not the more general. The choice involving

$$-P^{int+}(t_k)D(V^+)(t_k) - P^{int-}(t_k)D(V^-)(t_k),$$

does not give much more opportunities, [112].

Let us note that $\mathcal{T}_{int}(t_1,t_2,U,V) = 0$ for any rigid system velocity, i.e., for $D(V) = (V_1 - V_2) = 0$.

The virtual work of the acceleration forces is

$$\mathcal{T}_{acc}(t_1,t_2,U,V) = \sum_{i=1}^{2}\left\{\int_{t_1}^{t_2} m_i \frac{\mathrm{d}U_i}{\mathrm{d}t}(\tau)V_i(\tau)d\tau\right\} + \sum_{i=1}^{2}\left\{\sum_{t_k \in S(U,t_1,t_2)\cup S(V,t_1,t_2)} \left\{ m_i[U_i(t_k)].\frac{V_i^+(t_k)+V_i^-(t_k)}{2}\right\}\right\},$$

where $[U(t_k)] = U^+(t_k) - U^-(t_k)$ is the velocity discontinuity. It is clear that this work is a linear function of $V$ when $S(V,t_1,t_2)$ is fixed.

#### 10.1.3.1 The Theorem of Expanded Energy

Let us compute the actual work of the acceleration forces

$$
\begin{aligned}
&\mathcal{T}_{acc}(t_1,t_2,U,U) \\
&= \sum_{i=1}^{2}\left\{\int_{t_1}^{t_2} m_i\frac{\mathrm{d}U_i}{\mathrm{d}t}(\tau)U_i(\tau)d\tau + \sum_{t_k\in S(U,t_1,t_2)}\left\{m_i\left[U_i(t_k)\right].\frac{U_i^+(t_k)+U_i^-(t_k)}{2}\right\}\right\} \\
&= \sum_{i=1}^{2}\left\{\int_{t_1}^{t_2} m_i\frac{\mathrm{d}U_i}{\mathrm{d}t}(\tau)U_i(\tau)d\tau + \sum_{t_k\in S(U,t_1,t_2)}\frac{m_i}{2}\left[\left(U_i(t_k)\right)^2\right]\right\} \\
&= \sum_{i=1}^{2}\left\{\frac{m_i}{2}\left(U_i^-(t_2)\right)^2 - \frac{m_i}{2}\left(U_i^+(t_1)\right)^2\right\} = \mathcal{K}(U^-(t_2)) - \mathcal{K}(U^+(t_1)) \\
&= \mathcal{K}^-(t_2) - \mathcal{K}^+(t_1),
\end{aligned}
$$

which is equal to the variation of kinetic energy $\mathcal{K}$ between times $t_1$ and $t_2$. At any time we have kinetic energies $\mathcal{K}^-$ and $\mathcal{K}^+$ in case there is a collision.

The virtual work of the exterior forces is a linear function of the virtual velocities $V$ when $S(V,t_1,t_2)$ is fixed, we choose

$$
\begin{aligned}
&\mathcal{T}_{ext}(t_1,t_2,U,V) \\
&= \sum_{i=1}^{2}\left\{\int_{t_1}^{t_2} F_i(\tau)V_i(\tau)d\tau + \sum_{t_k\in S(U,t_1,t_2)\cup S(V,t_1,t_2)}\left\{P_i^{ext}(t_k)\frac{V_i^+(t_k)+V_i^-(t_k)}{2}\right\}\right\}.
\end{aligned}
$$

Exterior force $F_i$ may be the gravity force. Exterior percussion $P_i^{ext}$ may represent hammer blows applied to the points.

*Remark 10.4.* It is not common that an exterior percussion is applied to a mechanical system at the same time a collision occurs. An example is in a pin-ball machine where a collision may produce an external electrical blow.

### *10.1.4 The Equations of Motion*

We assume that the interval $]t_1,t_2[$ is split in distinct intervals $]t_k,t_l[$ where there is no discontinuity of $U$, the two ends of the intervals being discontinuity times. By choosing smooth virtual velocities with compact support in interval $]t_k,t_l[$, we get

$$
\sum_{i=1}^{2}\int_{t_k}^{t_l} m_i\frac{\mathrm{d}U_i}{\mathrm{d}t}(\tau)V_i(\tau)d\tau = -\int_{t_k}^{t_l} R^{int}(\tau)D(V)(\tau)d\tau + \sum_{i=1}^{2}\int_{t_k}^{t_l} F_i(\tau)V_i(\tau)d\tau.
$$

The fundamental lemma of variation calculus gives

$$m_1 \frac{dU_1}{dt} = -R^{int} + F_1, \ a.e. \ in \ ]t_k, t_l[,$$
$$m_2 \frac{dU_2}{dt} = R^{int} + F_2, \ a.e. \ in \ ]t_k, t_l[. \tag{10.1}$$

*Remark 10.5.* *a.e.*means almost everywhere or almost always in this context.

Because points $t_i$ are numerable, relationships (10.1) are satisfied almost everywhere in $]t_1, t_2[$. Thus the principle becomes

$$\sum_{i=1}^{2} \sum_{t_k \in S(U,t_1,t_2) \cup S(V,t_1,t_2)} \left\{ m_i \left[U_i(t_k)\right] \frac{V_i^+(t_k) + V_i^-(t_k)}{2} \right\}$$
$$= - \sum_{t_k \in S(U,t_1,t_2) \cup S(V,t_1,t_2)} \left\{ P^{int}(t_k) D\left(\frac{V^+(t_k) + V^-(t_k)}{2}\right) \right\}$$
$$+ \sum_{i=1}^{2} \sum_{t_k \in S(U,t_1,t_2) \cup S(V,t_1,t_2)} \left\{ P_i^{ext}(t_k) \frac{V_i^+(t_k) + V_i^-(t_k)}{2} \right\}.$$

At time $t_j$ between intervals $]t_k, t_j[$ and $]t_j, t_l[$, we have by choosing virtual velocities with compact support in $]t_k, t_l[$

$$\sum_{i=1}^{2} m_i \left[U_i(t_j)\right] \frac{V_i^+(t_j) + V_i^-(t_j)}{2}$$
$$= -P^{int}(t_j) D\left(\frac{V^+(t_k) + V^-(t_k)}{2}\right) + \sum_{i=1}^{2} P_i^{ext}(t_j) \frac{V_i^+(t_j) + V_i^-(t_j)}{2},$$

which gives immediately

$$m_1 \left[U_1(t_j)\right] = -P^{int}(t_j) + P_1^{ext}(t_j), \tag{10.2}$$
$$m_2 \left[U_2(t_j)\right] = P^{int}(t_j) + P_2^{ext}(t_j). \tag{10.3}$$

At time $t \in ]t_k, t_l[$ by choosing virtual velocities with support in $]t_k, t_l[$ having a unique discontinuity point at time $t$, we get easily that relationships (10.2) and (10.3) are satisfied at time $t$. Thus, they are satisfied at any time

$$m_1 \left[U_1(t_j)\right] = -P^{int}(t_j) + P_1^{ext}(t_j), \ \forall t \in ]t_1, t_2[,$$
$$m_2 \left[U_2(t_j)\right] = P^{int}(t_j) + P_2^{ext}(t_j), \ \forall t \in ]t_1, t_2[.$$

A detailed derivation of the equations of motion is given in [113].

*Remark 10.6.* The equations of motion are a simplified version of the equations given in Sect. 3.6 for two solids when the velocities and temperatures are discontinuous on the whole domains occupied by the two solids at time of collision.

### *10.1.5 An Example of Collisions: The Equations for $U^+$*

The constitutive laws, given in a general setting in next section, define the interior percussion $P^{int}$ with a pseudopotential of dissipation $\Phi$ depending on $D = D((U^+ + U^-)/2)$ and on parameter $\chi = D(U^-/2)$ which depends on the past. The non-interpenetration condition of the two balls is an internal constraint

$$U_1^+ - U_2^+ = D(U^+) \leq 0.$$

It is taken into account by the pseudopotential of dissipation

$$\Phi(D,\chi) = kD^2 + I_-(D-\chi), \tag{10.4}$$

where $I_-$ is the indicator function of the set of the non positive real numbers.

In order to solve the equations of motion, we rewrite them in a way adapted to numerical computation. We denote

$$D(U) = E\hat{U},$$

with deformation matrix $E$

$$E = \begin{bmatrix} 1 & -1 \end{bmatrix},$$

and

$$\hat{U} = \begin{bmatrix} U_1 \\ U_2 \end{bmatrix}.$$

Constitutive law is

$$\begin{aligned} P^{int} &= P^d + P^{reac} \in \partial\Phi\left(D\left(\frac{U^+ + U^-}{2}\right), D\left(\frac{U^-}{2}\right)\right), \\ P^d &= 2kD\left(\frac{U^+ + U^-}{2}\right) = kD(U^+ + U^-), \\ P^{reac} &\in \partial I_-(D\left(\frac{U^+ + U^-}{2}\right) - D(\frac{U^-}{2})) = \partial I_-(U_1^+ - U_2^+), \end{aligned} \tag{10.5}$$

where subdifferential $\partial\Phi$ of $\Phi(D,\chi)$ is computed with respect to $D$, see Appendix A. Equations of motion (10.2) and (10.3) are

$$M[U] + E^T P^{int} = 0, \tag{10.6}$$

where matrix $E^T$ is the transposed matrix of $E$ and the mass matrix is

$$M = \begin{bmatrix} m_1 & 0 \\ 0 & m_2 \end{bmatrix}.$$

Equations (10.5) and (10.6) give

$$M[U] + E^T \partial \Phi \left( D\left(\frac{U^+ + U^-}{2}\right), D\left(\frac{U^-}{2}\right)\right) \ni 0,$$

or

$$M[U] + E^T \partial \Phi \left( E\left(\frac{U^+ + U^-}{2}\right), E\left(\frac{U^-}{2}\right)\right) \ni 0.$$

By letting $X = U^+ + U^-$, the equation becomes

$$MX + E^T \partial \Phi \left( E\left(\frac{X}{2}\right), E\left(\frac{U^-}{2}\right)\right) \ni 2MU^-. \tag{10.7}$$

Note that to find $U^+$ when $U^-$ is known, is equivalent to find $X$. Equation (10.7) has good properties:

**Theorem 10.1.** *If dissipation coefficient $k$ is positive, (10.7) has one and only one solution.*

*Proof.* Operator

$$Y \to E^T \partial \Phi \left( E\left(\frac{Y}{2}\right), E\left(\frac{U^-}{2}\right)\right),$$

is monotone and maximal from $\mathbb{R}^2$ into $\mathbb{R}^2$ because $\Phi$ is convex. Operator

$$Y \to MY + E^T \partial \Phi \left( E\left(\frac{Y}{2}\right), E\left(\frac{U^-}{2}\right)\right),$$

is strictly monotone, maximal and surjective from $\mathbb{R}^2$ onto $\mathbb{R}^2$. It results (10.7) has one and only one solution. □

This result is general. It applies to collisions of a system involving any number of rigid bodies, [112]. The equations are used to perform numerical computations for engineering purposes, [83].

#### 10.1.5.1 Computation of the Velocities $U^+$: Computation of the Dissipated Work

Let us compute the velocities $U_1^+$ and $U_2^+$ after a collision assuming the velocities before the collision $U_1^-$ and $U_2^-$, which satisfy $U_1^- - U_2^- > 0$, are known. Depending on the values of the $m_i$ and $k$, there are two possible evolutions. The equations giving $U_1^+$, $U_2^+$ depending on $U_1^-$, $U_2^-$ and satisfying the impenetrability condition $U_1^- - U_2^- \geq 0$, are

$$\begin{aligned} m_1(U_1^+ - U_1^-) &= -P^{int} = -k(U_1^+ - U_2^+ + U_1^- - U_2^-) - P^{reac}, \\ P^{reac} &\in \partial I_-(U_1^+ - U_2^+), \\ m_2(U_2^+ - U_2^-) &= P^{int}, \end{aligned}$$

where $P^{reac}$ is the impenetrability reaction ($\partial I_-$ is the subdifferential set of function $I_-$ (see Appendix A), $P_{reac}$ is 0 if contact is not maintained after collision ($U_1^+ - U_2^+ < 0$), it is non negative if contact is maintained after collision ($U_1^+ - U_2^+ = 0$). We get by solving the previous equations or (10.7):

1. If $m_1 m_2 - k(m_1 + m_2) \leq 0$, i.e., if the balls are not heavy or if system is very dissipative, they rebound with velocities

$$\begin{aligned} U_1^+ &= \frac{(m_1 m_2 + k m_1 - k m_2)U_1^- + 2k m_2 U_2^-}{m_1 m_2 + k m_1 + k m_2}, \\ U_2^+ &= \frac{(m_1 m_2 + k m_2 - k m_1)U_2^- + 2k m_1 U_1^-}{m_1 m_2 + k m_1 + k m_2}, \end{aligned} \tag{10.8}$$

$$P_{reac} = 0, \; U_1^+ - U_2^+ \leq 0.$$

The dissipated work $P^{int} D\left(\frac{U^+ + U^-}{2}\right)$ which intervene in the thermal phenomenon described in next section is

$$\begin{aligned} P^{int} D\left(\frac{U^+ + U^-}{2}\right) &= \frac{k}{2}(U_1^+ - U_2^+ + U_1^- - U_2^-)^2 \\ &= \frac{2k m_1^2 m_2^2}{(m_1 m_2 + k m_1 + k m_2)^2}\left(U_1^- - U_2^-\right)^2. \end{aligned}$$

2. If $m_1 m_2 - k(m_1 + m_2) \geq 0$, i.e., if the balls are heavy or if system is not very dissipative, they do not rebound. Their common velocity is

$$U_1^+ = U_2^+ = \frac{m_1 U_1^- + m_2 U_2^-}{m_1 + m_2}.$$

The impenetrability reaction is

$$P_{reac} = \frac{m_1 m_2 - k(m_1 + m_2)}{m_1 + m_2}(U_1^- - U_2^-) \geq 0,$$

because $U_1^+ - U_2^+ = 0$. The dissipated work is

$$P^{int} D\left(\frac{U^+ + U^-}{2}\right) = \frac{k}{2}(U_1^+ - U_2^+ + U_1^- - U_2^-)^2 + P_{reac}\frac{U_1^- - U_2^-}{2}$$
$$= \frac{m_1 m_2}{2(m_1+m_2)}(U_1^- - U_2^-)^2.$$

3. If $m_1 m_2 - k(m_1+m_2) = 0$, case 1 and 2 solutions are valid with

$$U_1^+ - U_2^+ = 0, \; P_{reac} = 0.$$

#### 10.1.5.2 The Measure of Dissipation Parameter $k$

In case the two balls rebound, we have

$$U_1^+ - U_2^+ = \frac{m_1 m_2 - k(m_1+m_2)}{m_1 m_2 + k m_1 + k m_2}\left(U_1^- - U_2^-\right).$$

Experiments give easily coefficient $k$ by plotting the relative velocity after collision versus the relative velocity before.

In case the balls do not rebound, positive exterior percussion $P_2^{ext}$ may be applied at ball number 2 and negative exterior percussion $P_1^{ext}$ may be applied at ball 1, concomitant to collision, in such a way that

$$U_1^- + \frac{P_1^{ext}}{m_1} - (U_2^- + \frac{P_2^{ext}}{m_2}) < 0,$$

which gives $U_1^+ \neq U_2^+$ and

$$U_1^+ - U_2^+ = \frac{m_1 m_2 - k(m_1+m_2)}{m_1 m_2 + k m_1 + k m_2}(U_1^- + \frac{P_1^{ext}}{m_1} - (U_2^- + \frac{P_2^{ext}}{m_2})),$$

allowing to measure $k$. In practice, experiments may be performed with the balls at rest before collision. In this situation, parameter $k$ quantifies the difficulty to separate the balls in contact. It accounts for adhesion effects. The larger it is, the more difficult is the separation. Other way to describe adhesion are given in [83].

This example, shows how to identify more sophisticated constitutive laws by assuming that $k$ depends on $D = \left(U_1^+ - U_2^+ + U_1^- - U_2^-\right)/2$ and $\chi = U_1^- - U_2^-$.

### *10.1.6 An Other Example: The Impenetrability Condition*

The impenetrability condition is a physical property. There are case when it is conditional: let consider for instance, the solid number 1 to be a steel ball and solid number 2 to be a paper sheet which is hold by somebody. We let the steel ball fall on the sheet of paper (on Fig. 10.1, the downward vertical direction is toward the right). When the steel ball collides the sheet of paper, depending on its weight and velocity, it goes through the sheet or it is stopped. In this situation, the impenetrability depends on the circumstances. For the sake of simplicity, we assume that the phenomenon is unidimensional and that the sheet of paper is firmly hold in such a way its velocity remains constant, $U_2^- = U_2^+ = 0$. When colliding, the velocity of ball number 1 is $U_1^- > 0$.

We choose the pseudopotential of dissipation describing the collision properties of the system (steel ball)∪(sheet of paper) to be

$$\Phi(D,\chi) = kD^2 + k_0 pp(D-\chi),$$

where $pp(x) = \sup(x,0)$ is the positive part of $x$ and

$$D = \frac{U^+ + U^-}{2}, \ \chi = \frac{U^-}{2},$$

denoting $U = U_1$. Dissipative parameters $k$ and $k_0$ are positive. The function $D \to \Phi(D,\chi)$ is a pseudopotential of dissipation because

$$\Phi(0,\chi) = \Phi(0,\frac{U^-}{2}) = pp(-\frac{U}{2}) = 0,$$

because $U^- > 0$. The interior percussion is

$$p^{int} \in \partial\Phi\left(\frac{U^+ + U^-}{2}\right) = k\left(U^+ + U^-\right) + k_0 H\left(U^+\right),$$

where the graph $H$ is

$$H(x) = \begin{cases} 1, & \text{if } x > 0, \\ [0,1], & \text{if } x = 0, \\ 0, & \text{if } x < 0. \end{cases}$$

The equation to find the velocity of the steel ball $U^+$ after the collision with the sheet of paper is

$$m(U^+ - U^-) = -p^{int} \in -k(U^+ + U^-) - k_0 H(U^+).$$

By letting $X = U^+ + U^-$, it becomes

$$mX + k_0 H(X - U_N^-) + kX \ni 2mU_N^-.$$

Applying a theorem analog to Theorem 10.1, the preceding equation gives one and only one velocity $U^+$ when velocity $U^-$ is known. We have:

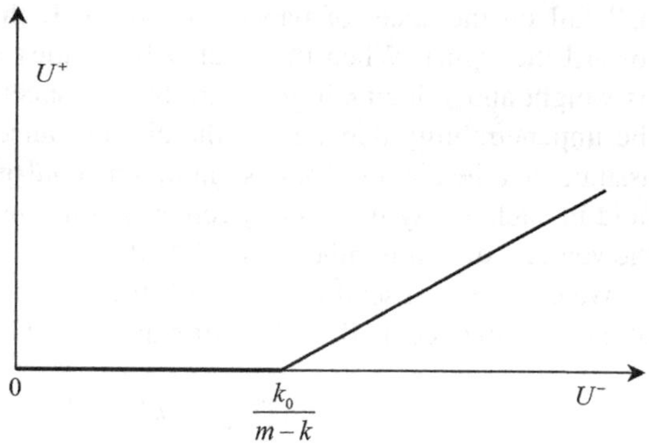


**Fig. 10.2** The velocity $U^+$ of the steel ball versus its velocity $U^-$ before colliding the sheet of paper. When the falling velocity is low, $U^- \leq k_0/(m-k)$, the ball is stopped by the sheet of paper. When it is large, $U^- \geq k_0/(m-k)$, the steel ball goes through the sheet of paper

1. If $m-k \leq 0$, i.e., if the steel ball is very light or the system (steel ball)$\cup$(sheet of paper) is very dissipative, the ball rebounds with velocity

$$U^+ = \frac{m-k}{m+k} U^-.$$

2. If $m-k \geq 0$, i.e., if the steel ball is heavy or the system is not very dissipative, we get, Fig. 10.2:

   a. If the falling velocity is small

   $$U^- \leq \frac{k_0}{m-k},$$

   the steel ball does not bounce. It is stopped by the sheet of paper

   $$U^+ = 0.$$

   b. If the falling velocity is large

   $$U^- \geq \frac{k_0}{m-k},$$

   the steel ball goes through the paper sheet

   $$U^+ = \frac{(m-k)U^- - k_0}{m+k},$$

   $$U^- > U^+ \geq 0.$$

   After collision, the ball is still falling but it has been slowed down when going through the paper sheet.

In Case 2 the dissipated work is

$$P^{int}\frac{U^+ + U^-}{2} = \frac{m}{2}(U^-)^2,$$

if the falling velocity is slow, and

$$P^{int}\frac{U^+ + U^-}{2} = \frac{m}{2}\left\{\frac{4km(U^-)^2 + 2k_0(m-k)U^- - k_0^2}{(m+k)^2}\right\},$$

if the falling velocity is high.

This example is due Eric Dimnet in order to describe rock avalanches in mountain forests. The rocks may break trees and continue their way down the slope, [84].

*Remark 10.7.* In both examples, the dissipated work increases with the relative velocity $U_1^- - U_2^-$ or $U^-$.

## 10.2 Collisions of Two Balls: The Thermal Phenomenon

Four temperatures appear in collisions : $T_i^+$ and $T_i^-$ temperatures of the points after and before collision. We denote

$$[T_1] = T_1^+ - T_1^-,\ [T_2] = T_2^+ - T_2^-.$$

These differences are analog to the temperature time derivative $\mathrm{d}T/\mathrm{d}t$ when the temperature is smooth. Let us recall that there is no dissipation with respect to this quantity because of Helmholtz relationship (see formula (3.8))

$$s = -\frac{\partial \Psi}{\partial T}(T,\eta),$$

defining the entropy $s$ when $E = (T,\eta)$ (see Sect. 3.2.3). We denote some average temperatures

$$\bar{T}_1 = \frac{T_1^+ + T_1^-}{2},\ \bar{T}_2 = \frac{T_2^+ + T_2^-}{2},\ \bar{\Theta} = \frac{T_1^+ + T_1^- + T_2^+ + T_2^-}{4} = \frac{\bar{T}_1 + \bar{T}_2}{2}.$$

The average temperature difference

$$\delta\bar{T} = \bar{T}_2 - \bar{T}_1,$$

is the analog of the spatial thermal gradient in a smooth situation. Let us also recall that there is dissipation with respect to this quantity: the almost universal example is the Fourier law. In the following investigation of the thermal effects of collisions, analogous choices are made remembering that they sum up what occur during the duration of collisions when Fourier law is assumed and Helmholtz relationship is valid.

In the sequel, we consider sums which are transformed in the following way

$$T^{+}\mathscr{B}^{+}+T^{-}\mathscr{B}^{-}=\bar{T}\Sigma(\mathscr{B})+[T]\Delta(\mathscr{B}),$$

with

$$\bar{T}=\frac{T^{+}+T^{-}}{2},\ [T]=T^{+}-T^{-},\ \Sigma(\mathscr{B})=\mathscr{B}^{+}+\mathscr{B}^{-},\ \Delta(\mathscr{B})=\frac{\mathscr{B}^{+}-\mathscr{B}^{-}}{2}.$$

### 10.2.1 First Law of Thermodynamics for a Point

The laws of thermodynamics are the same for the two points. For point 1, the first law is

$$[\mathscr{E}_1]+[\mathscr{K}_1]=\mathscr{T}_{ext1}(U)+\mathscr{C}_1,$$

where $\mathscr{T}_{ext1}(U)$ is the actual work of the percussions which are exterior to point 1 and $\mathscr{C}_1$ is the amount of heat received by the point in the collision. This heat involves the heats $T_1\mathscr{B}_1$ received from the exterior to the system and the heats $T_1\mathscr{B}_{12}$ received from the interior of the system, i.e., from the other point. We assume that these heats are received either at temperature $T_1^{+}$ or at temperature $T_1^{-}$

$$\begin{aligned}\mathscr{C}_1&=T_1^{+}\left(\mathscr{B}_1^{+}+\mathscr{B}_{12}^{+}\right)+T_1^{-}\left(\mathscr{B}_1^{-}+\mathscr{B}_{12}^{-}\right)\\&=\bar{T}_1\left(\Sigma(\mathscr{B}_1)+\Sigma(\mathscr{B}_{12})\right)+[T_1]\left(\Delta(\mathscr{B}_1)+\Delta(\mathscr{B}_{12})\right).\end{aligned}$$

The theorem of expanded energy for point 1 is

$$[\mathscr{K}_1]=\mathscr{T}_{acc1}(U)=\mathscr{T}_{ext1}(U).$$

It gives with the energy balance

$$[\mathscr{E}_1]=\mathscr{C}_1=\bar{T}_1\left(\Sigma(\mathscr{B}_1)+\Sigma(\mathscr{B}_{12})\right)+[T_1]\left(\Delta(\mathscr{B}_1)+\Delta(\mathscr{B}_{12})\right). \tag{10.9}$$

In the right hand side, there is no mechanical quantity because a point has no interior force, the work of which could contribute to the internal energy evolution.

### 10.2.2 Second Law of Thermodynamics for a Point

It is

$$[\mathscr{S}_1]\geq\mathscr{B}_1^{+}+\mathscr{B}_1^{-}+\mathscr{B}_{12}^{+}+\mathscr{B}_{12}^{-}=\Sigma(\mathscr{B}_1)+\Sigma(\mathscr{B}_{12}).$$

### 10.2.3 A Useful Inequality for a Point

Second law and relationship (10.9) give

$$[\mathcal{E}_1] - \bar{T}_1 [\mathcal{S}_1] \leq [T_1] (\Delta (\mathcal{B}_1) + \Delta (\mathcal{B}_{12})),$$

or by introducing $\Psi = \mathcal{E} - T\mathcal{S}$

$$[\Psi_1] + \overline{\mathcal{S}}_1 [T_1] \leq [T_1] \Delta (\mathcal{B}_1) + [T_1] \Delta (\mathcal{B}_{12}).$$

We denote

$$[\Psi_1] + \overline{\mathcal{S}}_1 [T_1] = -S_1 [T_1],$$

remembering that when $[T_1] \to 0$ we have $[\Psi_1] / [T_1] + \overline{\mathcal{S}}_1 \to 0$ because $\partial\Psi/\partial T + s = 0$. Then

$$0 \leq [T_1] (\Delta (\mathcal{B}_1) + \Delta (\mathcal{B}_{12}) + S_1).$$

Due to Helmholtz relationship, it is reasonable to assume no dissipation with respect to $[T_1]$ which is, as already said, the analog to $\mathrm{d}T/\mathrm{d}t$

$$\Delta (\mathcal{B}_1) + \Delta (\mathcal{B}_{12}) + S_1 = 0.$$

We deduce

$$[\mathcal{S}_1] = \Sigma (\mathcal{B}_1) + \Sigma (\mathcal{B}_{12}). \tag{10.10}$$

### 10.2.4 The First Law for the System

The internal energy and entropy of the system are the sums of the internal energies and entropies of its components $\mathcal{E}_1$, $\mathcal{S}_1$ and $\mathcal{E}_2$, $\mathcal{S}_2$ to which interaction internal energy and entropy $\mathcal{E}^{int}$, $\mathcal{S}^{int}$ are added:

$$\mathcal{E} = \mathcal{E}_1 + \mathcal{E}_2 + \mathcal{E}^{int},$$
$$\mathcal{S} = \mathcal{S}_1 + \mathcal{S}_2 + \mathcal{S}^{int}.$$

The first law for the system is

$$[\mathcal{E}] + [\mathcal{K}] = \mathcal{T}_{ext}(U) + \mathcal{C},$$

where $\mathcal{C}$ is the heat received by the system in collision

$$\mathcal{C} = T_1^+ \mathcal{B}_1^+ + T_1^- \mathcal{B}_1^- + T_2^+ \mathcal{B}_2^+ + T_2^- \mathcal{B}_2^-.$$

The theorem of expanded energy is

$$[\mathscr{K}] = \mathscr{T}_{acc}(U) = \mathscr{T}_{int}(D(U)) + \mathscr{T}_{ext}(U).$$

It gives with the first law

$$[\mathscr{E}] = -\mathscr{T}_{int}(D(U)) + \mathscr{C}. \tag{10.11}$$

The first laws for each point (10.9) and for the system (10.11) give

$$\begin{aligned}
[\mathscr{E}] &= [\mathscr{E}_1] + [\mathscr{E}_2] + \left[\mathscr{E}^{int}\right] \\
&= \left[\mathscr{E}^{int}\right] + \bar{T}_1\left(\Sigma\left(\mathscr{B}_1\right) + \Sigma\left(\mathscr{B}_{12}\right)\right) + [T_1]\left(\Delta\left(\mathscr{B}_1\right) + \Delta\left(\mathscr{B}_{12}\right)\right) \\
&\quad + \bar{T}_2\left(\Sigma\left(\mathscr{B}_2\right) + \Sigma\left(\mathscr{B}_{21}\right)\right) + [T_2]\left(\Delta\left(\mathscr{B}_2\right) + \Delta\left(\mathscr{B}_{21}\right)\right) \\
&= P^{int} D\left(\frac{U^+ + U^-}{2}\right) + T_1^+ \mathscr{B}_1^+ + T_1^- \mathscr{B}_1^- + T_2^+ \mathscr{B}_2^+ + T_2^- \mathscr{B}_2^- \\
&= P^{int} D\left(\frac{U^+ + U^-}{2}\right) + \bar{T}_1 \Sigma\left(\mathscr{B}_1\right) + [T_1]\Delta\left(\mathscr{B}_1\right) + \bar{T}_2 \Sigma\left(\mathscr{B}_2\right) + [T_2]\Delta\left(\mathscr{B}_2\right).
\end{aligned}$$

Then

$$\begin{aligned}
\left[\mathscr{E}^{int}\right] &= P^{int} D\left(\frac{U^+ + U^-}{2}\right) \\
&\quad - \bar{T}_1 \Sigma\left(\mathscr{B}_{12}\right) - [T_1]\Delta\left(\mathscr{B}_{12}\right) - \bar{T}_2 \Sigma\left(\mathscr{B}_{21}\right) - [T_2]\Delta\left(\mathscr{B}_{21}\right).
\end{aligned} \tag{10.12}$$

This relationship is interesting because only interior quantities intervene.

### *10.2.5 The Second Law for the System*

It is

$$[\mathscr{S}] \geq \mathscr{B}_1^+ + \mathscr{B}_1^- + \mathscr{B}_2^+ + \mathscr{B}_2^- = \Sigma\left(\mathscr{B}_1\right) + \Sigma\left(\mathscr{B}_2\right),$$

where

$$[\mathscr{S}] = [\mathscr{S}_1] + [\mathscr{S}_2] + \left[\mathscr{S}^{int}\right] \geq \Sigma\left(\mathscr{B}_1\right) + \Sigma\left(\mathscr{B}_2\right),$$

which gives with (10.10)

$$\left[\mathscr{S}^{int}\right] \geq -\Sigma\left(\mathscr{B}_{12}\right) - \Sigma\left(\mathscr{B}_{21}\right). \tag{10.13}$$

In this relationship only interior heat exchanges intervene. Together with the first law (10.12), it gives an inequality coupling the mechanical and thermal dissipations.

### 10.2.6 A Useful Inequality for the System

We have with (10.12) and (10.13)

$$\left[\mathscr{E}^{int}\right] - \bar{\Theta}\left[\mathscr{S}^{int}\right] \leq P^{int} D\left(\frac{U^+ + U^-}{2}\right) - \bar{T}_1 \Sigma\left(\mathscr{B}_{12}\right) - [T_1]\Delta\left(\mathscr{B}_{12}\right)$$
$$-\bar{T}_2 \Sigma\left(\mathscr{B}_{21}\right) - [T_2]\Delta\left(\mathscr{B}_{21}\right) + \bar{\Theta}\left(\Sigma\left(\mathscr{B}_{12}\right) + \Sigma\left(\mathscr{B}_{21}\right)\right).$$

We define the interaction free energy: $\Psi^{int} = \mathscr{E}^{int} - \Theta \mathscr{S}^{int}$. It gives with the preceding relationship

$$\left[\Psi^{int}\right] + \overline{\mathscr{S}^{int}}[\Theta] \leq P^{int} D\left(\frac{U^+ + U^-}{2}\right)$$
$$-\bar{T}_1 \Sigma\left(\mathscr{B}_{12}\right) - [T_1]\Delta\left(\mathscr{B}_{12}\right) - \bar{T}_2 \Sigma\left(\mathscr{B}_{21}\right) - [T_2]\Delta\left(\mathscr{B}_{21}\right)$$
$$+\bar{\Theta}\left(\Sigma\left(\mathscr{B}_{12}\right) + \Sigma\left(\mathscr{B}_{21}\right)\right),$$

with

$$\overline{\mathscr{S}^{int}} = \frac{\mathscr{S}^{int+} + \mathscr{S}^{int-}}{2}.$$

Remembering that we have denoted

$$[\Theta] = \frac{1}{2}\left([T_1] + [T_2]\right),$$

we define

$$\left[\Psi^{int}\right] + \overline{\mathscr{S}^{int}}[\Theta] = -S_{int}[\Theta].$$

We obtain

$$0 \leq P^{int} D\left(\frac{U^+ + U^-}{2}\right) - \bar{T}_1 \Sigma\left(\mathscr{B}_{12}\right) - [T_1]\left(\Delta\left(\mathscr{B}_{12}\right) - \frac{S_{int}}{2}\right)$$
$$-\bar{T}_2 \Sigma\left(\mathscr{B}_{21}\right) - [T_2]\left(\Delta\left(\mathscr{B}_{21}\right) - \frac{S_{int}}{2}\right) + \bar{\Theta}\left(\Sigma(\mathscr{B}_{12}) + \Sigma(\mathscr{B}_{21})\right).$$

It is reasonable not to have dissipation with respect to $[T_1]$ and $[T_2]$ which are temperature variations with respect to time and not with respect to space

$$\Delta\left(\mathscr{B}_{12}\right) - \frac{S_{int}}{2} = 0, \; \Delta\left(\mathscr{B}_{21}\right) - \frac{S_{int}}{2} = 0. \tag{10.14}$$

Then

$$0 \leq P^{int} D\left(\frac{U^+ + U^-}{2}\right) - \bar{T}_1 \Sigma\left(\mathscr{B}_{12}\right) - \bar{T}_2 \Sigma\left(\mathscr{B}_{21}\right) + \bar{\Theta}\left(\Sigma\left(\mathscr{B}_{12}\right) + \Sigma\left(\mathscr{B}_{21}\right)\right),$$

but

$$\bar{T}_1\Sigma(\mathscr{B}_{12})+\bar{T}_2\Sigma(\mathscr{B}_{21})=\bar{\Theta}\left(\Sigma(\mathscr{B}_{12})+\Sigma(\mathscr{B}_{21})\right)+\delta\bar{T}\frac{\Sigma(\mathscr{B}_{21})-\Sigma(\mathscr{B}_{12})}{2},$$

recalling that $\delta\bar{T}=\bar{T}_2-\bar{T}_1$. Thus we get

$$0\leq P^{int}D\left(\frac{U^{+}+U^{-}}{2}\right)-\delta\bar{T}\frac{\Sigma(\mathscr{B}_{21})-\Sigma(\mathscr{B}_{12})}{2}. \tag{10.15}$$

This relationship is important due to the quantities which intervene : the one which characterizes the mechanical evolution $D((U^{+}+U^{-})/2)$, and the other one which characterizes the spatial thermal heterogeneity $\delta\bar{T}$. Let us stress its analogy with the spatial thermal gradient in smooth situation. Relationship (10.15) is a guide to choose the quantities which are to be related to the interior forces. When dealing with experimental results relationship (10.15) is useful to choose the quantities which are to be plotted ones versus the others: $P^{int}$ versus $D(U^{+}+U^{-})$ and $\Sigma(\mathscr{B}_{21})-\Sigma(\mathscr{B}_{12})$ versus $\delta\bar{T}$.

Let us recall that on our way we have assumed no dissipation with respect to the temperature time variation.

### 10.2.7 The Constitutive Laws

Quantities which describe the mechanical evolution and thermal heterogeneity are $D((U^{+}+U^{-})/2)$ and $\delta\bar{T}$. We define the constitutive laws with a pseudopotential of dissipation

$$\hat{\Phi}\left(D\left(\frac{U^{+}+U^{-}}{2}\right),\delta\bar{T},\chi\right).$$

The quantity $\chi=D(U^{-}/2)$ which depends on the past is used to take into account the non-interpenetration of the points. Then the constitutive laws are

$$\left(P^{int},\frac{\Sigma(\mathscr{B}_{21})-\Sigma(\mathscr{B}_{12})}{2}\right)\in\partial\hat{\Phi}\left(D\left(\frac{U^{+}+U^{-}}{2}\right),\delta\bar{T},D\left(\frac{U^{-}}{2}\right)\right),$$

where the subdifferential set is computed with respect to the two first variables. It is easy to show that inequality (10.15) is satisfied. The choice of the pseudopotential of dissipation is to be guided by experiments.

### 10.2.8 An Example of Thermal Effects Due to Collisions

We choose a pseudopotential which does not couple the two first variables insuring that the mechanical problem is decoupled from the thermal one, for instance

$$\hat{\Phi}\left(D\left(\frac{U^{+}+U^{-}}{2}\right),\delta\bar{T},D\left(\frac{U^{-}}{2}\right)\right)=\hat{\Phi}_{meca}\left(D\left(\frac{U^{+}+U^{-}}{2}\right),D\left(\frac{U^{-}}{2}\right)\right)$$
$$+\frac{\lambda}{4}\left(\delta\bar{T}\right)^{2}.$$

The pseudopotential $\hat{\Phi}_{meca}$ gives the percussion $P^{int}$. The mechanical problem is not investigated in this Section but two examples have been given in Sect. 10.1. Complete investigation and results may be found in [113]. The thermal constitutive law is

$$\Sigma\left(\mathscr{B}_{21}\right)-\Sigma\left(\mathscr{B}_{12}\right)=-\lambda\delta\bar{T}.$$

We choose the same simple free energy for the points $\Psi=-CT\ln T$ and a zero interaction free energy, $\Psi^{int}=0$. This choice gives

$$S_{int}=0,$$

and with (10.14)

$$\Delta\left(\mathscr{B}_{12}\right)=0,\ \Delta\left(\mathscr{B}_{21}\right)=0.$$

Let us assume that the mechanical problem is solved, i.e., the work $P^{int}D((U^{+}+U^{-})/2)$ is known. The thermal equations are

$$[\mathscr{S}_1]=C\ln\frac{T_1^{+}}{T_1^{-}}=\Sigma\left(\mathscr{B}_1\right)+\Sigma\left(\mathscr{B}_{12}\right),$$

$$[\mathscr{S}_2]=C\ln\frac{T_2^{+}}{T_2^{-}}=\Sigma\left(\mathscr{B}_2\right)+\Sigma\left(\mathscr{B}_{21}\right),$$

and

$$[\mathscr{E}_1]=C\left[T_1\right]=\bar{T}_1\left(\Sigma\left(\mathscr{B}_1\right)+\Sigma\left(\mathscr{B}_{12}\right)\right)+\left[T_1\right]\Delta\left(\mathscr{B}_1\right),$$
$$[\mathscr{E}_2]=C\left[T_2\right]=\bar{T}_2\left(\Sigma\left(\mathscr{B}_2\right)+\Sigma\left(\mathscr{B}_{21}\right)\right)+\left[T_2\right]\Delta\left(\mathscr{B}_2\right),$$

$$0=\left[\mathscr{E}^{int}\right]=P^{int}D\left(\frac{U^{+}+U^{-}}{2}\right)-\bar{T}_1\Sigma\left(\mathscr{B}_{12}\right)-\bar{T}_2\Sigma\left(\mathscr{B}_{21}\right).$$

These equations are coupled with the description of the thermal exchanges with the exterior of the system. For instance, collision may be either isentropic

$$\Sigma\left(\mathscr{B}_1\right)=\Sigma\left(\mathscr{B}_2\right)=0,$$

or adiabatic

$$T_1^+ \mathscr{B}_1^+ + T_1^- \mathscr{B}_1^- = \bar{T}_1 \Sigma(\mathscr{B}_1) + [T_1] \Delta(\mathscr{B}_1) = 0,$$
$$\bar{T}_2 \Sigma(\mathscr{B}_2) + [T_2] \Delta(\mathscr{B}_2) = 0.$$

The solution of these equations gives the temperatures $T_i^+$ and the different entropy fluxes $\mathscr{B}_i$ and $\mathscr{B}_{ij}$.

#### 10.2.8.1 The Isentropic Case

The equations give

$$C \ln \frac{T_1^+}{T_1^-} = \Sigma(\mathscr{B}_{12}), \; C \ln \frac{T_2^+}{T_2^-} = \Sigma(\mathscr{B}_{21}),$$
$$C[T_1] = \bar{T}_1 \Sigma(\mathscr{B}_{12}) + [T_1] \Delta(\mathscr{B}_1), \; C[T_2] = \bar{T}_2 \Sigma(\mathscr{B}_{21}) + [T_2] \Delta(\mathscr{B}_2),$$
$$P^{int} D\left(\frac{U^+ + U^-}{2}\right) = \bar{T}_1 \Sigma(\mathscr{B}_{12}) + \bar{T}_2 \Sigma(\mathscr{B}_{21}),$$
$$\Sigma(\mathscr{B}_{21}) - \Sigma(\mathscr{B}_{12}) = -\lambda \delta \bar{T}.$$

Then we have

$$C[T_1] = \bar{T}_1 \Sigma(\mathscr{B}_{12}) + [T_1] \Delta(\mathscr{B}_1),$$
$$C[T_2] = \bar{T}_2 \Sigma(\mathscr{B}_{21}) + [T_2] \Delta(\mathscr{B}_2),$$
$$P^{int} D\left(\frac{U^+ + U^-}{2}\right) = \bar{T}_1 C \ln \frac{T_1^+}{T_1^-} + \bar{T}_2 C \ln \frac{T_2^+}{T_2^-},$$
$$C \ln \frac{T_2^+}{T_2^-} + \lambda \bar{T}_2 = C \ln \frac{T_1^+}{T_1^-} + \lambda \bar{T}_1.$$

The two last equations give the temperatures $T_1^+$ and $T_2^+$ after a collision. The first two equations give the thermal exchanges with the exterior : they are computed with $\Delta(\mathscr{B}_1)$ and $\Delta(\mathscr{B}_2)$.

#### 10.2.8.2 The Adiabatic Case

The equations give

$$[\mathscr{S}_1] = C \ln \frac{T_1^+}{T_1^-} = \Sigma(\mathscr{B}_1) + \Sigma(\mathscr{B}_{12}),$$
$$[\mathscr{S}_2] = C \ln \frac{T_2^+}{T_2^-} = \Sigma(\mathscr{B}_2) + \Sigma(\mathscr{B}_{21}),$$

and

$$C[T_1] = \bar{T}_1 \Sigma(\mathscr{B}_{12}),\ C[T_2] = \bar{T}_2 \Sigma(\mathscr{B}_{21}),$$

and

$$P^{int} D\left(\frac{U^+ + U^-}{2}\right) = C[T_1] + C[T_2],$$

$$C\frac{[T_2]}{\bar{T}_2} - C\frac{[T_1]}{\bar{T}_1} = -\lambda \delta \bar{T}.$$

The two last equations give the temperatures $T_1^+$ and $T_2^+$ after a collision. The first two give the entropic exchanges $\Sigma(\mathscr{B}_1)$ and $\Sigma(\mathscr{B}_2)$ with the exterior.

#### 10.2.8.3 A Computation

Computations of temperatures $T_1^+$ and $T_2^+$ in the two preceding situations may be performed with numerics or with closed form solutions which are sophisticated. A case is easily solved: we assume that the collision occurs at the ambient temperature and that the temperature variations are negligible compared with the ambient temperature. We may say that we assume small perturbations. This is the case for an every day experiment. In the isentropic and adiabatic case, we have

$$\ln \frac{T^+}{T^-} \simeq \frac{[T]}{T^-},\ \frac{[T]}{\bar{T}} \simeq \frac{[T]}{T^-}.$$

In both cases we have

$$P^{int} D\left(\frac{U^+ + U^-}{2}\right) = C[T_1] + C[T_2],$$

$$C\frac{[T_2]}{T_2^-} - C\frac{[T_1]}{T_1^-} = -\lambda \delta \bar{T}$$

$$= -\lambda(T_2^- - T_1^- + \frac{[T_2]}{2} - \frac{[T_1]}{2}).$$

These relationships give

$$P^{int} D\left(\frac{U^+ + U^-}{2}\right) = C[T_1] + C[T_2],$$

$$\left(\frac{C}{T_2^-} + \frac{\lambda}{2}\right)[T_2] - \left(\frac{C}{T_1^-} + \frac{\lambda}{2}\right)[T_1] = -\lambda(T_2^- - T_1^-).$$

To simplify again, let us assume that $\lambda$ is small compared to $C/T$. It results

$$P^{int}D\left(\frac{U^{+}+U^{-}}{2}\right)=C[T_1]+C[T_2],$$

$$\frac{C}{T_2^{-}}[T_2]-\frac{C}{T_1^{-}}[T_1]=-\lambda(T_2^{-}-T_1^{-}),$$

which gives

$$C[T_1]=\frac{T_1^{-}}{T_2^{-}+T_1^{-}}\left(P^{int}D\left(\frac{U^{+}+U^{-}}{2}\right)+\lambda T_2^{-}(T_2^{-}-T_1^{-})\right),$$

$$C[T_2]=\frac{T_2^{-}}{T_2^{-}+T_1^{-}}\left(P^{int}D\left(\frac{U^{+}+U^{-}}{2}\right)-\lambda T_1^{-}(T_2^{-}-T_1^{-})\right). \qquad (10.16)$$

These formulas show that the mechanical work

$$\frac{T_i^{-}}{T_2^{-}+T_1^{-}}P^{int}D\left(\frac{U^{+}+U^{-}}{2}\right)\geq 0,$$

tends to warm up the two points and that the conduction effect

$$\frac{T_i^{-}}{T_2^{-}+T_1^{-}}\left(\pm\lambda T_j^{-}(T_2^{-}-T_1^{-})\right),$$

tends to equalize their temperatures.

## 10.3 Experimental Results

As far as we know, there are almost no thermal measurements in collisions. Nevertheless some results are available. For instance, temperature of the back face of a metallic plate collided by a steel ball has been measured with an infrared camera, [188, 189]. A temperature increase up to $7^{o}C$ in less than a hundredth of a second is reported in Fig. 10.3. At the engineering scale, temperature is actually discontinuous, even if at a finer time scale it is continuous.

If we assume the plate and the steel ball have the same temperature before collision, the formulas (10.16) within the small perturbation assumption give

$$C[\theta]=P^{int}D\left(\frac{U^{+}+U^{-}}{2}\right),$$

where $C$ is the heat capacity of the steel ball, see also the examples of the following example . By assuming the ball has the same temperature than the plate, this result may be used to have information on the mechanical constitutive law.

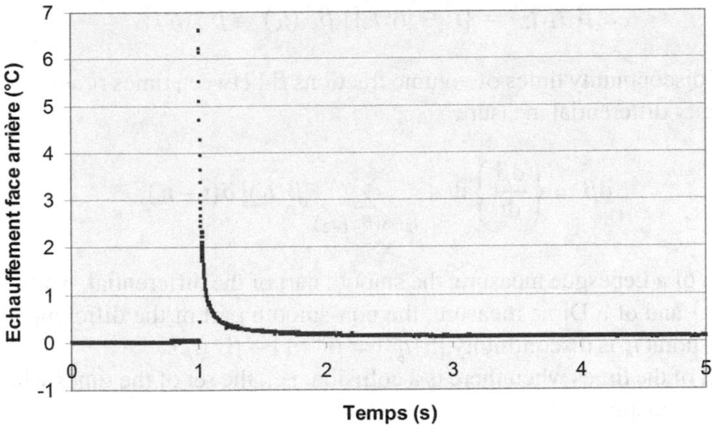


**Fig. 10.3** A metallic ball with 3 mm diameter collides a metallic plate with 0.85 mm thickness. The kinetic energy of the ball is 0.27 J. Temperature of the plate back face versus time is discontinuous at the time scale of the measurements, [189]

## 10.4 Collision of Two Balls: The Thermomechanical Theory Involving Phase Change

Let us come back to the problem of warm rain falling on frozen ground or to the problem of a hailstone falling on the ground, i.e., to the thermal consequences of collisions of materials which can change phase. The heat source due to the collision can produce a very fast phase change concomitant to the collision, for instance the hailstone may melt. The volume fraction state quantities, $\beta_i$, are involve. Let us recall that their evolution is described by an equation of motion. We assume the collision is instantaneous, thus it is wise to assume the phase change is also instantaneous: the very fast evolution of the volume fraction is replaced by a discontinuity. The functions $\beta_i(t)$ are bounded variations functions of time $t$. The discontinuities $[\beta_i]$ are to be described by non-smooth equation of motion accounting for the microscopic motions.

### *10.4.1 The Equations of Motion*

The equations of motion are derived from the principle of virtual work which introduces different contributions. The actual and virtual velocities, $U$ and $V$, are bounded variation functions of time $t$. The actual and virtual volume fractions, $\beta = (\beta_1, \beta_2)$ and $b = (b_1, b_2)$, are also bounded variation functions of time $t$. We denote

$$S(\beta,t_1,t_2) = \{t_k \in ]t_1,t_2[\,|\, \beta^+(t_k) \neq \beta^-(t_k)\},$$

the set of discontinuity times of volume fractions $\beta$ between times $t_1$ and $t_2$ ($t_1 < t_2$). The velocity differential measure

$$\mathrm{d}\beta = \left\{\frac{\mathrm{d}\beta}{\mathrm{d}t}\right\}\mathrm{d}t + \sum_{t_k\in S(\beta,t_1,t_2)} [\beta(t_k)]\,\delta(t-t_k),$$

is the sum of a Lebesgue measure, the smooth part of the differential, whose density is $\{\mathrm{d}\beta/\mathrm{d}t\}$ and of a Dirac measure, the non-smooth part of the differential, whose density at point $t_k$ is discontinuity $[\beta(t_k)] = \beta^+(t_k) - \beta^-(t_k)$.

The set of the times when there is a collision, i.e., the set of the times when either $U$ or $\beta$ is discontinuous), is

$$S(U,\beta,t_1,t_2) = S(U,t_1,t_2) \cup S(\beta,t_1,t_2).$$

The virtual work of the acceleration forces has already been given at the beginning of the chapter

$$\begin{aligned}\mathscr{T}_{acc}(t_1,t_2,U,V,\beta,b) &= \sum_{i=1}^{2}\int_{t_1}^{t_2} m_i\frac{\mathrm{d}U_i}{\mathrm{d}t}(\tau)V_i(\tau)d\tau \\ &+\sum_{i=1}^{2}\left\{\sum_{t_k\in S(U,\beta,t_1,t_2)\cup S(V,b,t_1,t_2)} m_i\,[U_i(t_k)]\,\frac{V_i^+(t_k)+V_i^-(t_k)}{2}\right\},\end{aligned}$$

where $m_1$, $m_2$ are the masses of the two balls.

The virtual work of the interior forces is defined by

$$\begin{aligned}&\mathscr{T}_{int}(t_1,t_2,U,V,\beta,b) \\ &= -\int_{t_1}^{t_2} R^{int}(\tau)D(V)(\tau) + B_i(\tau)b_i(\tau)d\tau \\ &\quad - \sum_{t_k\in S(U,\beta,t_1,t_2)\cup S(V,b,t_1,t_2)} P^{int}(t_k)D\left(\frac{V^+(t_k)+V^-(t_k)}{2}\right) \\ &\quad -\sum_{i=1}^{2}\left\{\sum_{t_k\in S(U,\beta,t_1,t_2)\cup S(V,b,t_1,t_2)} B_{pi}(t_k)\,[b_i(t_k)]\right\},\end{aligned}$$

where $b = (b_1,b_2)$ is the virtual velocity of phase change, $B = (B_1,B_2)$ and $B_p = (B_{p1},B_{p2})$ are the interior microscopic works and percussion works.

For the sake of completeness, we assume exterior work, $A_i(t)$, and percussion work, $A_{ip}$, with respect to the microscopic motion. The virtual work of the exterior forces is

$$\mathscr{T}_{ext}(t_1,t_2,U,V,\beta,b)$$
$$= \sum_{i=1}^{2}\left\{\int_{t_1}^{t_2} F_i(\tau)V_i(\tau)+A_i(\tau)b_i(\tau)d\tau\right\}$$
$$+\sum_{i=1}^{2}\left\{\sum_{t_k\in S(U,\beta,t_1,t_2)\cup S(V,b,t_1,t_2)}\left\{P_i^{ext}(t_k)\frac{V_i^+(t_k)+V_i^-(t_k)}{2}+A_{pi}(t_k)\left[b_i(t_k)\right]\right\}\right\},$$

with forces $F_i$, percussion $P_i^{ext}$, work $A_i$ and percussion work $A_{pi}$.

The principle of virtual work, gives easily the equations of motion at any time $t$

$$m_1\left[U_1(t)\right] = -P^{int}(t)+P_1^{ext}(t),$$
$$m_2\left[U_2(t)\right] = P^{int}(t)+P_2^{ext}(t),$$
$$-B_{p1}(t)+A_{p1}(t) = 0, \tag{10.17}$$
$$-B_{p2}(t)+A_{p2}(t) = 0, \tag{10.18}$$

and almost everywhere

$$m_1\frac{dU_1}{dt} = -R^{int}+F_1,\ m_2\frac{dU_2}{dt} = R^{int}+F_2,$$
$$-B_1+A_1=0,\ -B_2+A_2=0.$$

*Remark 10.8.* Let us note that in the virtual works, the virtual macroscopic velocity intervene with $(V^+ + V^-)/2$ whereas the virtual microscopic velocities intervene with $[b(t_k)]$, because velocity $U(t)$ is a discontinuous function of time and velocity $d\beta/dt$ is the derivative with respect to time of a discontinuous function of time. All the virtual works depend on the velocities. Set $S(\beta,t_1,t_2)$ actually depends on velocity differential measure $d\beta$.

## 10.4.2 The Laws of Thermodynamics for Each Ball

In the sequel we focus on the collisions and do not describe the smooth evolutions. Thus the equations are written for a unique collision.

### 10.4.2.1 The First Law

The first law or energy balance for a ball, say ball number 1, is

$$[\mathscr{E}_1]+[\mathscr{K}_1] = \mathscr{T}_{ext1}(t,t,U,U,\beta,\beta)+\mathscr{C}_1, \tag{10.19}$$

where

$$\mathscr{T}_{ext1}(t,t,U,V,\beta,b) = \left(-P^{int}(t) + P_1^{ext}(t)\right) \frac{V_1^+(t)+V_1^-(t)}{2} + A_{p1}(t)\,[b_i(t)],$$

is the virtual work of the exterior percussions applied to ball 1 and,

$$\begin{aligned}
\mathscr{C}_1 &= T_1^+\left(\mathscr{B}_1^+ + \mathscr{B}_{12}^+\right) + T_1^-\left(\mathscr{B}_1^- + \mathscr{B}_{12}^-\right) \\
&= \bar{T}_1\left(\Sigma(\mathscr{B}_1) + \Sigma(\mathscr{B}_{12})\right) + [T_1]\left(\Delta(\mathscr{B}_1) + \Delta(\mathscr{B}_{12})\right),
\end{aligned}$$

is the heat impulse provided to the ball at the time of collision. This quantity includes the heat received from outside the system, $T_1\mathscr{B}_1$, and the heat received from the inside, that is from the other ball, $T_1\mathscr{B}_{12}$.

*Remark 10.9.* The bold letter $B$ represents an interior percussion work of damage whereas the calligraphic letter $\mathscr{B}$ represents an entropy impulse (une bouffée de chaleur in French).

The theorem of expended energy for ball 1,

$$[\mathscr{K}_1] = \mathscr{T}_{acc1}(t,t,U,U,\beta,\beta) = \mathscr{T}_{int1}(t,t,U,U,\beta,\beta) + \mathscr{T}_{ext1}(t,t,U,U,\beta,\beta),$$

where

$$\mathscr{T}_{int1}(t,t,U,U,\beta,b) = -B_{p1}(t)\,[b_i(t)].$$

With first law (10.19) we get

$$\begin{aligned}
[\mathscr{E}_1] &= \mathscr{C}_1 - \mathscr{T}_{int1}(t,t,U,U,\beta,\beta) \\
&= \bar{T}_1\left(\Sigma(\mathscr{B}_1) + \Sigma(\mathscr{B}_{12})\right) + [T_1]\left(\Delta(\mathscr{B}_1) + \Delta(\mathscr{B}_{12})\right) + B_{p1}\,[\beta_1], \quad (10.20)
\end{aligned}$$

with

$$\bar{T} = \frac{T^+ + T^-}{2},\ \Sigma(\mathscr{B}) = \mathscr{B}^+ + \mathscr{B}^-,\ \Delta(\mathscr{B}) = \frac{\mathscr{B}^+ - \mathscr{B}^-}{2}.$$

#### 10.4.2.2 The Second Law

Let be $\mathscr{S}_1$ the ball number 1 entropy, and recall the second law is

$$[\mathscr{S}_1] \geq \mathscr{B}_1^+ + \mathscr{B}_1^- + \mathscr{B}_{12}^+ + \mathscr{B}_{12}^- = \Sigma(\mathscr{B}_1) + \Sigma(\mathscr{B}_{12}). \quad (10.21)$$

By means of (10.20), the second law of thermodynamics gives

$$[\mathscr{E}_1] - \bar{T}_1\,[\mathscr{S}_1] \leq [T_1]\left(\Delta(\mathscr{B}_1) + \Delta(\mathscr{B}_{12})\right) + B_{p1}\,[\beta_1],$$

or, introducing the free energy $\Psi = \mathscr{E} - T\mathscr{S}$,

$$[\Psi_1] + \overline{\mathscr{S}}_1 [T_1] \leq [T_1] (\Delta (\mathscr{B}_1) + \Delta (\mathscr{B}_{12})) + B_{p1} [\beta_1]. \tag{10.22}$$

Free energy of ball 1 is

$$\Psi_1(T_1, \beta_1) = -C_1 T_1 \ln T_1 - \beta_1 \frac{L_1}{T_0}(T_1 - T_0) + I(\beta_1),$$

where $C_1$ is the heat capacity, $L_1$ is the latent heat at the phase change temperature $T_0$ and the indicator function $I$ of interval $[0,1]$ takes into account the internal constraint on the volume fraction

$$0 \leq \beta_1 \leq 1.$$

We define

$$\hat{\Psi}_1(T_1, \beta_1) = -\beta_1 \frac{L_1}{T_0}(T_1 - T_0) + I(\beta_1).$$

Then

$$[\Psi_1] = [-C_1 T_1 \ln T_1] + \left[\hat{\Psi}_1\right], \tag{10.23}$$

where

$$\begin{aligned}\left[\hat{\Psi}_1\right] &= \hat{\Psi}_1(T_1^+, \beta_1^+) - \hat{\Psi}_1(T_1^-, \beta_1^-)\\ &= \hat{\Psi}_1(T_1^+, \beta_1^+) - \hat{\Psi}_1(T_1^+, \beta_1^-) + \hat{\Psi}_1(T_1^+, \beta_1^-) - \hat{\Psi}_1(T_1^-, \beta_1^-),\end{aligned}$$

and

$$\hat{\Psi}_1(T_1^+, \beta_1^-) - \hat{\Psi}_1(T_1^-, \beta_1^-) = -\beta_1^- \frac{L_1}{T_0} [T_1]. \tag{10.24}$$

In view of (10.23) and (10.24), inequality (10.22) transforms into

$$\begin{aligned}&[\Psi_1] + \overline{\mathscr{S}}_1 [T_1]\\ &= [-C_1 T_1 \ln T_1] + \hat{\Psi}_1(T_1^+, \beta_1^+) - \hat{\Psi}_1(T_1^+, \beta_1^-) - \beta_1^- \frac{L_1}{T_0} [T_1] + \overline{\mathscr{S}}_1 [T_1]\\ &\leq [T_1] (\Delta (\mathscr{B}_1) + \Delta (\mathscr{B}_{12})) + B_{p1} [\beta_1].\end{aligned} \tag{10.25}$$

Since

$$\frac{[-C_1 T_1 \ln T_1]}{[T_1]},$$

has a limit when $[T_1] \longrightarrow 0$, we note

$$[-C_1 T_1 \ln T_1] + \left(\overline{\mathscr{S}}_1 - \beta_1^- \frac{L_1}{T_0}\right) [T_1] = -S_1 [T_1]. \tag{10.26}$$

From (10.25), we have

$$\hat{\Psi}_1(T_1^+,\beta_1^+) - \hat{\Psi}_1(T_1^+,\beta_1^-) \leq [T_1]\left(\Delta\left(\mathscr{B}_1\right) + \Delta\left(\mathscr{B}_{12}\right) + S_1\right) + B_{p1}\left[\beta_1\right]. \quad (10.27)$$

It is reasonable to assume that there is no dissipation with respect to $[T_1]$

$$\Delta\left(\mathscr{B}_1\right) + \Delta\left(\mathscr{B}_{12}\right) + S_1 = 0, \quad (10.28)$$

then, inequality (10.27) becomes

$$\hat{\Psi}_1(T_1^+,\beta_1^+) - \hat{\Psi}_1(T_1^+,\beta_1^-) \leq B_{p1}\left[\beta_1\right]. \quad (10.29)$$

## 10.4.3 The Laws of Thermodynamics for the System

### 10.4.3.1 The First Law

The energy and entropy of the system are

$$\mathscr{E} = \mathscr{E}_1 + \mathscr{E}_2 + \mathscr{E}^{int},$$
$$\mathscr{S} = \mathscr{S}_1 + \mathscr{S}_2 + \mathscr{S}^{int},$$

where $\mathscr{E}^{int}$ and $\mathscr{S}^{int}$ are the interaction energy and entropy. The first law for the system is

$$[\mathscr{E}] + [\mathscr{K}] = \mathscr{T}_{ext}(t,t,U,U,\beta,\beta) + \mathscr{C},$$

where $\mathscr{C}$ is the heat impulse provided to the system at the time of collision

$$\mathscr{C} = T_1^+\mathscr{B}_1^+ + T_1^-\mathscr{B}_1^- + T_2^+\mathscr{B}_2^+ + T_2^-\mathscr{B}_2^-.$$

The theorem of expended energy,

$$\begin{aligned}[\mathscr{K}] &= \mathscr{T}_{acc}(t,t,U,U,\beta,\beta)\\ &= \mathscr{T}_{int}(t,t,U,U,\beta,\beta) + \mathscr{T}_{ext}(t,t,U,U,\beta,\beta),\end{aligned}$$

and the first law give

$$[\mathscr{E}] = \mathscr{C} - \mathscr{T}_{int}(t,t,U,U,\beta,\beta). \quad (10.30)$$

The first law for every ball (10.20) and for the system (10.30) yield

$$
\begin{aligned}
[\mathscr{E}] &= [\mathscr{E}_1] + [\mathscr{E}_2] + \left[\mathscr{E}^{int}\right] \\
&= \bar{T}_1 \left(\Sigma\left(\mathscr{B}_1\right) + \Sigma\left(\mathscr{B}_{12}\right)\right) + [T_1]\left(\Delta\left(\mathscr{B}_1\right) + \Delta\left(\mathscr{B}_{12}\right)\right) + B_{p1}[\beta_1] \\
&\quad + \bar{T}_2 \left(\Sigma\left(\mathscr{B}_2\right) + \Sigma\left(\mathscr{B}_{21}\right)\right) + [T_2]\left(\Delta\left(\mathscr{B}_2\right) + \Delta\left(\mathscr{B}_{21}\right)\right) + B_{p2}[\beta_2] + \left[\mathscr{E}^{int}\right] \\
&= T_1^+ \mathscr{B}_1^+ + T_1^- \mathscr{B}_1^- + T_2^+ \mathscr{B}_2^+ + T_2^- \mathscr{B}_2^- \\
&\quad + P^{int} D\left(\frac{U^+ + U^-}{2}\right) + B_{p1}[\beta_1] + B_{p2}[\beta_2] \\
&= \bar{T}_1 \Sigma\left(\mathscr{B}_1\right) + [T_1]\Delta\left(\mathscr{B}_1\right) + \bar{T}_2 \Sigma\left(\mathscr{B}_2\right) + [T_2]\Delta\left(\mathscr{B}_2\right) \\
&\quad + P^{int} D\left(\frac{U^+ + U^-}{2}\right) + B_{p1}[\beta_1] + B_{p2}[\beta_2].
\end{aligned}
$$

It follows that

$$
\begin{aligned}
\left[\mathscr{E}^{int}\right] &= P^{int} D\left(\frac{U^+ + U^-}{2}\right) \\
&\quad - \bar{T}_1 \Sigma\left(\mathscr{B}_{12}\right) - [T_1]\Delta\left(\mathscr{B}_{12}\right) - \bar{T}_2 \Sigma\left(\mathscr{B}_{21}\right) - [T_2]\Delta\left(\mathscr{B}_{21}\right). \qquad (10.31)
\end{aligned}
$$

#### 10.4.3.2 The Second Law

The second law of thermodynamics is

$$
\begin{aligned}
[\mathscr{S}] &= [\mathscr{S}_1] + [\mathscr{S}_2] + \left[\mathscr{S}^{int}\right] \\
&\geq \mathscr{B}_1^+ + \mathscr{B}_1^- + \mathscr{B}_2^+ + \mathscr{B}_2^- = \Sigma\left(\mathscr{B}_1\right) + \Sigma\left(\mathscr{B}_2\right). \qquad (10.32)
\end{aligned}
$$

#### 10.4.3.3 A Useful Inequality

Let us note that if relationships

$$\left[\mathscr{S}^{int}\right] \geq -\Sigma\left(\mathscr{B}_{12}\right) - \Sigma\left(\mathscr{B}_{21}\right), \qquad (10.33)$$

and (10.21) are satisfied, the second law (10.32) is satisfied.

Let us get an inequality equivalent to (10.33). To get this new inequality, we assume (10.33) is satisfied and transform relationship (10.31). Let be

$$\bar{\Theta} = \frac{\bar{T}_1 + \bar{T}_2}{2};$$

from (10.31) and (10.33), we have

$$\left[\mathscr{E}^{int}\right]-\bar{\Theta}\left[\mathscr{S}^{int}\right]\leq P^{int}D\left(\frac{U^{+}+U^{-}}{2}\right)-\bar{T}_{1}\Sigma\left(\mathscr{B}_{12}\right)-\left[T_{1}\right]\Delta\left(\mathscr{B}_{12}\right)$$

$$-\bar{T}_{2}\Sigma\left(\mathscr{B}_{21}\right)-\left[T_{2}\right]\Delta\left(\mathscr{B}_{21}\right)+\bar{\Theta}\left(\Sigma\left(\mathscr{B}_{12}\right)+\Sigma\left(\mathscr{B}_{21}\right)\right). \tag{10.34}$$

Introducing the free energy of interaction $\Psi^{int}=\mathscr{E}^{int}-\Theta\mathscr{S}^{int}$, inequality (10.34) can be written as

$$\left[\Psi^{int}\right]+\overline{\mathscr{S}^{int}}\left[\Theta\right]\leq P^{int}D\left(\frac{U^{+}+U^{-}}{2}\right)-\bar{T}_{1}\Sigma\left(\mathscr{B}_{12}\right)-\left[T_{1}\right]\Delta\left(\mathscr{B}_{12}\right)$$

$$-\bar{T}_{2}\Sigma\left(\mathscr{B}_{21}\right)-\left[T_{2}\right]\Delta\left(\mathscr{B}_{21}\right)+\bar{\Theta}\left(\Sigma\left(\mathscr{B}_{12}\right)+\Sigma\left(\mathscr{B}_{21}\right)\right), \tag{10.35}$$

where

$$\overline{\mathscr{S}^{int}}=\frac{\left(\mathscr{S}^{int}\right)^{+}+\left(\mathscr{S}^{int}\right)^{-}}{2},\ [\Theta]=\frac{[T_1]+[T_2]}{2}.$$

Assuming $\Psi^{int}$ to be a smooth function of $\Theta$, we have

$$\lim_{[\Theta]\to 0}\frac{\left[\Psi^{int}\right]}{[\Theta]}+\overline{\mathscr{S}^{int}}=0.$$

Therefore, we can put

$$\left[\Psi^{int}\right]+\overline{\mathscr{S}^{int}}\left[\Theta\right]=-S^{int}\left[\Theta\right], \tag{10.36}$$

and, from (10.35), we find

$$0\leq P^{int}D\left(\frac{U^{+}+U^{-}}{2}\right)-\bar{T}_{1}\Sigma\left(\mathscr{B}_{12}\right)-\left[T_{1}\right]\left(\Delta\left(\mathscr{B}_{12}\right)-\frac{S^{int}}{2}\right)$$

$$-\bar{T}_{2}\Sigma\left(\mathscr{B}_{21}\right)-\left[T_{2}\right]\left(\Delta\left(\mathscr{B}_{21}\right)-\frac{S^{int}}{2}\right)+\bar{\Theta}\left(\Sigma\left(\mathscr{B}_{12}\right)+\Sigma\left(\mathscr{B}_{21}\right)\right). \tag{10.37}$$

It is reasonable to assume that there is no dissipation with respect to $[T_1]$ and $[T_2]$

$$\Delta\left(\mathscr{B}_{12}\right)-\frac{S^{int}}{2}=0,\ \Delta\left(\mathscr{B}_{21}\right)-\frac{S^{int}}{2}=0\,; \tag{10.38}$$

then

$$0\leq P^{int}D\left(\frac{U^{+}+U^{-}}{2}\right)-\bar{T}_{1}\Sigma\left(\mathscr{B}_{12}\right)$$

$$-\bar{T}_{2}\Sigma\left(\mathscr{B}_{21}\right)+\bar{\Theta}\left(\Sigma\left(\mathscr{B}_{12}\right)+\Sigma\left(\mathscr{B}_{21}\right)\right). \tag{10.39}$$

But

$$\bar{T}_1\Sigma(\mathscr{B}_{12})+\bar{T}_2\Sigma(\mathscr{B}_{21})$$
$$=\bar{\Theta}\left(\Sigma(\mathscr{B}_{12})+\Sigma(\mathscr{B}_{21})\right)+\delta\bar{T}\frac{\Sigma(\mathscr{B}_{21})-\Sigma(\mathscr{B}_{12})}{2}, \tag{10.40}$$

where the difference of temperature, $\delta\bar{T}$, is defined by

$$\delta\bar{T}=\bar{T}_2-\bar{T}_1.$$

Finally, inequality (10.39) yields

$$0\leq P^{int}D\left(\frac{U^{+}+U^{-}}{2}\right)-\delta\bar{T}\frac{\Sigma(\mathscr{B}_{21})-\Sigma(\mathscr{B}_{12})}{2}. \tag{10.41}$$

This last relationship links mechanical and thermal dissipations.

We have proved that inequality (10.33) implies inequality (10.41). Let us prove the converse.

**Theorem 10.2.** *Let $\Psi^{int}$ be a smooth function of $\Theta$. We assume no dissipation with respect to $[T_1]$ and $[T_2]$, (relationship (10.38)). If relationship (10.31) is satisfied, then (10.33) and (10.41) are equivalent.*

*Proof.* We assume (10.41) is satisfied. By means of (10.40) and supposing no dissipation with respect to $[T_1]$ and $[T_2]$, it follows that (10.37) is satisfied; thanks to (10.36), inequality (10.37) yields

$$\left[\Psi^{int}\right]\leq P^{int}D\left(\frac{U^{+}+U^{-}}{2}\right)-\bar{T}_1\Sigma(\mathscr{B}_{12})-[T_1]\Delta(\mathscr{B}_{12})$$
$$-\bar{T}_2\Sigma(\mathscr{B}_{21})-[T_2]\Delta(\mathscr{B}_{21})+\bar{\Theta}\left(\Sigma(\mathscr{B}_{12})+\Sigma(\mathscr{B}_{21})\right)-\overline{\mathscr{S}^{int}}[\Theta]. \tag{10.42}$$

Substituting the following identity

$$\left[\Psi^{int}\right]=\left[\mathscr{E}^{int}\right]-[\Theta]\overline{\mathscr{S}^{int}}-\bar{\Theta}\left[\mathscr{S}^{int}\right],$$

into (10.42), we have

$$-\overline{\Theta}\left[\mathscr{S}^{int}\right]\leq-\left[\mathscr{E}^{int}\right]+P^{int}D\left(\frac{U^{+}+U^{-}}{2}\right)$$
$$-\bar{T}_1\Sigma(\mathscr{B}_{12})-[T_1]\Delta(\mathscr{B}_{12})-\bar{T}_2\Sigma(\mathscr{B}_{21})-[T_2]\Delta(\mathscr{B}_{21})$$
$$+\overline{\Theta}\left(\Sigma(\mathscr{B}_{12})+\Sigma(\mathscr{B}_{21})\right)\;; \tag{10.43}$$

finally, by virtue of (10.31), (10.43) entails

$$-\overline{\Theta}\left[\mathscr{S}^{int}\right] \leq \overline{\Theta}\left(\Sigma\left(\mathscr{B}_{12}\right)+\Sigma\left(\mathscr{B}_{21}\right)\right),$$

that is, inequality (10.33) because temperatures are positive. □

### 10.4.4 The Constitutive Laws

For ball 1, we split the interior force $B_{p1}$ into a possibly dissipative part due to the free energy, indexed by $^{fe}$, and a dissipative part, indexed by $^{d}$

$$B_{p1} = B_{p1}^{fe} + B_{p1}^{d}.$$

The interior force, $B_1^{fe}$, is defined by free energy $\Psi_1(T_1, \beta_1)$, as follows

$$B_{p1}^{fe} \in \partial\Psi_1(T_1^+, \beta_1^+) = -\frac{L_1}{T_0}(T_1^+ - T_0) + \partial I(\beta_1^+), \tag{10.44}$$

where subdifferential $\partial\Psi_1$, is computed with respect to $\beta_1^+$. Dissipative interior force, $B_{p1}^d$, is defined by

$$B_{p1}^d \in \partial\Phi_1\left([\beta_1]\right), \tag{10.45}$$

and satisfies

$$B_{p1}^d\left[\beta_1\right] \geq 0, \tag{10.46}$$

where $\Phi_1\left([\beta_1]\right)$ is a pseudopotential of dissipation and the subdifferential is computed with respect to $[\beta_1]$.

For the system, we choose the following pseudopotential of dissipation

$$\Phi\left(D\left(\frac{U^+ + U^-}{2}\right), \delta\bar{T}, \chi\right) = \Phi\left(D\left(\frac{U^+ + U^-}{2}\right), \delta\bar{T}\right) + I_+(D\left(U^+\right)),$$

which depends on the velocity $D((U^+ + U^-)/2)$, on the thermal heterogeneity $\delta\bar{T}$ and on the quantity $\chi = D(U^-/2)$, which ensures the non-interpenetration condition, for instance, $D(U^+) = U_1^+ - U_2^+ \geq 0$, in case ball 1 is on the right of ball 2.

The constitutive laws are relationships (10.44), (10.45) and

$$\left(P^{int}, \frac{\Sigma\left(\mathscr{B}_{21}\right) - \Sigma\left(\mathscr{B}_{12}\right)}{2}\right) \in \partial\Phi\left(D\left(\frac{U^+ + U^-}{2}\right), \delta\bar{T}, \chi\right), \tag{10.47}$$

where the subdifferential set of $\Phi$ is computed with respect to the first two variables.

*Remark 10.10.* From (10.47), applying the classical properties of pseudopotentials, we see that inequality (10.41) is satisfied.

Moreover, the following theorems prove that the constitutive laws are such that the internal constraints and the second law are satisfied.

**Theorem 10.3.** *If the constitutive laws (10.44) and (10.45) are satisfied, then inequality (10.29) holds and the internal constraint* $0 \leq \beta^+ \leq 1$ *is verified.*

*Proof.* If

$$B_{p1}^{fe} \in \partial \Psi_1(T_1^+, \beta_1^+),$$

then, using the definition of subdifferential, we have

$$\hat{\Psi}_1(T_1^+, \beta_1^+) - \hat{\Psi}_1(T_1^+, \beta_1^-) \leq B_{p1}^{fe}(\beta_1^+ - \beta_1^-) = B_{p1}^{fe}[\beta_1];$$

with (10.46) resulting from (10.45), the previous inequality entails

$$\hat{\Psi}_1(T_1^+, \beta_1^+) - \hat{\Psi}_1(T_1^+, \beta_1^-) \leq B_{p1}^{fe}[\beta_1] + B_{p1}^{d}[\beta_1] = B_{p1}[\beta_1].$$

Since

$$\partial \Psi_1(T_1^+, \beta_1^+) \neq \emptyset,$$

the internal constraint $0 \leq \beta_1^+ \leq 1$ is verified. □

**Theorem 10.4.** *Assume no dissipation with respect to* $[T_1]$ *and* $[T_2]$ *(relationships (10.28) and (10.38)). If the first laws (10.20), (10.30) and the constitutive laws (10.44), (10.45) and (10.47) are verified, then the second law of thermodynamics holds for every ball and for the system.*

*Proof.* Due to Theorem 10.3, (10.23), (10.45) and (10.24), we have

$$\begin{aligned}[\Psi_1] &= [-C_1 T_1 \ln T_1] + \hat{\Psi}_1(T_1^+, \beta_1^+) - \hat{\Psi}_1(T_1^+, \beta_1^-) - \beta_1^- \frac{L_1}{T_0}[T_1] \\ &\leq [-C_1 T_1 \ln T_1] + B_{p1}[\beta_1] - \beta_1^- \frac{L_1}{T_0}[T_1]. \end{aligned} \tag{10.48}$$

We have

$$[\Psi_1] = [\mathscr{E}_1] - [T_1 \mathscr{S}_1] = [\mathscr{E}_1] - [T_1]\overline{\mathscr{S}}_1 - \bar{T}_1[\mathscr{S}_1], \tag{10.49}$$

from (10.20), (10.48) and (10.49), it follows that

$$\begin{aligned}
\bar{T}_1\left[\mathscr{S}_1\right] &\geq \left[\mathscr{E}_1\right]-\left[T_1\right]\overline{\mathscr{S}}_1+\left[C_1 T_1 \ln T_1\right]-B_{p1}\left[\beta_1\right]+\beta_1^{-}\frac{L_1}{T_0}\left[T_1\right] \\
&= \bar{T}_1\left(\Sigma\left(\mathscr{B}_1\right)+\Sigma\left(\mathscr{B}_{12}\right)\right)+\left[T_1\right]\left(\Delta\left(\mathscr{B}_1\right)+\Delta\left(\mathscr{B}_{12}\right)\right)+B_{p1}\left[\beta_1\right] \\
&\quad -\left[T_1\right]\overline{\mathscr{S}}_1+\left[C_1 T_1 \ln T_1\right]-B_{p1}\left[\beta_1\right]+\beta_1^{-}\frac{L_1}{T_0}\left[T_1\right] \\
&= \bar{T}_1\left(\Sigma\left(\mathscr{B}_1\right)+\Sigma\left(\mathscr{B}_{12}\right)\right)+\left[T_1\right]\left(\Delta\left(\mathscr{B}_1\right)+\Delta\left(\mathscr{B}_{12}\right)+S_1\right),
\end{aligned}$$

with $S_1$ defined as in (10.26). From the previous inequality, assuming no dissipation with respect to $[T_1]$ (relationship (10.28)), we obtain

$$\left[\mathscr{S}_1\right] \geq \Sigma\left(\mathscr{B}_1\right)+\Sigma\left(\mathscr{B}_{12}\right), \tag{10.50}$$

that is, the second law of thermodynamics for ball 1.

Due to constitutive laws (10.47) and Theorem 10.2, inequality (10.33) is satisfied. It results from (10.21) for each ball and from (10.33) that

$$\begin{aligned}
[\mathscr{S}] &= \left[\mathscr{S}_1\right]+\left[\mathscr{S}_2\right]+\left[\mathscr{S}^{int}\right] \\
&\geq \Sigma\left(\mathscr{B}_1\right)+\Sigma\left(\mathscr{B}_{12}\right)+\Sigma\left(\mathscr{B}_2\right)+\Sigma\left(\mathscr{B}_{21}\right)-\Sigma\left(\mathscr{B}_{12}\right)-\Sigma\left(\mathscr{B}_{21}\right) \\
&= \Sigma\left(\mathscr{B}_1\right)+\Sigma\left(\mathscr{B}_2\right),
\end{aligned}$$

which is the second law for the system. □

*Remark 10.11.* The reaction to the internal constraint $0 \leq \beta^+ \leq 1$ is dissipative or non workless because

$$\partial I(\beta^+)(\beta^+ - \beta^-),$$

may be different from zero, in fact positive. Thus, the reaction is dissipative, as we may expect. This property is not true in a smooth evolution where the internal constraint is workless or non-dissipative because

$$\partial I(\beta)\frac{\mathrm{d}\beta}{\mathrm{d}t} = 0, \ a.e..$$

This is an example of the properties of the power of the reaction forces $\mathscr{P}^{reac}$ given in Sect. 3.4.1.

## 10.5 Examples of Thermal Effects with Phase Changes

Most of the following examples have been investigated by Anna Maria Caucci, [61]. We choose a pseudopotential of dissipation $\Phi$, without non-diagonal terms, for example

$$\Phi\left(D\left(\frac{U^+ + U^-}{2}\right), \delta\bar{T}, D\left(\frac{U^-}{2}\right)\right)$$
$$= \Phi^{mech}\left(D\left(\frac{U^+ + U^-}{2}\right), D\left(\frac{U^-}{2}\right)\right) + \frac{\lambda}{4}(\delta\bar{T})^2.$$

With this choice, the mechanical problem is split up with the thermal one. The thermal constitutive law is

$$\Sigma(\mathscr{B}_{21}) - \Sigma(\mathscr{B}_{12}) = -\lambda\,\delta\bar{T}. \tag{10.51}$$

Let us recall that the free energies chosen for the balls are

$$\Psi_i(T_i, \beta_i) = -C_i T_i \ln T_i - \beta_i \frac{L_i}{T_0}(T_i - T_0) + I(\beta_i), \quad i = 1, 2.$$

We choose the free energy of interaction $\Psi^{int} = 0$ and have

$$S^{int} = 0.$$

and, from (10.38)

$$\Delta(\mathscr{B}_{12}) = 0, \ \Delta(\mathscr{B}_{21}) = 0.$$

It results from (10.28) that

$$\Delta(\mathscr{B}_1) + S_1 = 0, \ \Delta(\mathscr{B}_2) + S_2 = 0.$$

Now, we suppose that the macroscopic mechanical problem is solved, that is, we know the quantity $P^{int} D((U^+ + U^-)/2)$ and we assume no external percussion work $A_1 = A_2 = 0$. Due to equations of motion (10.17), (10.18), the thermal equations are

$$[\mathscr{S}_1] = [C_1 \ln T_1] + \frac{L_1}{T_0}[\beta_1] = C_1 \ln \frac{T_1^+}{T_1^-} + \frac{L_1}{T_0}[\beta_1] = \Sigma(\mathscr{B}_1) + \Sigma(\mathscr{B}_{12}),$$

$$[\mathscr{S}_2] = [C_2 \ln T_2] + \frac{L_2}{T_0}[\beta_2] = C_2 \ln \frac{T_2^+}{T_2^-} + \frac{L_2}{T_0}[\beta_2] = \Sigma(\mathscr{B}_2) + \Sigma(\mathscr{B}_{21}),$$

$$[\mathscr{E}_1] = C_1[T_1] + L_1[\beta_1] = \bar{T}_1(\Sigma(\mathscr{B}_1) + \Sigma(\mathscr{B}_{12})) + [T_1]\Delta(\mathscr{B}_1),$$

$$[\mathscr{E}_2] = C_2[T_2] + L_2[\beta_2] = \bar{T}_2(\Sigma(\mathscr{B}_2) + \Sigma(\mathscr{B}_{21})) + [T_2]\Delta(\mathscr{B}_2),$$

$$[\mathscr{E}^{int}] = P^{int} D\left(\frac{U^+ + U^-}{2}\right) - \bar{T}_1 \Sigma(\mathscr{B}_{12}) - \bar{T}_2 \Sigma(\mathscr{B}_{21}) = 0.$$

These equations are completed by the description of thermal relationships between the system and the outside and by the equations of microscopic motion

$$0 = B_{pi} = B_{pi}^{fe} + B_{pi}^{d} \in -\frac{L_i}{T_0}(T_i^+ - T_0) + \partial I(\beta_i^+) + \partial \Phi_i([\beta_i]).$$

### 10.5.1 The Thermal Equations in the Isentropic Case

If the collision is isentropic

$$\Sigma(\mathscr{B}_1) = \Sigma(\mathscr{B}_2) = 0,$$

and the thermal equations become

$$\Delta(\mathscr{B}_{12}) = 0,\ \Delta(\mathscr{B}_{21}) = 0,$$

$$\Delta(\mathscr{B}_1) + S_1 = 0,\ \Delta(\mathscr{B}_2) + S_2 = 0, \tag{10.52}$$

$$C_1 \ln\frac{T_1^+}{T_1^-} + \frac{L_1}{T_0}[\beta_1] = \Sigma(\mathscr{B}_{12}),$$

$$C_2 \ln\frac{T_2^+}{T_2^-} + \frac{L_2}{T_0}[\beta_2] = \Sigma(\mathscr{B}_{21}),$$

$$C_1[T_1] + L[\beta_1] = \bar{T}_1\Sigma(\mathscr{B}_{12}) + [T_1]\Delta(\mathscr{B}_1), \tag{10.53}$$

$$C_2[T_2] + L[\beta_2] = \bar{T}_2\Sigma(\mathscr{B}_{21}) + [T_2]\Delta(\mathscr{B}_2), \tag{10.54}$$

$$P^{int}D\left(\frac{U^+ + U^-}{2}\right) = \bar{T}_1\Sigma(\mathscr{B}_{12}) + \bar{T}_2\Sigma(\mathscr{B}_{21}). \tag{10.55}$$

Substituting the expressions of $\Sigma(\mathscr{B}_{12})$ and $\Sigma(\mathscr{B}_{21})$ in (10.51) and (10.55), we obtain

$$C_2 \ln\frac{T_2^+}{T_2^-} + \frac{L_1}{T_0}[\beta_2] + \lambda\bar{T}_2 = C_1 \ln\frac{T_1^+}{T_1^-} + \frac{L_2}{T_0}[\beta_1] + \lambda\delta\bar{T}_1, \tag{10.56}$$

$$P^{int}D\left(\frac{U^+ + U^-}{2}\right) = \bar{T}_1\left(C_1 \ln\frac{T_1^+}{T_1^-} + \frac{L_1}{T_0}[\beta_1]\right) + \bar{T}_2\left(C_2 \ln\frac{T_2^+}{T_2^-} + \frac{L_2}{T_0}[\beta_2]\right). \tag{10.57}$$

Equations (10.53) and (10.54) give the heat exchanges, $\Delta(\mathscr{B}_1)$ and $\Delta(\mathscr{B}_2)$, with the outside, while (10.56) and (10.57) give the temperatures $T_1^+$ and $T_2^+$ after the collision. The splitting of the heat exchanges with the exterior, the $\mathscr{B}^+$ and $\mathscr{B}^-$ is given by (10.52).

### 10.5.2 The Thermal Equations in the Adiabatic Case

If the collision is adiabatic

$$T_1^+\mathscr{B}_1^+ + T_1^-\mathscr{B}_1^- = \bar{T}_1\Sigma(\mathscr{B}_1) + [T_1]\Delta(\mathscr{B}_1) = 0,$$
$$T_2^+\mathscr{B}_2^+ + T_2^-\mathscr{B}_2^- = \bar{T}_2\Sigma(\mathscr{B}_2) + [T_2]\Delta(\mathscr{B}_2) = 0,$$

and we have the following equations

$$\Delta(\mathscr{B}_{12}) = 0,\ \Delta(\mathscr{B}_{21}) = 0,$$

$$\Delta(\mathscr{B}_1) + S_1 = 0,\ \Delta(\mathscr{B}_2) + S_2 = 0,$$

$$[\mathscr{S}_1] = C_1 \ln\frac{T_1^+}{T_1^-} + \frac{L_1}{T_0}[\beta_1] = \Sigma(\mathscr{B}_1) + \Sigma(\mathscr{B}_{12}),$$

$$[\mathscr{S}_2] = C_2 \ln\frac{T_2^+}{T_2^-} + \frac{L_2}{T_0}[\beta_2] = \Sigma(\mathscr{B}_2) + \Sigma(\mathscr{B}_{21}),$$

$$C_1[T_1] + L_1[\beta_1] = \bar{T}_1\Sigma(\mathscr{B}_{12}),$$
$$C_2[T_2] + L_2[\beta_2] = \bar{T}_2\Sigma(\mathscr{B}_{21}),$$
$$P^{int}D\left(\frac{U^+ + U^-}{2}\right) = C_1[T_1] + L_1[\beta_1] + C_2[T_2] + L_2[\beta_2], \quad (10.58)$$

$$C_2\frac{[T_2]}{\bar{T}_2} + \frac{L_2}{\bar{T}_2}[\beta_2] - C_1\frac{[T_1]}{\bar{T}_1} - \frac{L_1}{\bar{T}_1}[\beta_1] = -\lambda\delta\bar{T}. \quad (10.59)$$

The last two equations give the temperatures $T_1^+$ and $T_2^+$ after the collision, the first give the entropy exchanges, $\Sigma(\mathscr{B}_1)$ and $\Sigma(\mathscr{B}_2)$, with the outside.

### 10.5.3 The Adiabatic Situation with the Small Perturbation Assumption

We consider the adiabatic case and we assume small perturbations, that is

$$T_i^\pm = T_0 + \theta_i^\pm,\ |\theta_i^\pm| \ll T_0.$$

We can write (10.58) and (10.59) as follows

$$P^{int} D\left(\frac{U^{+}+U^{-}}{2}\right)=C_1\left[\theta_1\right]+L_1\left[\beta_1\right]+C_2\left[\theta_2\right]+L_2\left[\beta_2\right], \tag{10.60}$$

$$C_2\frac{[\theta_2]}{T_0}+\frac{L_1}{T_0}[\beta_2]-C_1\frac{[\theta_1]}{T_0}-\frac{L_2}{T_0}[\beta_1]=-\lambda\left(\theta_2^{-}-\theta_1^{-}+\frac{[\theta_2]}{2}-\frac{[\theta_1]}{2}\right), \tag{10.61}$$

and get the system

$$\begin{aligned} 2C_1\left[\theta_1\right] &= P^{int} D\left(\frac{U^{+}+U^{-}}{2}\right)-2L_1\left[\beta_1\right] \\ &\quad +\lambda T_0\left(\theta_2^{-}-\theta_1^{-}+\frac{[\theta_2]}{2}-\frac{[\theta_1]}{2}\right), \end{aligned} \tag{10.62}$$

$$\begin{aligned} 2C_2\left[\theta_2\right] &= P^{int} D\left(\frac{U^{+}+U^{-}}{2}\right)-2L_2\left[\beta_2\right] \\ &\quad -\lambda T_0\left(\theta_2^{-}-\theta_1^{-}+\frac{[\theta_2]}{2}-\frac{[\theta_1]}{2}\right), \end{aligned} \tag{10.63}$$

where the volume fractions $\beta_1$ and $\beta_2$ satisfy the equations of microscopic motion

$$0 \in -\frac{L_1}{T_0}(T_1^{+}-T_0)+\partial I(\beta_1^{+})+\partial\Phi_1\left([\beta_1]\right), \tag{10.64}$$

$$0 \in -\frac{L_2}{T_0}(T_2^{+}-T_0)+\partial I(\beta_2^{+})+\partial\Phi_2\left([\beta_2]\right), \tag{10.65}$$

which are equivalent to

$$\theta_1^{+} \in \partial I(\beta_1^{+})+\frac{T_0}{L_1}\partial\Phi_1\left([\beta_1]\right), \tag{10.66}$$

$$\theta_2^{+} \in \partial I(\beta_2^{+})+\frac{T_0}{L_2}\partial\Phi_2\left([\beta_2]\right). \tag{10.67}$$

It is easy to prove that (10.62), (10.66) and (10.63), (10.67) giving the temperatures and phase fractions $\theta_1^{+}$, $\theta_2^{+}$ and $\beta_1^{+}$, $\beta_2^{+}$ have unique solutions.

### 10.5.4 *Identical Balls at the Same Temperature Before the Collision*

The temperatures after the collision are equal. Equations (10.62) and (10.66) show that they do not depend on $\lambda$.

#### 10.5.4.1 Case Where the Phase Change Is Not Dissipative

We consider two identical balls of ice, $C_1 = C_2 = C$, $L_1 = L_2 = L$ at the same temperature, $\theta^-$, before the collision. The balls have the same temperature and volume fraction after the collision, $\theta^+$, $\beta^+$

$$\theta_1^- = \theta_2^- = \theta^-, \; \theta_1^+ = \theta_2^+ = \theta^+, \tag{10.68}$$

$$\beta_1^- = \beta_2^- = 0, \; \beta_1^+ = \beta_2^+ = \beta^+. \tag{10.69}$$

We assume no dissipation with respect to the volume fractions discontinuities $[\beta_1]$ and $[\beta_2]$

$$\Phi_1([\beta_1]) = \Phi_2([\beta_2]) = 0. \tag{10.70}$$

Equations (10.62) and (10.63) give

$$C[\theta] = \frac{1}{2}(\mathcal{T} - 2L\beta^+), \tag{10.71}$$

where

$$\mathcal{T} = P^{int} D\left(\frac{U^+ + U^-}{2}\right).$$

Because of (10.70), we find easily from the equations of microscopic motion (10.66), (10.67)

$$\theta^+ \in \partial I(\beta^+). \tag{10.72}$$

The evolution of the balls is described in the following theorem

**Theorem 10.5.** *We have*

1 *if $\mathcal{T} \leq -2C\theta^-$, then $\beta^+ = 0$: after collision there is ice with temperature given by (10.71);*
2 *if $\mathcal{T} \geq 2(L - C\theta^-)$, then $\beta^+ = 1$: after collision there is liquid water with temperature given by (10.71);*
3 *if $-2C\theta^- < \mathcal{T} < 2(L - C\theta^-)$, then $0 < \beta^+ < 1$: after collision there is a mixture of ice and liquid water with temperature $\theta^+ = 0$. Volume fraction $\beta^+$ is given by (10.71).*

*Proof.* 1 If $\mathscr{T} \leq -2C\theta^-$, then, from (10.71), it follows that

$$C[\theta] + L\beta^+ \leq -C\theta^-.$$

The previous inequality entails

$$\theta^+ \leq -\frac{L}{C}\beta^+ \leq 0;$$

therefore, in view of (10.72), we obtain

$$\beta^+ = 0.$$

2 Let be $\mathscr{T} \geq 2(L - C\theta^-)$. We get

$$C[\theta] + L\beta^+ \geq L - C\theta^-,$$

that is

$$\theta^+ \geq \frac{L}{C}(1-\beta^+) \geq 0.$$

On the other hand, because $\theta^+ \in \partial I(\beta^+)$, we get

$$\beta^+ = 1.$$

3 If $-2C\theta^- < \mathscr{T} < 2(L - C\theta^-)$, then

$$-\frac{L}{C}\beta^+ < \theta^+ < \frac{L}{C}(1-\beta^+). \tag{10.73}$$

Now, if $\theta^+ > 0$, because of (10.72), we have $\beta^+ = 1$ and, substituting into (10.73)

$$\theta^+ < 0;$$

analogously, if $\theta^+ < 0$, we find $\beta^+ = 0$ and, from (10.73)

$$\theta^+ > 0.$$

Finally, if we suppose $\theta^+ = 0$, then (10.73) is satisfied and constitutive law (10.72) implies

$$0 < \beta^+ < 1.$$

Volume fraction is given by (10.71)

$$\beta^+ = \frac{\mathscr{T} + 2C\theta^-}{2L}.$$

The result of the theorem agrees with what is expected: a violent collision produces a phase change whereas a non violent collision does not. Violent means dissipative, i.e., $\mathscr{T}$ large.

#### 10.5.4.2 Case Where the Phase Change Is Dissipative

For every ball, let us choose the pseudopotential of dissipation as

$$\Phi_i([\beta_i]) = \frac{c}{2}[\beta_i]^2, \; i = 1,2,$$

where $c$ is a positive constant. Thus, equations of motion (10.64) and (10.65) give

$$\frac{L}{T_0}\theta_1^+ \in \partial I(\beta_1^+) + c[\beta_1],$$

$$\frac{L}{T_0}\theta_2^+ \in \partial I(\beta_2^+) + c[\beta_2],$$

and we have the equations

$$C[\theta] = \frac{1}{2}(\mathscr{T} - 2L\beta^+),$$

$$\frac{L}{T_0}\theta^+ \in \partial I(\beta^+) + c\beta^+.$$

The following result is easily proved

**Theorem 10.6.** *We have*

1 *if* $\mathscr{T} \leq -2C\theta^-$, *then* $\beta^+ = 0$*: after collision there is solid with* $\theta^+$*;*
2 *if* $\mathscr{T} \geq 2(L - C\theta^- + cCT_0/L)$, *then* $\beta^+ = 1$*: after collision there is liquid with* $\theta^+ \geq (cT_0)/L$*;*
3 *if* $-2C\theta^- < \mathscr{T} < 2(L - C\theta^- + cCT_0/L)$, *then* $0 < \beta^+ < 1$*: after collision there is a mixture of solid and liquid with* $\theta^+ = (cT_0\beta^+)/L$*.*

When there is dissipation the collision has to be more violent to melt the solid balls. The phase change occurs with temperature slightly above $T_0$, as it is the case for dissipative phase changes, see Sect. 4.7.

### *10.5.5 Identical Balls at Different Temperatures Before the Collision: Collision of Two Pieces of Ice*

When two pieces of ice at different temperature collide, the dissipation due to the collision may be large enough to melt the warmest of them. We look for conditions on the state quantities before the collision and on the dissipated work, such that this phenomenon occurs. We expect that the temperatures before collision cannot be very cold and that the dissipated work has to be large.

We assume that there is no dissipation with respect to $[\beta_1]$ and $[\beta_2]$

$$\Phi_1([\beta_1]) = \Phi_2([\beta_2]) = 0,$$

thus the equations of microscopic motion are (10.66), (10.67)

$$\theta_1^+ \in \partial I(\beta_1^+),\ \theta_2^+ \in \partial I(\beta_1^+). \tag{10.74}$$

The two identical pieces of ice before collision satisfy

$$\beta_1^- = \beta_2^- = 0,\ \theta_1^- \leq 0,\ \theta_2^- \leq 0.$$

We look for conditions such that ball 1 melts and ball 2 remains frozen

$$\beta_1^+ = 1,\ \beta_2^+ = 0. \tag{10.75}$$

Thus from (10.74), we have

$$\theta_1^+ \geq 0,\ \theta_2^+ \leq 0. \tag{10.76}$$

#### 10.5.5.1 Case with $\lambda$ Small with Respect to $C/T_0$

Equations (10.19) and (10.27) give

$$p^{int} D\left(\frac{U^+ + U^-}{2}\right) = C_1[\theta_1] + L[\beta_1] + C_2[\theta_2] + L[\beta_2],$$

$$\left(\frac{C_2}{T_0} + \frac{\lambda}{2}\right)[\theta_2] + \frac{L}{T_0}[\beta_2] - \left(\frac{C_1}{T_0} + \frac{\lambda}{2}\right)[\theta_1] - \frac{L}{T_0}[\beta_1] = -\lambda\left(\theta_2^- - \theta_1^-\right).$$

We assume $\lambda$ small with respect to $C_i/T_0$, $i = 1,2$ (see Remark 10.12) and get

$$p^{int} D\left(\frac{U^+ + U^-}{2}\right) = C_1[\theta_1] + L[\beta_1] + C_2[\theta_2] + L[\beta_2],$$

$$\frac{C_2}{T_0}[\theta_2] + \frac{L}{T_0}[\beta_2] - \frac{C_1}{T_0}[\theta_1] - \frac{L}{T_0}[\beta_1] = -\lambda\left(\theta_2^- - \theta_1^-\right),$$

which gives the system

$$C_1\left[\theta_1\right]=\frac{1}{2}\left(P^{int}D\left(\frac{U^{+}+U^{-}}{2}\right)-2L\left[\beta_1\right]+\lambda T_0\left(\theta_2^{-}-\theta_1^{-}\right)\right), \qquad (10.77)$$

$$C_2\left[\theta_2\right]=\frac{1}{2}\left(P^{int}D\left(\frac{U^{+}+U^{-}}{2}\right)-2L\left[\beta_2\right]-\lambda T_0\left(\theta_2^{-}-\theta_1^{-}\right)\right), \qquad (10.78)$$

where the volume fractions $\beta_1$ and $\beta_2$ satisfy the equations of microscopic motion (10.74).

Equations (10.77), (10.78) give with $C_1=C_2=C$, and (10.75)

$$C\left[\theta_1\right]=\frac{1}{2}\left(\mathcal{T}-2L+\lambda T_0\left(\theta_2^{-}-\theta_1^{-}\right)\right), \qquad (10.79)$$

$$C\left[\theta_2\right]=\frac{1}{2}\left(\mathcal{T}-\lambda T_0\left(\theta_2^{-}-\theta_1^{-}\right)\right). \qquad (10.80)$$

By means of (10.79), (10.80), conditions (10.76) are satisfied if and only if

$$\begin{aligned}(2C-\lambda T_0)\,\theta_1^{-}+\lambda T_0\theta_2^{-}+\mathcal{T}-2L\geq 0,\\ \lambda T_0\theta_1^{-}+(2C-\lambda T_0)\,\theta_2^{-}+\mathcal{T}\leq 0,\end{aligned} \qquad (10.81)$$

with $\theta_1^{-}\leq 0,\ \theta_2^{-}\leq 0$. Because of our hypothesis on $\lambda$, we have

$$2C-\lambda T_0>0. \qquad (10.82)$$

If $\mathcal{T}<2L$, inequality $(10.81)_1$ and (10.82) show that it is impossible to satisfy system (10.81) with $\theta_1^{-}\leq 0, \theta_2^{-}\leq 0$. Thus if the dissipation is small, it is impossible to melt one piece of ice. Both of them remain frozen.

If $\mathcal{T}\geq 2L$, it is possible to find $(\theta_1^{-}, \theta_2^{-})$ satisfying system (10.81) with $\theta_1^{-}\leq 0$, $\theta_1^{-}\simeq 0$ ($\theta_1^{-}=0$ if $\mathcal{T}=2L$) and $\theta_2^{-}<\theta_1^{-}$. Thus if the dissipation is large, one piece of ice melts, the other one remains frozen. The temperature of the coldest piece of ice has to be sufficiently cold. Examples are given on Figs. 10.4 and 10.5 for $\lambda=0\ \mathrm{J/K^2}$ and $\lambda=1000\ \mathrm{J/K^2}$, $\mathcal{T}=10\,L$, with $L=3.33\times 10^5$ J, $C=10^6$ J/K and $T_0=273$ K.

#### 10.5.5.2 Case When $\lambda$ Is Not Small with Respect to $C/T_0$

We have

$$[\beta_1]=1,\ [\beta_2]=0,$$

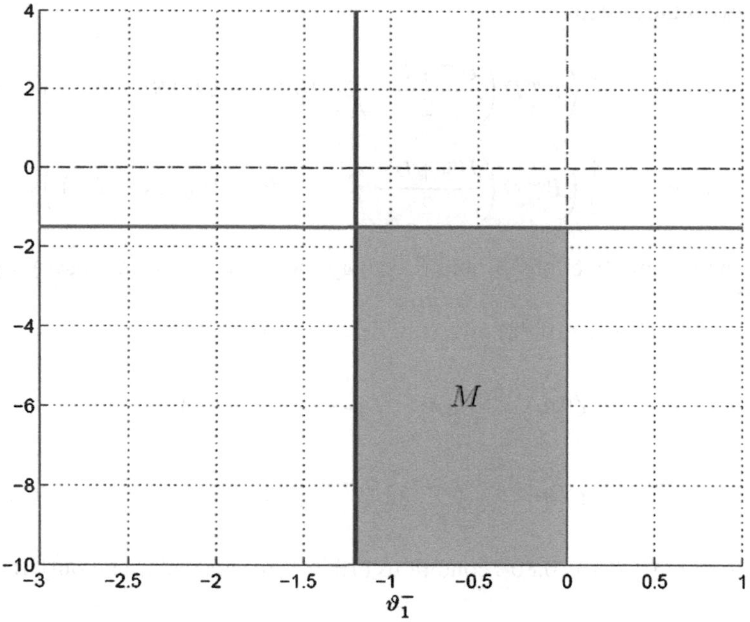


**Fig. 10.4** The case $\lambda = 0$ J/K$^2$ and $\mathscr{T} = 10\,L$, with $L = 3.33 \times 10^5$ J, $C = 10^6$ J/K. The inequalities (10.81) have solutions if $\theta_1^-$ is negative and satisfies $\theta_1^- \geq -(\mathscr{T} - 2L)/2C$, equality on the blue line and $\theta_2^- \leq -\mathscr{T}/2C$, equality on the green line. The point $(\theta_1^-, \theta_2^-)$ has to belong to set $M$

Equations (10.60), (10.61) give

$$\mathscr{T} = C[\theta_1] + L + C[\theta_2], \tag{10.83}$$

$$C\frac{[\theta_2]}{T_0} - C\frac{[\theta_1]}{T_0} - \frac{L}{T_0} = -\lambda\left(\theta_2^- - \theta_1^- + \frac{[\theta_2]}{2} - \frac{[\theta_1]}{2}\right). \tag{10.84}$$

From (10.83), we get

$$[\theta_2] = \frac{1}{C}(\mathscr{T} - C[\theta_1] - L)\ ; \tag{10.85}$$

substituting (10.85) into (10.84), we obtain

$$\theta_1^+ = \frac{2C}{2C + \lambda T_0}\theta_1^- + \frac{\lambda T_0}{2C + \lambda T_0}\theta_2^- + \frac{\mathscr{T} - L}{2C} - \frac{L}{2C + \lambda T_0},$$

$$\theta_2^+ = \frac{2C}{2C + \lambda T_0}\theta_2^- + \frac{\lambda T_0}{2C + \lambda T_0}\theta_1^- + \frac{\mathscr{T} - L}{2C} + \frac{L}{2C + \lambda T_0}. \tag{10.86}$$

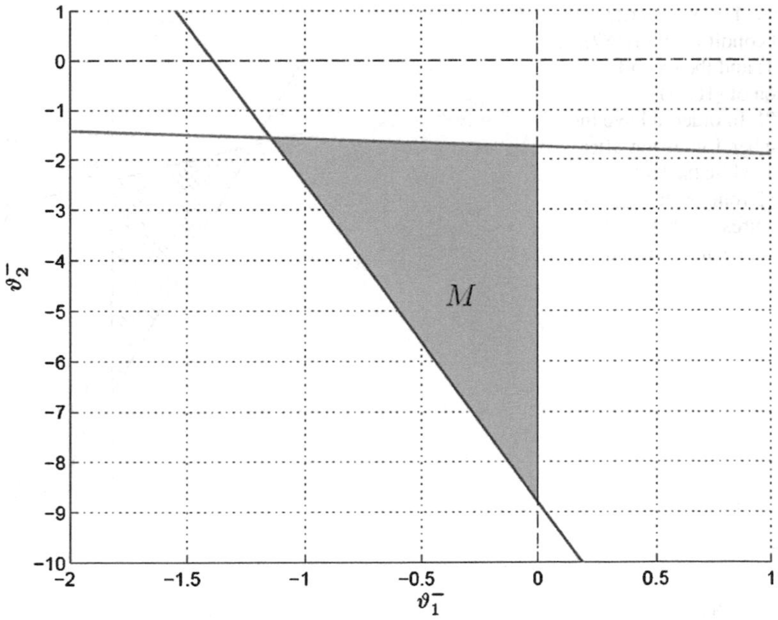


**Fig. 10.5** The case $\lambda = 1000\ \mathrm{J/K^2}$ and $\mathcal{T} = 10\,L$, with $L = 3.33 \times 10^5$ J, $C = 10^6$ J/K. The inequalities (10.81) have solutions if point $(\theta_1^-, \theta_2^-)$ belongs to set $M$ which is defined by the blue and green lines of (10.81) and $\theta_1^- \le 0$, $\theta_2^- \le 0$

Conditions (10.76) are satisfied if and only if

$$\frac{2C}{2C+\lambda T_0}\theta_1^- + \frac{\lambda T_0}{2C+\lambda T_0}\theta_2^- + \frac{\mathcal{T}-L}{2C} - \frac{L}{2C+\lambda T_0} \ge 0,$$
$$\frac{2C}{2C+\lambda T_0}\theta_2^- + \frac{\lambda T_0}{2C+\lambda T_0}\theta_1^- + \frac{\mathcal{T}-L}{2C} + \frac{L}{2C+\lambda T_0} \le 0. \qquad (10.87)$$

Because $\theta_1^- \le 0$, $\theta_2^- \le 0$, to satisfy $(10.87)_1$ a necessary condition is

$$\frac{\mathcal{T}-L}{2C} \ge \frac{L}{2C+\lambda T_0}.$$

Thus the dissipation has to be large in order to melt one of the pieces of ice.

If $\lambda < 2C/T_0$, then system (10.87) has solutions $(\theta_1^-, \theta_2^-)$ such that $\theta_1^- \le 0$, $\theta_2^- \le 0$, if

$$\frac{\mathcal{T}-L}{2C} \ge \frac{L}{2C-\lambda T_0} > \frac{L}{2C+\lambda T_0},$$

see Figs. 10.6 and 10.7. This relationship is valid when $\lambda$ is negligible. If $\lambda$ is small, only the mechanical effect warms the balls whereas the conduction has a negligible effect. We have

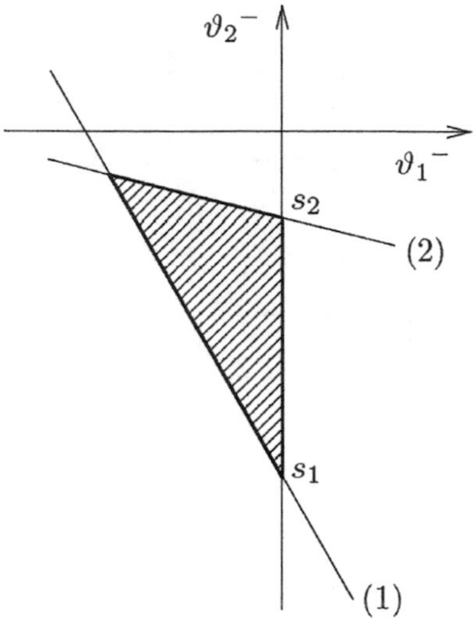


**Fig. 10.6** For $\lambda < C/(2T_0)$, the first condition of (10.87), curve (1) and the second condition of (10.87), curve (2). In order to have the ball number 1 unfrozen after collision while the ball number 2 remains frozen, the temperatures $\theta_1^-$ and $\theta_2^-$ have to be in the hatchered triangle

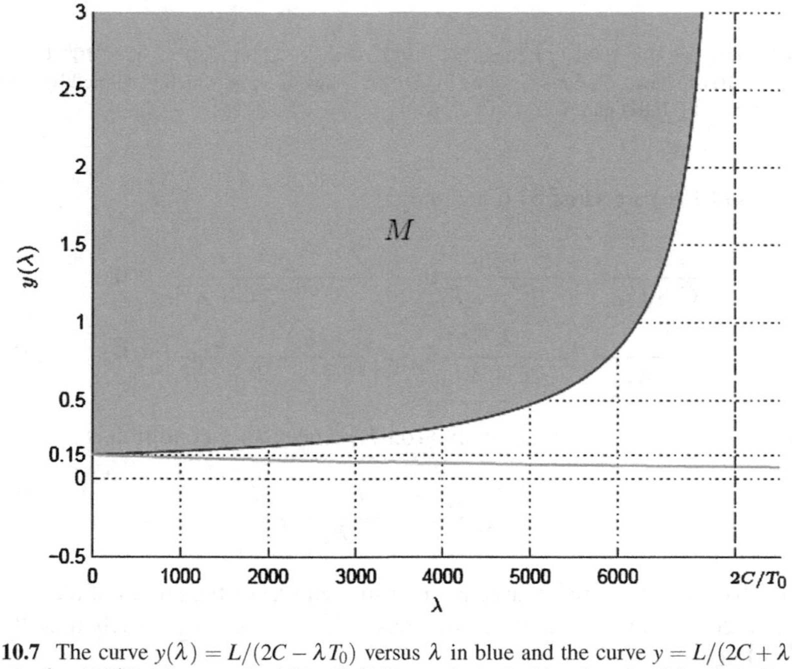


**Fig. 10.7** The curve $y(\lambda) = L/(2C - \lambda T_0)$ versus $\lambda$ in blue and the curve $y = L/(2C + \lambda T_0)$ in green for $\lambda < 2C/T_0$. The quantity $(\mathscr{T} - L)/2C$ has to be in the blue domain $M$ for the ball number 1 to melt. Thus the collision has to be violent enough, i.e., $\mathscr{T}$ large enough for the ball number 1 to melt

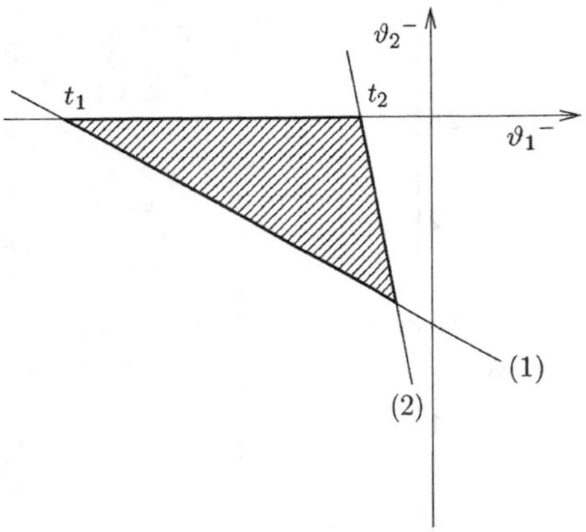


**Fig. 10.8** For $\lambda > C/(2T_0)$, the first condition (10.87), curve (1) and the second condition of (10.87), curve (2). In order to have the ball number 1 unfrozen after collision while the ball number 2 remains frozen, the temperatures $\theta_1^-$ and $\theta_2^-$ have to be in the hatchered triangles

$$\theta_2^- \leq \theta_1^- \leq 0, \text{ and } \theta_2^+ \leq 0 \leq \theta_1^+.$$

The warmest piece of ice melts in the collision and the coldest remains frozen.

If $\lambda > 2C/T_0$, then system (10.87) has solutions such that $\theta_1^- \leq 0$, $\theta_2^- \leq 0$ if

$$\frac{\mathcal{T} - L}{2C} \geq -\frac{L}{2C - \lambda T_0} > \frac{L}{2C + \lambda T_0},$$

see Figs. 10.8 and 10.9. If $C$ is small, i.e., if the heat capacity is negligible, it is difficult for the system to store energy (the only possibility to store energy is with change phase). Because we have assumed the system to be adiabatic, the heat has to remain in the system and very large temperature variations occur: the effect of conduction is added to the mechanical effect and it increases the temperature of the coldest ball. Therefore, the coldest ball before collision becomes the warmest and *vice versa*. We have

$$\theta_1^- \leq \theta_2^- \leq 0, \text{ and } \theta_2^+ \leq 0 \leq \theta_1^+.$$

In the extreme situation where $\lambda = \infty$, formulas (10.86) show that

$$\theta_1^+ = \theta_2^-, \; \theta_2^+ = \theta_1^-.$$

When the thermal dissipation is very large the temperatures exchange. Let us note the same result holds for the velocities with a quadratic mechanical pseudopotential of dissipation (10.4). When the mechanical dissipation parameter $k = \infty$, formulas (10.8) show that the velocities exchange

$$U_1^+ = U_2^-, \; U_2^+ = U_1^-,$$

in case $m_1 = m_2$.

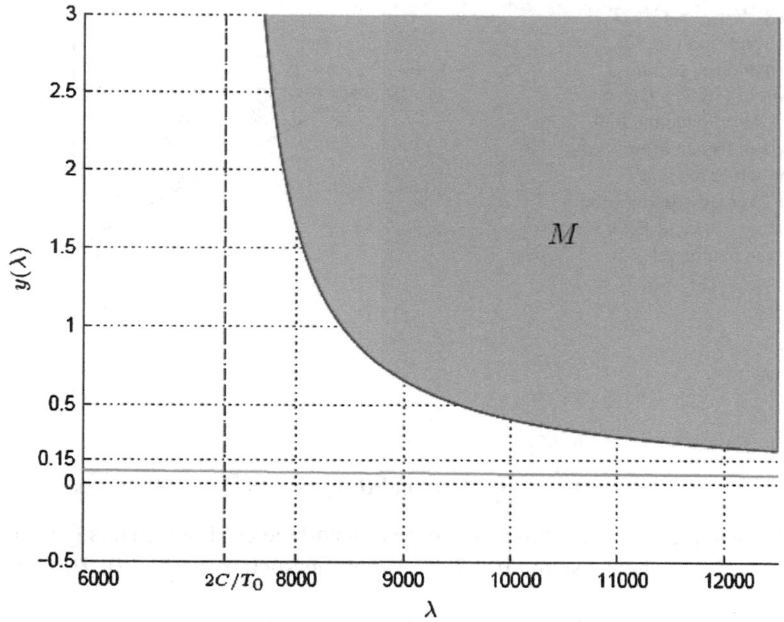


**Fig. 10.9** The curve $y(\lambda) = -L/(2C - \lambda T_0)$ versus $\lambda$ in red and the curve $y = L/(2C + \lambda T_0)$ in green for $\lambda > 2C/T_0$. The quantity $(\mathscr{T} - L)/2C$ has to be in the blue domain $M$ for the ball number 1 to melt. Thus the collision has to be violent enough, i.e., $\mathscr{T}$ large enough for the ball number 1 to melt

In the extreme situation where the heat capacity is zero, $C = 0$, the system cannot store energy except by changing phase. The energy $L$ which is needed to melt the piece of ice has to be equal to the dissipated work $\mathscr{T}$ and the discontinuities of temperature are opposite due to (10.83) extended by continuity.

*Remark 10.12.* The assumption $\lambda > 2C/T_0$ is not often realistic because $\lambda$ is proportional to the contact surface of the two pieces of ice and $C$ is proportional to the volume of the pieces of ice. To have $\lambda > 2C/T_0$, the contact surface has to be large compared to the volume. For instance the two pieces of ice are planes with small thickness. But in this case the contact surface with the atmosphere is also large in contradiction with the adiabatic assumption in collision. Thus the assumption $\lambda > 2C/T_0$ is not coherent when assuming adiabatic collisions, i.e., collisions without heat exchange with the exterior.

If $\lambda = 2C/T_0$, then, from (10.84), we have

$$\left(\frac{C}{T_0} + \frac{\lambda}{2}\right)(\theta_2^+ - \theta_1^+) = \frac{L}{T_0} + \left(\frac{C}{T_0} - \frac{\lambda}{2}\right)(\theta_2^- - \theta_1^-) = \frac{L}{T_0};$$

this entails

$$\theta_2^+ > \theta_1^+,$$

which forbids to have $\theta_1^+ \geq 0 \geq \theta_2^+$. In this situation, the effect of conduction and heat storage have opposite effects and cancel each other.

### 10.5.6 *Identical Balls at Different Temperatures Before the Collision: Collision of a Droplet of Water with a Piece of Ice*

We assume that $\lambda$ is small with respect to $C/T_0$ and that there is no dissipation with respect to $[\beta_1]$ and $[\beta_2]$

$$\Phi_1([\beta_1]) = \Phi_2([\beta_2]) = 0.$$

A droplet of rain and a piece of ice collide. Before collision we have

$$\beta_1^- = 1,\ \beta_2^- = 0,\ \theta_1^- \geq 0,\ \theta_2^- \leq 0.$$

We look for conditions such that the droplet of rain freezes and the ice remains frozen

$$\beta_1^+ = 0,\ \beta_2^+ = 0. \tag{10.88}$$

Thus from (10.74), we have

$$\theta_1^+ \leq 0,\ \theta_2^+ \leq 0. \tag{10.89}$$

We assume that the collision is adiabatic and to be coherent that $\lambda$ is small with respect to $C/T_0$. Equations (10.62), (10.63) give with $C_1 = C_2 = C$, using and (10.88)

$$C[\theta_1] = \frac{1}{2}\left(\mathcal{T} + 2L + \lambda T_0\left(\theta_2^- - \theta_1^-\right)\right),$$

$$C[\theta_2] = \frac{1}{2}\left(\mathcal{T} - \lambda T_0\left(\theta_2^- - \theta_1^-\right)\right).$$

By means of (10.79), (10.80), conditions (10.89) are satisfied if and only if

$$(2C - \lambda T_0)\,\theta_1^- + \lambda T_0 \theta_2^- + \mathcal{T} + 2L \leq 0,$$
$$\lambda T_0 \theta_1^- + (2C - \lambda T_0)\,\theta_2^- + \mathcal{T} \leq 0, \tag{10.90}$$

with $\theta_1^- \geq 0$, $\theta_2^- \leq 0$. Because of the hypothesis $2C - \lambda T_0 > 0$, it is always possible to satisfy conditions (10.90) by having $\theta_2^-$ negative enough, i.e., by having the ice piece cold enough. The maximum value of $\theta_2^-$ is given by $\theta_1^+ = 0$. It is

$$\theta_2^- = -\frac{\mathcal{T}+2L}{\lambda T_0} - \frac{2C-\lambda T_0}{\lambda T_0},$$

or

$$\theta_2^- = -\frac{\mathcal{T}+2L}{\lambda T_0} - \frac{2C}{\lambda T_0}\theta_1^-,$$

because $\lambda$ is small with respect to $C/T_0$. As it may be expected, the maximum value of $\theta_2^-$ is negative and decreasing when the dissipated $\mathcal{T}$ work is increasing and when $\theta_1^-$ is increasing.

These results have been reported in [62].

### 10.5.7 A Droplet of Rain Falls on a Frozen Ground: Does It Give Black Ice?

Let us consider a droplet of rain falling on a frozen ground and keep the assumption that the two solid colliding have uniform temperatures. Because the heat capacities of the media are proportional to their volumes, it is wise to have $C_2 = \infty$ (the soil is solid number 2). It results from (10.63) that indeed the temperature of the ground is not modified by the collision of the water droplet. It remains frozen

$$\theta_2^+ = \theta_2^-,\ \beta_2^+ = \beta_2^-.$$

The two remaining equations are

$$C_1[\theta_1] = \frac{1}{2}\left(\mathcal{T} - 2L_1[\beta_1] + \lambda T_0\left(\theta_2^- - \theta_1^- - \frac{[\theta_1]}{2}\right)\right),$$
$$\theta_1^+ \in \partial I(\beta_1^+), \tag{10.91}$$

with

$$\theta_1^- > 0,\ \beta_1^- = 1,\ \theta_2^- < 0.$$

They give

$$L_1\beta_1^+ + \partial I(\beta_1^+) \ni \frac{1}{2}(\mathcal{T} + 2L_1 + \lambda T_0(\theta_2^- - \theta_1^-)) + (C_1 + \frac{\lambda T_0}{4})\theta_1^- = \mathcal{A}.$$

The solution of the system (10.91) is:

- If $\mathcal{A} \geq L_1$, $\beta_1^+ = 1$. The droplet remains liquid and there is no black ice. This situation requires that

$$\frac{1}{2}(\mathcal{T} + \lambda T_0(\theta_2^- - \theta_1^-)) + (C_1 + \frac{\lambda T_0}{4})\theta_1^- \geq 0,$$

or

$$\frac{\mathscr{T}}{2}+\frac{\lambda T_0}{2}\theta_2^-+(C_1-\frac{\lambda T_0}{4})\theta_1^-\geq 0.$$

Assuming $C_1-\lambda T_0/4\geq 0$, the only quantity which is negative is $(\lambda T_0/4)\theta_2^-$. Thus the temperature of the frozen ground has to be not too cold to avoid black ice. In the same way, warm droplets and very dissipative collisions avoid black ice. After colliding the soil, the liquid water with temperature

$$\theta_1^+=\theta_1^-+\frac{1}{2C_1+\frac{\lambda T_0}{2}}\left(\mathscr{T}+\lambda T_0\left(\theta_2^--\theta_1^-\right)\right)\geq 0,$$

is in contact with a cold layer. The equations of Chap. 9 are to be used to describe its evolution.

- If $\mathscr{A}\leq 0$, $\beta_1^+=0$. The droplet freezes and there is black ice. This situation requires that

$$\frac{1}{2}\left(\mathscr{T}+2L_1+\lambda T_0\left(\theta_2^--\theta_1^-\right)\right)+(C_1+\frac{\lambda T_0}{4})\theta_1^-\leq 0,$$

or

$$\frac{\mathscr{T}}{2}+L_1+\frac{\lambda T_0}{2}\theta_2^-+(C_1-\frac{\lambda T_0}{4})\theta_1^-\leq 0.$$

The temperature of the ground has to be cold enough to get black ice with temperature

$$\theta_1^+=\theta_1^-+\frac{1}{2C_1+\frac{\lambda T_0}{2}}\left(\mathscr{T}+2L_1+\lambda T_0\left(\theta_2^--\theta_1^-\right)\right)\leq 0.$$

- If $0<\mathscr{A}<L_1$, $0<\beta_1^+<1$. The droplet freezes partially and its temperature is $\theta_1^+=0$. The unfrozen water volume fraction is

$$\beta_1^+=\frac{\mathscr{A}}{L_1}.$$

After colliding the soil, the evolution of the mixture of ice and water is described by the equations of Chap. 9.

# Chapter 11
# Collisions of Deformable Bodies and Phase Change

Let us consider again a piece of ice colliding with a warm soil. Work is mostly dissipated on the contact surface and in its neighbourhood. Thus the temperature discontinuity is more important in this zone and some melting may occur there and not elsewhere. In this chapter, collisions of deformable solids with volume discontinuities of velocity are investigated together with the resulting phase changes. We introduce percussions related to volume velocity discontinuities and to phase volume fraction discontinuities. We assume the contact surface of the colliding solids is conformal, i.e., it has a nonzero Lebesgue measure.

## 11.1 The Principle of Virtual Work

The virtual work of the interior percussion forces, is

$$\begin{aligned}\mathcal{T}_{int}(\Omega_1 \cup \Omega_2, \mathbf{V}, \gamma) = &-\int_{\Omega_1} \Sigma_1 : \mathsf{D}(\frac{\mathbf{V}_1^+ + \mathbf{V}_1^-}{2}) d\Omega_1 - \int_{\Omega_2} \Sigma_2 : \mathsf{D}(\frac{\mathbf{V}_2^+ + \mathbf{V}_2^-}{2}) d\Omega_2 \\ &- \int_{\partial\Omega_1 \cap \partial\Omega_2} \mathbf{R}_p \cdot \mathbf{D}_s(\frac{\mathbf{V}^+ + \mathbf{V}^-}{2}) d\Gamma \\ &- \int_{\Omega_1} B_{p1}[\gamma_1] + \mathbf{H}_{p1} \cdot \operatorname{grad}[\gamma_1] d\Omega_1 \\ &- \int_{\Omega_2} B_{p2}[\gamma_2] + \mathbf{H}_{p2} \cdot \operatorname{grad}[\gamma_2] d\Omega_2,\end{aligned}$$

where the virtual velocities are $\mathbf{V} = (\mathbf{V}_1, \mathbf{V}_2)$, $\gamma = (\gamma_1, \gamma_2)$, the $\mathsf{D}(\mathbf{V}_1)$, $\mathsf{D}(\mathbf{V}_2)$ are the usual strain rates and $\mathbf{D}_s(\mathbf{V}) = \mathbf{V}_2 - \mathbf{V}_1$ is the gap velocity. The domains occupied by the two solids at time $t$ are $\Omega_1(t) = \Omega_1$ and $\Omega_2(t) = \Omega_2$. It appears, contact percussions $\mathbf{R}_p$, percussion stresses $\Sigma$ and, percussion works $B$, percussion work flux vectors $\mathbf{H}_p$. The virtual work of the acceleration forces is

M. Frémond, *Phase Change in Mechanics*, Lecture Notes of the Unione Matematica Italiana 13, DOI 10.1007/978-3-642-24609-8_11,

$$\mathcal{T}_{acc}(\Omega_1 \cup \Omega_2, \mathbf{V}, \gamma) = \int_{\Omega_1} \rho_1 [\mathbf{U}_1] \cdot \frac{\mathbf{V}_1^+ + \mathbf{V}_1^-}{2} d\Omega_1 + \int_{\Omega_2} \rho_2 [\mathbf{U}_2] \cdot \frac{\mathbf{V}_2^+ + \mathbf{V}_2^-}{2} d\Omega_2.$$

*Remark 11.1.* The mass balance equations are

$$[\rho] = 0.$$

When the material are incompressible, the incompressibility conditions are taken into account by the pseudopotentials of dissipation introducing percussion pressures, [100, 115].

The exterior forces have the virtual work

$$\begin{aligned}
\mathcal{T}_{ext}(\Omega_1 \cup \Omega_2, \mathbf{V}, \gamma) = & \int_{\Omega_1} \mathbf{F}_{p1} \cdot \frac{\mathbf{V}_1^+ + \mathbf{V}_1^-}{2} d\Omega_1 + \int_{\Omega_2} \mathbf{F}_{p2} \cdot \frac{\mathbf{V}_2^+ + \mathbf{V}_2^-}{2} d\Omega_2 \\
& + \int_{\partial\Omega_1} \mathbf{T}_{p1} \cdot \frac{\mathbf{V}_1^+ + \mathbf{V}_1^-}{2} d\Gamma_1 + \int_{\partial\Omega_2} \mathbf{T}_{p2} \cdot \frac{\mathbf{V}_2^+ + \mathbf{V}_2^-}{2} d\Gamma_2 \\
& + \int_{\Omega_1} A_1 [\gamma_1] d\Omega_1 + \int_{\Omega_2} A_2 [\gamma_2] d\Omega_2 \\
& + \int_{\partial\Omega_1} a_1 [\gamma_1] d\Gamma_1 + \int_{\partial\Omega_2} a_2 [\gamma_2] d\Gamma_2.
\end{aligned}$$

We assume that the surface exterior percussions $\mathbf{T}_p$ are applied to the whole boundary of each solid (they can be equal to zero in some parts of the boundaries). The $\mathbf{F}_p$ are the volume exterior percussions. The $A$ and $a$ are the volume and surface percussion work provided by the exterior by electrical, radiative,... actions. The equations of motion results from the principle of virtual work

$$\forall \mathbf{V}, \forall \gamma, \mathcal{T}_{acc}(\Omega_1 \cup \Omega_2, \mathbf{V}, \gamma) = \mathcal{T}_{int}(\Omega_1 \cup \Omega_2, \mathbf{V}, \gamma) + \mathcal{T}_{ext}(\Omega_1 \cup \Omega_2, \mathbf{V}, \gamma).$$

Different choices of the virtual velocities $\mathbf{V} = (\mathbf{V}_1, \mathbf{V}_2)$ and $\gamma = (\gamma_1, \gamma_2)$ give

$$\rho_1 [\mathbf{U}_1] = \operatorname{div} \Sigma_1 + \mathbf{F}_{p1}, \ -B_{p1} + \operatorname{div} \mathbf{H}_{p1} + A_1 = 0, \ in\ \Omega_1, \tag{11.1}$$

$$\rho_2 [\mathbf{U}_2] = \operatorname{div} \Sigma_2 + \mathbf{F}_{p2}, \ -B_{p2} + \operatorname{div} \mathbf{H}_{p2} + A_2 = 0, \ in\ \Omega_2, \tag{11.2}$$

$$\begin{aligned}
\Sigma_1 \mathbf{N}_1 &= \mathbf{R}_p + \mathbf{T}_{p1}, \ \mathbf{H}_{p1} \cdot \mathbf{N}_1 = a_1, \ on\ \partial\Omega_1 \cap \partial\Omega_2, \\
\Sigma_2 \mathbf{N}_2 &= -\mathbf{R}_p + \mathbf{T}_{p2}, \ \mathbf{H}_{p2} \cdot \mathbf{N}_2 = a_2, \ on\ \partial\Omega_1 \cap \partial\Omega_2, \\
\Sigma_1 \mathbf{N}_1 &= \mathbf{T}_{p1}, \ \mathbf{H}_{p1} \cdot \mathbf{N}_1 = a_1, \ on\ \partial\Omega_1 \backslash (\partial\Omega_1 \cap \partial\Omega_2), \\
\Sigma_2 \mathbf{N}_2 &= \mathbf{T}_{p2}, \ \mathbf{H}_{p2} \cdot \mathbf{N}_2 = a_2, \ on\ \partial\Omega_2 \backslash (\partial\Omega_1 \cap \partial\Omega_2).
\end{aligned}$$

## 11.2 The First Law of Thermodynamics

As in the previous chapter, we define

$$[T_2] = T_2^+ - T_2^-, \; [T_1] = T_1^+ - T_1^- \;;$$

those two quantities are analogous to the velocity $dT/dt$ in a smooth evolution, and

$$\underline{T}_2 = \frac{T_2^+ + T_2^-}{2}, \; \underline{T}_1 = \frac{T_1^+ + T_1^-}{2},$$

in $\Omega_1$ and $\Omega_2$,

$$\delta \underline{T} = \underline{T}_2 - \underline{T}_1, \; \Theta = \frac{T_2 + T_1}{2}, \; [\Theta] = \Theta^+ - \Theta^- = \frac{[T_1] + [T_2]}{2},$$

$$\underline{T}_m = \frac{\underline{T}_2 + \underline{T}_1}{2} = \frac{\Theta^+ + \Theta^-}{2} = \frac{1}{4}\left(T_2^+ + T_1^+ + T_2^- + T_1^-\right),$$

in $\partial\Omega_1 \cap \partial\Omega_2$.

Quantity $\delta \underline{T}$ at point $\mathbf{x}_1 = \mathbf{x}_2 = \mathbf{x} \in \partial\Omega_1 \cap \partial\Omega_2$ is analogous to $\mathrm{grad}\, T$ in a smooth situation. The states quantities of the solids are $E_1 = (T_1, \beta_1, \mathrm{grad}\, \beta_1, \varepsilon_1)$ and $E_2 = (T_2, \beta_2, \mathrm{grad}\, \beta_2, \varepsilon_2)$, where the deformation $\varepsilon$ are continuous in the collision. The state quantities of the contact surface are $E^{\pm} = \Theta$ on $\partial\Omega_1 \cap \partial\Omega_2$.

The evolutions of the two solids are described by

$$\Delta E_1^{\pm} = (\mathsf{D}(\frac{\mathbf{U}_1^+ + \mathbf{U}_1^-}{2}), [\beta_1], \mathrm{grad}\,[\beta_1], \mathrm{grad}\, \underline{T}_1) \; in \; \Omega_1,$$

$$\Delta E_2^{\pm} = (\mathsf{D}(\frac{\mathbf{U}_2^+ + \mathbf{U}_2^-}{2}), [\beta_2], \mathrm{grad}\,[\beta_2], \mathrm{grad}\, \underline{T}_2) \; in \; \Omega_2.$$

The evolution of the contact surface is describe by

$$\delta E^{\pm} = (\mathbf{D}_s(\frac{\mathbf{U}^+ + \mathbf{U}^-}{2}), \delta \underline{T}) \; on \; \partial\Omega_1 \cap \partial\Omega_2.$$

The free energies of the two solids are $\Psi_1(E_1)$ and $\Psi_2(E_2)$. The structure of the expression of the work of the interior percussions leads us to introduce surface free energies, $\Psi_{1s}$, $\Psi_{2s}$.

Now let us consider the thermal effects. We focus on the basic problem, the collision of two solids. Thus the equations which are considered are relative to a collision which occurs at time $t$. For the sake of simplicity, time $t$ is deleted from the formulas. Following our rules, the heat intakes are equal to the entropy intakes multiplied by the temperatures at which the entropies are received, either at temperature $T^+$ or at temperature $T^-$. Let $\mathcal{D}_1 \cup \mathcal{D}_2$ be a subdomain of $\Omega_1 \cup \Omega_2$. Let us denote $O_1 = \partial\Omega_1 \cap \partial\mathscr{D}_1$, $O_2 = \partial\Omega_2 \cap \mathscr{D}_2$, and $O = O_1 \cap O_2$. When the solids collide, we assume that each solid receives heat impulses from the other through local actions. The heat impulse received by subdomain $\mathscr{D}_1 \cup \mathscr{D}_2$ is

$$\begin{aligned}
\mathscr{Q}(\mathscr{D}_1 \cup \mathscr{D}_2) &= \int_{\mathscr{D}_1} \Sigma(T_1\mathscr{B}_1)d\Omega_1 + \int_{\mathscr{D}_2} \Sigma(T_2\mathscr{B}_2)d\Omega_2 \\
&\quad + \int_{\partial\mathscr{D}_1\setminus O} \Sigma(T_1Q_{p1})d\Gamma_1 + \int_{\partial\mathscr{D}_1} \Sigma(T_1\mathscr{B}_{s1})d\Gamma_1 \\
&\quad + \int_{\partial\mathscr{D}_2\setminus O} \Sigma(T_2Q_{p2})d\Gamma_2 + \int_{\partial\mathscr{D}_2} \Sigma(T_2\mathscr{B}_{s2})d\Gamma_2 \\
&= \int_{\mathscr{D}_1} \Sigma(T_1\mathscr{B}_1)d\Omega_1 + \int_{\mathscr{D}_2} \Sigma(T_2\mathscr{B}_2)d\Omega_2 \\
&\quad + \int_{\partial\mathscr{D}_1} \Sigma\left(T_1\left(Q_{p1} + \mathscr{B}_{s1}\right)\right)d\Gamma_1 + \int_{\partial\mathscr{D}_2} \Sigma\left(T_2\left(Q_{p2} + \mathscr{B}_{s2}\right)\right)d\Gamma_2 \\
&\quad - \int_O \Sigma\left(T_1Q_{p1} + T_2Q_{p2}\right)d\Gamma,
\end{aligned}$$

where $\Sigma(X) = X^+ + X^-$. The $TQ_p$'s are the heat impulses supplied by contact action from the exterior of the system to the surfaces, and the $T\mathscr{B}_s$'s are the exterior surface heat impulses, for instance, sources due to a chemical rapid reaction (the heat $T\mathscr{B}_s$ is present everywhere in the preceding expression, whereas the heat supplied by the exterior $TQ_p$ is absent from part $\partial\mathscr{D}_1 \cap \partial\mathscr{D}_2$ where there is no heat supplied from the exterior by contact). The $T\mathscr{B}$'s are the volume heat impulses received from the exterior. The functions $TQ_p$ are defined on $\partial\Omega_1 \cap \partial\Omega_2$: they are the heat supplied by one solid to another one by contact (this property can be seen by choosing one of the subdomains $\mathscr{D}$ empty).

The structure of the expression of the work of the interior forces and the structure of the expression of the received heat lead us to assume surface internal energies $e_s$ besides the volume internal energies. The energy balance of the subdomain $\mathscr{D}_1 \cup \mathscr{D}_2$ at any time $t$, is

$$\int_{\mathscr{D}_1\cup\mathscr{D}_2} [e]d\Omega + \int_{O_1} [e_{s1}]d\Gamma_1 + \int_{O_2} [e_{s2}]d\Gamma_2 = -\mathscr{T}_{int}(\mathscr{D}_1 \cup \mathscr{D}_2, \mathbf{U}, \beta) + \mathscr{Q}(\mathscr{D}_1 \cup \mathscr{D}_2),$$

where $\mathscr{T}_{int}\ (\mathscr{D}_1 \cup \mathscr{D}_2, \mathbf{U}, \beta)$ is the actual work of the percussions interior to the subdomain $\mathscr{D}_1 \cup \mathscr{D}_2$. We get

$$\begin{aligned}
\int_{\mathscr{D}_1\cup\mathscr{D}_2} [e]d\Omega + \int_{O_1} [e_{s1}]d\Gamma_1 + \int_{O_2} [e_{s2}]d\Gamma_2 &= \int_O \mathbf{R}_p \cdot \mathbf{D}_s\Big(\frac{\mathbf{U}^+ + \mathbf{U}^-}{2}\Big)d\Gamma \\
&\quad + \int_{\mathscr{D}_1} \Sigma_1 : \mathsf{D}\Big(\frac{\mathbf{U}_1^+ + \mathbf{U}_1^-}{2}\Big) \\
&\quad + B_{p1}\left[\beta_1\right] + \mathbf{H}_{p1} \cdot \mathrm{grad}\left[\beta_1\right] d\Omega_1
\end{aligned}$$

$$+\int_{\mathcal{D}_2} \Sigma_2 : \mathsf{D}(\frac{\mathbf{U}_2^+ + \mathbf{U}_2^-}{2})$$
$$+B_{p2}[\beta_2] + \mathbf{H}_{p2} \cdot \operatorname{grad}[\beta_2]\, d\Omega_2$$
$$+\mathcal{Q}(\mathcal{D}_1 \cup \mathcal{D}_2).$$

The energy balance inside the solids is

$$\begin{aligned}\int_{\mathcal{D}_1\cup\mathcal{D}_2} [e] d\Omega &= \int_{\mathcal{D}_1} \Sigma_1 : \mathsf{D}(\frac{\mathbf{U}_1^+ + \mathbf{U}_1^-}{2}) + B_{p1}[\beta_1] + \mathbf{H}_{p1} \cdot \operatorname{grad}[\beta_1]\, d\Omega_1 \\ &+ \int_{\mathcal{D}_2} \Sigma_2 : \mathsf{D}(\frac{\mathbf{U}_2^+ + \mathbf{U}_2^-}{2}) + B_{p2}[\beta_2] + \mathbf{H}_{p2} \cdot \operatorname{grad}[\beta_2]\, d\Omega_2 \\ &- \int_{\partial\mathcal{D}_1} \Sigma(T_1 \mathbf{Q}_{p1} \cdot \mathbf{N}_1) d\Gamma_1 - \int_{\partial\mathcal{D}_2} \Sigma(T_2 \mathbf{Q}_{p2} \cdot \mathbf{N}_2) d\Gamma_2 \\ &+ \int_{\mathcal{D}_1} \Sigma(T_1 \mathcal{B}_1) d\Omega_1 + \int_{\mathcal{D}_2} \Sigma(T_2 \mathcal{B}_2) d\Omega_2,\end{aligned}$$

where the subdomains $\mathcal{D}_1$ and $\mathcal{D}_2$ are interior to each solid and where $T_1\mathbf{Q}_{p1}$, $T_2\mathbf{Q}_{p2}$ are the heat flux vectors in $\Omega_1$ and in $\Omega_2$. With this relationship, we get

$$\begin{aligned}\int_{O_1} [e_{s1}] d\Gamma_1 + \int_{O_2} [e_{s2}] d\Gamma_2 &= \int_O \mathbf{R}_p \cdot \mathbf{D}_s(\frac{\mathbf{U}^+ + \mathbf{U}^-}{2}) d\Gamma \\ &+ \int_{\partial\mathcal{D}_1} \Sigma\left\{T_1\left(\mathbf{Q}_{p1} \cdot \mathbf{N}_1 + Q_{p1} + \mathcal{B}_{s1}\right)\right\} d\Gamma_1 \\ &+ \int_{\partial\mathcal{D}_2} \Sigma\left\{T_2\left(\mathbf{Q}_{p2} \cdot \mathbf{N}_2 + Q_{p2} + \mathcal{B}_{s2}\right)\right\} d\Gamma_2 \\ &- \int_O \Sigma\left(T_1 Q_{p1} + T_2 Q_{p2}\right) d\Gamma.\end{aligned}$$

This results in the energy balance in $\Omega_1$

$$\begin{aligned}[e_1] + \operatorname{div}\Sigma\left(T_1 \mathbf{Q}_{p1}\right) &= \Sigma_1 : \mathsf{D}(\frac{\mathbf{U}_1^+ + \mathbf{U}_1^-}{2}) + B_{p1}[\beta_1] + \mathbf{H}_{p1} \cdot \operatorname{grad}[\beta_1] + \Sigma(T_1 \mathcal{B}_1) \\ &= \Sigma_1 : \mathsf{D}(\frac{\mathbf{U}_1^+ + \mathbf{U}_1^-}{2}) + B_{p1}[\beta_1] + \mathbf{H}_{p1} \cdot \operatorname{grad}[\beta_1] + \underline{T}_1 \Sigma(\mathcal{B}_1) \\ &+ [T_1] \Delta(\mathcal{B}_1),\end{aligned}$$

or

$$\begin{aligned}&[e_1] + \operatorname{div}\left(\underline{T}_1\left(\Sigma(\mathbf{Q}_{p1})\right)\right) + \operatorname{div}([T_1] \Delta(\mathbf{Q}_{p1})) \\ &= [e_1] + \underline{T}_1 \operatorname{div}(\Sigma(\mathbf{Q}_{p1})) + \operatorname{grad}\underline{T}_1 \cdot \Sigma(\mathbf{Q}_{p1}) + [T_1] \operatorname{div}\Delta(\mathbf{Q}_{p1}) \\ &+ \operatorname{grad}[T_1] \cdot \Delta(\mathbf{Q}_{p1})\end{aligned}$$

$$= \Sigma_1 : \mathrm{D}(\frac{\mathbf{U}_1^+ + \mathbf{U}_1^-}{2}) + B_{p1}[\beta_1] + \mathbf{H}_{p1} \cdot \mathrm{grad}[\beta_1] + \underline{T}_1 \Sigma(\mathscr{B}_1) + [T_1] \Delta(\mathscr{B}_1),$$

where we denote as usual $\Delta(X) = (X^+ - X^-)/2$, $\Sigma(X) = X^+ + X^-$ and we use the relationship $\Sigma(XY) = \Sigma(X)\underline{Y} + \Delta(X)[Y]$, with $\underline{Y} = (Y^+ + Y^-)/2$ and $[Y] = Y^+ - Y^-$.

We choose the volume free energies as

$$\Psi_1 = -C_1 T_1 \ln T_1 + \check{\Psi}_1,$$

with

$$\check{\Psi}_1(T_1, \beta_1, \mathrm{grad}\,\beta_1) = -\beta_1 \frac{L_1}{T_0}(T_1 - T_0) + \frac{k_1}{2}(\mathrm{grad}\,\beta_1)^2 + I(\beta_1).$$

We have

$$\begin{aligned}[\check{\Psi}_1] &= \check{\Psi}_1(T_1^+, \beta_1^+, \mathrm{grad}\,\beta_1^+) - \check{\Psi}_1(T_1^-, \beta_1^-, \mathrm{grad}\,\beta_1^-) \\ &= \check{\Psi}_1(T_1^+, \beta_1^+, \mathrm{grad}\,\beta_1^+) - \check{\Psi}_1(T_1^+, \beta_1^-, \mathrm{grad}\,\beta_1^-) \\ &\quad + \check{\Psi}_1(T_1^+, \beta_1^-, \mathrm{grad}\,\beta_1^-) - \check{\Psi}_1(T_1^-, \beta_1^-, \mathrm{grad}\,\beta_1^-),\end{aligned}$$

and

$$\check{\Psi}_1(T_1^+, \beta_1^-) - \check{\Psi}_1(T_1^-, \beta_1^-) = -\beta_1^- \frac{L_1}{T_0}[T_1].$$

Thus

$$\begin{aligned}[\Psi_1] &= [-C_1 T_1 \ln T_1] + \check{\Psi}_1(T_1^+, \beta_1^+) - \check{\Psi}_1(T_1^+, \beta_1^-) - \beta_1^- \frac{L_1}{T_0}[T_1] \\ &\le [-C_1 T_1 \ln T_1] - \beta_1^- \frac{L_1}{T_0}[T_1] + B_{p1}^{fe}[\beta_1] + \mathbf{H}_{p1}^{fe} \cdot \mathrm{grad}[\beta_1],\end{aligned}$$

with

$$\left(B_{p1}^{fe}, \mathbf{H}_{p1}^{fe}\right) \in \partial \check{\Psi}_1(T_1^+, \beta_1^+, \mathrm{grad}\,\beta_1^+) = \left(-\frac{L_1}{T_0}(T_1^+ - T_0) + \partial I(\beta_1^+), k_1\,\mathrm{grad}\,\beta_1^+\right), \tag{11.3}$$

where the subdifferential is computed with respect to $\beta_1^+$ and $\mathrm{grad}\,\beta_1^+$. Moreover

$$\begin{aligned}[e_1] = [\Psi_1 + T_1 s_1] &= [\Psi_1] + \underline{s}_1[T_1] + \underline{T}_1[s_1] \\ &\le -S_1[T_1] + \underline{T}_1[s_1] + B_{p1}^{fe}[\beta_1] + \mathbf{H}_{p1}^{fe} \cdot \mathrm{grad}[\beta_1],\end{aligned}$$

by denoting

$$[-C_1 T_1 \ln T_1] - \beta_1^- \frac{L_1}{T_0}[T_1] + \underline{s}_1[T_1] = -S_1[T_1].$$

Thus the energy balance in $\Omega_1$ is

$$\underline{T}_1\,[s_1] + \underline{T}_1\,\mathrm{div}(\Sigma(\mathbf{Q}_{p1})) \geq \Sigma_1 : \mathsf{D}\left(\frac{\mathbf{U}_1^+ + \mathbf{U}_1^-}{2}\right)$$

$$+ (B_{p1} - B_{p1}^{fe})\,[\beta_1] + (\mathbf{H}_{p1} - \mathbf{H}_{p1}^{fe}) \cdot \mathrm{grad}\,[\beta_1] + \underline{T}_1 \Sigma(\mathscr{B}_1)$$

$$+ [T_1]\,(\Delta(\mathscr{B}_1) + S_1 - \mathrm{div}\,\Delta(\mathbf{Q}_{p1})) - \mathrm{grad}\,[T_1] \cdot \Delta(\mathbf{Q}_{p1}) - \mathrm{grad}\,\underline{T}_1 \cdot \Sigma(\mathbf{Q}_{p1}).$$

When the temperature discontinuity is zero, $S_1$ is zero since

$$\frac{\partial \Psi}{\partial T} + s = 0.$$

This relationship is extended by assuming that there is no dissipation with respect to $[T_1]$ and grad $[T_1]$. Then

$$\Delta(\mathscr{B}_1) + S_1 - \mathrm{div}\,\Delta(\mathbf{Q}_{p1}) = 0, \;\; \Delta(\mathbf{Q}_{p1}) = 0,$$

which give

$$\Delta(\mathscr{B}_1) + S_1 = 0, \; \mathbf{Q}_{p1}^+ - \mathbf{Q}_{p1}^- = \mathbf{Q}_{p1}. \tag{11.4}$$

The energy balance becomes

$$[s_1] + 2\,\mathrm{div}\,\mathbf{Q}_{p1} \geq \frac{1}{\underline{T}_1}\left\{\Sigma_1 : \mathsf{D}\left(\frac{\mathbf{U}_1^+ + \mathbf{U}_1^-}{2}\right)\right.$$

$$\left. + (B_{p1} - B_{p1}^{fe})\,[\beta_1] + (\mathbf{H}_{p1} - \mathbf{H}_{p1}^{fe}) \cdot \mathrm{grad}\,[\beta_1]\right\}$$

$$- \frac{2}{\underline{T}_1}\,\mathrm{grad}\,\underline{T}_1 \cdot \mathbf{Q}_{p1} + \Sigma(\mathscr{B}_1), \tag{11.5}$$

In the same way, the energy balance in $\Omega_2$ is

$$[s_2] + 2\,\mathrm{div}\,\mathbf{Q}_{p2} \geq \frac{1}{\underline{T}_2}\left\{\Sigma_2 : \mathsf{D}\left(\frac{\mathbf{U}_2^+ + \mathbf{U}_2^-}{2}\right)\right.$$

$$\left. + (B_{p2} - B_{p2}^{fe})\,[\beta_2] + (\mathbf{H}_{p2} - \mathbf{H}_{p2}^{fe}) \cdot \mathrm{grad}\,[\beta_2]\right\}$$

$$- \frac{2}{\underline{T}_2}\,\mathrm{grad}\,\underline{T}_2 \cdot \mathbf{Q}_{p2} + \Sigma(\mathscr{B}_2), \tag{11.6}$$

with

$$\Delta(\mathscr{B}_2) + S_2 = 0, \;\; \mathbf{Q}_{p2}^+ = \mathbf{Q}_{p2}^- = \mathbf{Q}_{p2}. \tag{11.7}$$

*Remark 11.2.* In this section the discontinuity of the free energy

$$\begin{aligned}[\Psi] &= \Psi(T^+,\beta^+,\operatorname{grad}\beta^+) - \Psi(T^-,\beta^-,\operatorname{grad}\beta^-) \\ &= \Psi(T^+,\beta^+,\operatorname{grad}\beta^+) - \Psi(T^+,\beta^-,\operatorname{grad}\beta^-) \\ &\quad + \Psi(T^+,\beta^-,\operatorname{grad}\beta^-) - \Psi(T^-,\beta^-,\operatorname{grad}\beta^-),\end{aligned}$$

is as in Sect. 3.6, split into two terms $\Psi(T^+,\beta^+,\operatorname{grad}\beta^+) - \Psi(T^+,\beta^-,\operatorname{grad}\beta^-)$ and $\Psi(T^+,\beta^-,\operatorname{grad}\beta^-) - \Psi(T^-,\beta^-,\operatorname{grad}\beta^-)$. The first one is treated as in Sect. 3.6 by using the convexity of function $(\beta,\operatorname{grad}\beta) \to \Psi(T^+,\beta,\operatorname{grad}\beta)$. The second one is replaced by

$$\left(\frac{[-C_1 T_1 \ln T_1]}{[T_1]} - \beta_1^- \frac{L_1}{T_0}\right)[T_1],$$

which is different from

$$\partial\Psi_T(T_1^-,\beta^-,\operatorname{grad}\beta^-)[T_1] = \left(-C_1(1+\ln T_1^-) - \beta_1^- \frac{L_1}{T_0}\right)[T_1].$$

This difference appears in (11.4) and (3.42) where the two $S$ which are slightly different, give $\Delta(\mathscr{B})$ or the allocation of the external entropy production $\mathscr{B}^+ + \mathscr{B}^-$ between temperatures $T^+$ and $T^-$. We think that the two way of doing are convenient: to use either the differential quotient or the derivative of $-C_1 T_1 \ln T_1$. This remark give a hint to the way to deal with non convex free energies: in case $(\beta,\operatorname{grad}\beta) \to \Psi(T^+,\beta,\operatorname{grad}\beta)$ is not convex, the discontinuities may be replaced by differential quotients. But it is a long and complicated task.

The energy balance on $\partial\Omega_1$ is

$$\begin{aligned}[e_{s1}] &= \underline{T}_1 2\mathbf{Q}_{p1}\cdot\mathbf{N}_1 + \Sigma\left\{T_1(Q_{p1}+\mathscr{B}_{s1})\right\} \\ &= \underline{T}_1 2\mathbf{Q}_{p1}\cdot\mathbf{N}_1 + \underline{T}_1\Sigma(Q_{p1}+\mathscr{B}_{s1}) \\ &\quad + \Delta(Q_{p1}+\mathscr{B}_{s1})[T_1].\end{aligned} \tag{11.8}$$

Moreover due to the relationship $e_{s1} = \Psi_{s1} + s_{s1}T_1$

$$[e_{s1}] = [\Psi_{s1}] + \underline{s}_{s1}[T_1] + \underline{T}_1[s_{s1}] = -S_{s1}[T_1] + \underline{T}_1[s_{s1}],$$

by denoting

$$[\Psi_{s1}] + \underline{s}_{s1}[T_1] = -S_{s1}[T_1].$$

Then the energy balance is

$$\underline{T}_1[s_{s1}] = \underline{T}_1 2\mathbf{Q}_{p1}\cdot\mathbf{N}_1 + \underline{T}_1\Sigma(Q_{p1}+\mathscr{B}_{s1}) + \left\{\Delta(Q_{p1}+\mathscr{B}_{s1}) + S_{s1}\right\}[T_1].$$

Assuming as usual no dissipation with respect to $[T_1]$

$$\Delta\left(Q_{p1}+\mathscr{B}_{s1}\right)+S_{s1}=0, \tag{11.9}$$

the energy balance becomes

$$\underline{T}_1\left[s_{s1}\right]=\underline{T}_1 2\mathbf{Q}_{p1}\cdot\mathbf{N}_1+\underline{T}_1\Sigma\left(Q_{p1}+\mathscr{B}_{s1}\right). \tag{11.10}$$

The different heat sources which produce the evolution of the internal energy are thermal. There is no mechanical source in (11.10).

The energy balance on $\partial\Omega_2$ is

$$[e_{s2}]=\underline{T}_2 2\mathbf{Q}_{p2}\cdot\mathbf{N}_2+\Sigma\left\{T_2(Q_{p2}+\mathscr{B}_{s2})\right\}. \tag{11.11}$$

Then

$$\underline{T}_2[s_{s2}]=\underline{T}_2 2\mathbf{Q}_{p2}\cdot\mathbf{N}_2+\underline{T}_2\Sigma\left(Q_{p2}+\mathscr{B}_{s2}\right), \tag{11.12}$$

with no dissipation with respect to $[T_2]$

$$\Delta\left(Q_{p2}+\mathscr{B}_{s2}\right)+S_{s2}=0, \tag{11.13}$$

where $S_{s2}$ is defined by

$$[\Psi_{s2}]+\underline{s}_{s2}[T_2]=-S_{s2}\left[T_2\right].$$

The energy balance on $\partial\Omega_1\cap\partial\Omega_2$ is

$$\begin{aligned}[e_{s1}]+[e_{s2}]=&\ \mathbf{R}_p\cdot\mathbf{D}_s\left(\frac{\mathbf{U}^++\mathbf{U}^-}{2}\right)\\&+\Sigma(T_1\mathscr{B}_{s1})+\Sigma\{T_2\mathscr{B}_{s2}\}+\underline{T}_1 2\mathbf{Q}_{p1}\cdot\mathbf{N}_1\\&+\underline{T}_2 2\mathbf{Q}_{p2}\cdot\mathbf{N}_2.\end{aligned} \tag{11.14}$$

We have

$$\begin{aligned}[e_{s1}]+[e_{s2}]=&\ \mathbf{R}_p\cdot\mathbf{D}_s\left(\frac{\mathbf{U}^++\mathbf{U}^-}{2}\right)\\&+[T_1]\,\Delta\left(\mathscr{B}_{s1}\right)+\underline{T}_1\Sigma\left(\mathscr{B}_{s1}\right)+[T_2]\,\Delta\left(\mathscr{B}_{s2}\right)\\&+\underline{T}_2\Sigma\left(\mathscr{B}_{s2}\right)\\&+\underline{T}_1 2\mathbf{Q}_{p1}\cdot\mathbf{N}_1+\underline{T}_2 2\mathbf{Q}_{p2}\cdot\mathbf{N}_2.\end{aligned}$$

By subtracting relationships (11.8) and (11.11), it results

$$0 = \mathbf{R}_p \cdot \mathbf{D}_s(\frac{\mathbf{U}^+ + \mathbf{U}^-}{2}) - [T_1]\Delta(Q_{p1}) - \underline{T}_1\Sigma(Q_{p1}) - [T_2]\Delta(Q_{p2}) - \underline{T}_2\Sigma(Q_{p2})$$
$$= \mathbf{R}_p \cdot \mathbf{D}_s(\frac{\mathbf{U}^+ + \mathbf{U}^-}{2}) - [T_1]\Delta(Q_{p1}) - [T_2]\Delta(Q_{p2}) - \frac{\delta \underline{T}}{2}(\Sigma(Q_{p2}) - \Sigma(Q_{p1}) - \underline{T}_m(\Sigma(Q_{p2}) + \Sigma(Q_{p1}).$$

Assuming as usual no dissipation with respect to the $[T]$'s, we let

$$\Delta(Q_{p1}) = 0,\ \Delta(Q_{p2}) = 0. \tag{11.15}$$

Then $Q^+_{p1} = Q^-_{p1} = Q_{p1}$, $Q^+_{p2} = Q^-_{p2} = Q_{p2}$ and

$$0 = \mathbf{R}_p \cdot \mathbf{D}_s(\frac{\mathbf{U}^+ + \mathbf{U}^-}{2}) - \delta \underline{T}(Q_{p2} - Q_{p1}) - 2\underline{T}_m(Q_{p2} + Q_{p1}). \tag{11.16}$$

*Remark 11.3.* On $\partial\Omega_1 \cap \partial\Omega_2$, the relationships (11.9) and (11.13) become

$$\Delta(\mathscr{B}_{s1}) + S_{s1} = 0,\ \Delta(\mathscr{B}_{s2}) + S_{s2} = 0,$$

due to (11.15).

## 11.3 The Second Law of Thermodynamics

Surface entropies $s_s$ are introduced. The second law of thermodynamics is

$$\int_{\mathscr{D}_1 \cup \mathscr{D}_2} [s] d\Omega + \int_{O_1} [s_{s1}] d\Gamma_1 + \int_{O_2} [s_{s2}] d\Gamma_2 \geq - \int_{\partial\mathscr{D}_1 \setminus O_1} 2\mathbf{Q}_{p1} \cdot \mathbf{N}_1 d\Gamma_1$$
$$- \int_{\partial\mathscr{D}_2 \setminus O_2} 2\mathbf{Q}_{p2} \cdot \mathbf{N}_2 d\Gamma_2$$
$$+ \int_{\mathscr{D}_1} \Sigma(\mathscr{B}_1) d\Omega_1 + \int_{\mathscr{D}_2} \Sigma(\mathscr{B}_2) d\Omega_2$$
$$+ \int_{O_1 \setminus O} \Sigma(Q_{p1}) + \Sigma(\mathscr{B}_{s1}) d\Gamma_1$$
$$+ \int_{O_2 \setminus O} \Sigma(Q_{p2}) + \Sigma(\mathscr{B}_{s2}) d\Gamma_2$$
$$+ \int_{O} \Sigma(\mathscr{B}_{s1}) d\Gamma_1 + \int_{O} \Sigma(\mathscr{B}_{s2}) d\Gamma_2.$$

By integrating by parts

$$\int_{\mathscr{D}_1} [s_1] + \mathrm{div}(2\mathbf{Q}_{p1}) - \Sigma(\mathscr{B}_{s1}) d\Omega_1 + \int_{\mathscr{D}_2} [s_2] + \mathrm{div}(2\mathbf{Q}_{p2}) - \Sigma(\mathscr{B}_{s2}) d\Omega_2$$
$$+ \int_{O_1} [s_{s1}] d\Gamma_1 + \int_{O_2} [s_{s2}] d\Gamma_2$$
$$\geq \int_{O_1 \backslash O} 2\mathbf{Q}_{p1} \cdot \mathbf{N}_1 + \Sigma(Q_{p1}) + \Sigma(\mathscr{B}_{s1}) d\Gamma_1$$
$$+ \int_{O_2 \backslash O} 2\mathbf{Q}_{p2} \cdot \mathbf{N}_2 + \Sigma(Q_{p2}) + \Sigma(\mathscr{B}_{s2}) d\Gamma_2$$
$$+ \int_{O} 2\mathbf{Q}_{p1} \cdot \mathbf{N}_1 + \Sigma(\mathscr{B}_{s1}) d\Gamma_1 + \int_{O} 2\mathbf{Q}_{p2} \cdot \mathbf{N}_2 + \Sigma(\mathscr{B}_{s2}) d\Gamma_2,$$

then

$$\int_{\mathscr{D}_1} [s_1] + \mathrm{div}(2\mathbf{Q}_{p1}) - \Sigma(\mathscr{B}_{s1}) d\Omega_1 + \int_{\mathscr{D}_2} [s_2] + \mathrm{div}(2\mathbf{Q}_{p2}) - \Sigma(\mathscr{B}_{s2}) d\Omega_2$$
$$+ \int_{O_1} [s_{s1}] d\Gamma_1 + \int_{O_2} [s_{s2}] d\Gamma_2$$
$$\geq \int_{O_1} \left\{ 2\mathbf{Q}_{p1} \cdot \mathbf{N}_1 + \Sigma(Q_{p1}) + \Sigma(\mathscr{B}_{s1}) \right\} d\Gamma_1$$
$$+ \int_{O_2} \left\{ 2\mathbf{Q}_{p2} \cdot \mathbf{N}_2 + \Sigma(Q_{p2}) + \Sigma(\mathscr{B}_{s2}) \right\} d\Gamma_2$$
$$+ \int_{O} \{ -\Sigma(Q_{p1}) - \Sigma(Q_{p2}) \} d\Gamma.$$

The preceding inequality gives

$$[s_1] + \mathrm{div}(2\mathbf{Q}_{p1}) - \Sigma(\mathscr{B}_1) \geq 0, \ in \ \Omega_1, \tag{11.17}$$

$$[s_2] + \mathrm{div}(2\mathbf{Q}_{p2}) - \Sigma(\mathscr{B}_2) \geq 0, \ in \ \Omega_2, \tag{11.18}$$

$$[s_{s1}] - \left\{ 2\mathbf{Q}_{p1} \cdot \mathbf{N}_1 + \Sigma(Q_{p1}) + \Sigma(\mathscr{B}_{s1}) \right\} \geq 0, \ on \ \partial\Omega_1, \tag{11.19}$$

$$[s_{s2}] - \left\{ 2\mathbf{Q}_{p2} \cdot \mathbf{N}_2 + \Sigma(Q_{p2}) + \Sigma(\mathscr{B}_{s2}) \right\} \geq 0, \ on \ \partial\Omega_2, \tag{11.20}$$

$$[s_{s1}] + [s_{s2}]$$
$$- \left\{ 2\mathbf{Q}_{p1} \cdot \mathbf{N}_1 + \Sigma(\mathscr{B}_{s1}) \right\} - \left\{ 2\mathbf{Q}_{p2} \cdot \mathbf{N}_2 + \Sigma(\mathscr{B}_{s2}) \right\} \geq 0, \ on \ \partial\Omega_1 \cap \partial\Omega_2. \tag{11.21}$$

The energy balances (11.10) and (11.12) imply the inequalities (11.19) and (11.20) are satisfied (they are equalities). It results that (11.21) is equivalent to

$$\Sigma(Q_{p1}) + \Sigma(Q_{p2}) = 2(Q_{p2} + Q_{p1}) \geq 0.$$

The energy balances (11.5), (11.6) and (11.16) induce to choose the basic assumptions to establish the constitutive laws:

$$\Sigma_1 : \mathsf{D}\left(\frac{\mathbf{U}_1^+ + \mathbf{U}_1^-}{2}\right) + (B_{p1} - B_{p1}^{fe})\,[\beta_1] + (\mathbf{H}_{p1} - \mathbf{H}_{p1}^{fe}) \cdot \mathrm{grad}\,[\beta_1]$$

$$-2\,\mathrm{grad}\,\underline{T}_1 \cdot \mathbf{Q}_{p1} \geq 0, \; in \; \Omega_1, \tag{11.22}$$

$$\Sigma_2 : \mathsf{D}\left(\frac{\mathbf{U}_2^+ + \mathbf{U}_2^-}{2}\right) + (B_{p2} - B_{p2}^{fe})\,[\beta_2] + (\mathbf{H}_{p2} - \mathbf{H}_{p2}^{fe}) \cdot \mathrm{grad}\,[\beta_2]$$

$$-2\,\mathrm{grad}\,\underline{T}_2 \cdot \mathbf{Q}_{p2} \geq 0, \; in \; \Omega_2, \tag{11.23}$$

$$\mathbf{R}_p \cdot \mathbf{D}_s\left(\frac{\mathbf{U}^+ + \mathbf{U}^-}{2}\right) - \delta\underline{T}(Q_{p2} - Q_{p1}) \geq 0, \; on \; \partial\Omega_1 \cap \partial\Omega_2. \tag{11.24}$$

## 11.4 The Constitutive Laws

We assume pseudopotential of dissipation

$$\Phi_1(\Delta E_1^{\pm}, E_1^{\pm}, \chi) = \Phi_1\left(\mathsf{D}\left(\frac{\mathbf{U}_1^+ + \mathbf{U}_1^-}{2}\right), [\beta_1], \mathrm{grad}\,[\beta_1], \mathrm{grad}\,\underline{T}_1, E_1^{\pm}, \chi\right),$$

$$\Phi_2(E_2^{\pm}, \Delta E_2^{\pm}, \chi) = \Phi_2\left(\mathsf{D}\left(\frac{\mathbf{U}_2^+ + \mathbf{U}_2^-}{2}\right)\right), [\beta_2], \mathrm{grad}\,[\beta_2], \mathrm{grad}\,\underline{T}_2, E_2^{\pm}, \chi\right),$$

$$\Phi_s(\delta E^{\pm}, E^{\pm}, \chi) = \Phi_s\left(\mathbf{D}_s(\frac{\mathbf{U}^+ + \mathbf{U}^-}{2}), \delta\underline{T}, E^{\pm}, \chi\right),$$

which imply relationships (11.22)–(11.24) are satisfied. The surface pseudopotential of dissipation $\Phi_s$ takes into account the impenetrability condition

$$\mathbf{D}_s(\mathbf{U}^+) \cdot \mathbf{N}_1 \geq 0,$$

with the indicator function

$$I_+(\mathbf{D}_s(\mathbf{U}^+) \cdot \mathbf{N}_1), \tag{11.25}$$

and we have

$$\Phi_s(\delta E^{\pm}, E^{\pm}, \chi) = \Phi_s(\delta E^{\pm}, E^{\pm}, \chi) + I_+(\mathbf{D}_s(\mathbf{U}^+) \cdot \mathbf{N}_1),$$

where quantities $\chi$ involve $\mathbf{D}_s(\mathbf{U}^-)$.

The constitutive laws are

$$(\Sigma_1, B_{p1} - B_{p1}^{fe}, \mathbf{H}_{p1} - \mathbf{H}_{p1}^{fe}, -2\mathbf{Q}_{p1}) \tag{11.26}$$
$$\in \partial\Phi_1\left(\mathsf{D}\left(\frac{\mathbf{U}_1^+ + \mathbf{U}_1^-}{2}\right), [\beta_1], \mathrm{grad}\,[\beta_1], \mathrm{grad}\,\underline{T}_1, E_1^{\pm}, \chi\right),$$

$$(\Sigma_2, B_{p2} - B_{p2}^{fe}, \mathbf{H}_{p2} - \mathbf{H}_{p2}^{fe}, -2\mathbf{Q}_{p2}) \tag{11.27}$$
$$\in \partial\Phi_2\left(\mathsf{D}\left(\frac{\mathbf{U}_2^+ + \mathbf{U}_2^-}{2}\right), [\beta_2], \mathrm{grad}\,[\beta_2], \mathrm{grad}\,\underline{T}_2, E_2^{\pm}, \chi\right),$$

$$(\mathbf{R}_p, -(Q_{p2} - Q_{p1})) \in \partial\Phi_s\left(\mathbf{D}_s\left(\frac{\mathbf{U}^+ + \mathbf{U}^-}{2}\right), \delta\underline{T}, E^{\pm}, \chi\right),$$

where the subdifferential are computed with respect to the first quantities and not with respect to the $E$'s and $\chi$. For example, we choose

$$\Phi_1\left(\mathsf{D}\left(\frac{\mathbf{U}_1^+ + \mathbf{U}_1^-}{2}\right), [\beta_1], \mathrm{grad}\,[\beta_1], \mathrm{grad}\,\underline{T}_1, E_1^{\pm}, \chi\right)$$
$$= \Phi_1\left(\mathsf{D}\left(\frac{\mathbf{U}_1^+ + \mathbf{U}_1^-}{2}\right), [\beta_1], \mathrm{grad}\,[\beta_1], \mathrm{grad}\,\underline{T}_1, \underline{T}_1\right)$$
$$= 2k_M\left(\mathsf{D}\left(\frac{\mathbf{U}_1^+ + \mathbf{U}_1^-}{2}\right)\right)^2 + \frac{k}{2\underline{T}_1}(\mathrm{grad}\,\underline{T}_1)^2 + \frac{c}{2}[\beta_1]^2$$
$$+ \frac{k_m}{2}(\mathrm{grad}\,[\beta_1])^2,$$

and

$$\Phi_s\left(\mathbf{D}_s\left(\frac{\mathbf{U}^+ + \mathbf{U}^-}{2}\right), \delta\underline{T}, E^{\pm}, \chi\right) = \Phi_s\left(\mathbf{D}_s\left(\frac{\mathbf{U}^+ + \mathbf{U}^-}{2}\right), \delta\underline{T}, \underline{T}_1, \underline{T}_2, \mathbf{D}_s(\mathbf{U}^+)\right)$$
$$= k_{ls}\left(\mathbf{D}_s\left(\frac{\mathbf{U}^+ + \mathbf{U}^-}{2}\right)\right)^2 + \frac{\underline{T}_1 + \underline{T}_2}{\underline{T}_1\underline{T}_2}\frac{k_{ps}^d}{2}(\delta\underline{T})^2 + I_+(\mathbf{D}_s(\mathbf{U}^+)\cdot\mathbf{N}_1),$$

where the $k$'s parameters and $c$ denote different thermal conductivities and dissipative parameters. For the sake of simplicity, we assume $\Phi_1 \equiv \Phi_2$. The pseudopotential of dissipation have been chosen in such a way that the constitutive laws are linear besides the compulsory impenetrability condition. We think that the basic physical phenomena are to be described by these laws. It is only when dealing with particular problems that sophisticated non linear constitutive laws are to be introduced.

*Remark 11.4.* The coefficient of $(\delta\underline{T})^2$ in pseudopotential $\Phi_s$ is

$$\frac{\underline{T}_1 + \underline{T}_2}{\underline{T}_1\underline{T}_2}\frac{k_{ps}^d}{2},$$

whereas it is

$$\frac{\lambda}{4},$$

in pseudopotential $\Phi$ of Chap. 10 for collisions of rigid balls. In the small perturbation assumption where the temperature have values close to $T_0$, we have

$$\lambda = \frac{4k_{ps}^d}{T_0},$$

if the rigid body collision is understood as an approximation of the deformable body collision. The coefficient in front of $k_{ps}^d$ is useful to have rather simple boundary conditions (for instance, (11.51)).

The constitutive laws are

$$\begin{aligned}\Sigma_1 &= 2k_M \mathsf{D}\left(\mathbf{U}_1^+ + \mathbf{U}_1^-\right),\\ B_{p1} - B_{p1}^{fe} &= c\,[\beta_1]\,,\ \mathbf{H}_{p1} - \mathbf{H}_{p1}^{fe} = k_m \operatorname{grad}[\beta_1]\,,\\ -\mathbf{Q}_{p1} &= \frac{k}{\underline{T}_1}\operatorname{grad}\underline{T}_1,\ in\ \Omega_1,\end{aligned}$$

$$\begin{aligned}\Sigma_2 &= 2k_M \mathsf{D}\left(\mathbf{U}_2^+ + \mathbf{U}_2^-\right),\\ B_{p2} - B_{p2}^{fe} &= c\,[\beta_2]\,,\ \mathbf{H}_{p2} - \mathbf{H}_{p2}^{fe} = k_m \operatorname{grad}[\beta_2]\,,\\ -\mathbf{Q}_{p2} &= \frac{k}{\underline{T}_2}\operatorname{grad}\underline{T}_2,\ in\ \Omega_2,\end{aligned}$$

and

$$\begin{aligned}\mathbf{R}_p &= k_{ls}\left(\mathbf{U}_2^+ - \mathbf{U}_1^+ + \mathbf{U}_2^- - \mathbf{U}_1^-\right) + P^{dr}\mathbf{N}_1,\\ -(Q_{p2} - Q_{p1}) &= \frac{\underline{T}_1 + \underline{T}_2}{\underline{T}_1\underline{T}_2} k_{ps}^d \delta\underline{T},\ on\ \partial\Omega_1 \cap \partial\Omega_2,\end{aligned} \tag{11.28}$$

where $P^{dr}$ is the impenetrability reaction

$$P^{dr} \in \partial I_+(\mathbf{D}_s(\mathbf{U}^+)\cdot\mathbf{N}_1) = \partial I_+((\mathbf{U}_2^+ - \mathbf{U}_1^+)\cdot\mathbf{N}_1).$$

*Remark 11.5.* The pseudopotentials $\Phi_i$ does not have constraint on $[\beta_i]$, because $\beta_i^+$ is between 0 and 1 due to the definition of $B_i^{nd}$ (see relationship (11.3) which involves $\partial I(\beta_1^+)$).

Note that the constitutive laws are only dissipative as are the constitutive laws for rigid bodies collisions. This property is characteristic of collisions where even the impenetrability reactions are dissipative in contrast with many other cases where the reactions to internal constraints are workless (see Sect. 3.4.1). The following theorem ensures that the second law of thermodynamics is satisfied with the constitutive laws that have just been chosen.

**Theorem 11.1.** *If the constitutive laws are satisfied and if the temperatures are positive, then:*

- *The impenetrability internal constraint is satisfied.*
- *The second law of thermodynamics is satisfied.*

*Proof.* The proof is straightforward. □

## 11.5 Evolution in a Collision

The data of this problem are the state and the velocities before the collision, as well as the exterior actions: the percussions $\mathbf{F}_p$, $\mathbf{T}_p$, $A$, $a$, and the heat impulse sources $\Sigma(T\mathscr{B})$ in $\Omega_1$ and $\Omega_2$, $\Sigma T(Q_p)$ on $\partial\Omega_1\backslash(\partial\Omega_1\cap\partial\Omega_2)$ and $\partial\Omega_2\backslash(\partial\Omega_1\cap\partial\Omega_2)$, and $\Sigma T(\mathscr{B}_s)$ in $\partial\Omega_1$ and $\partial\Omega_2$. The unknowns are the state quantities $(T,\beta)$ and the velocities after the collision. The mechanical equations are

$$\begin{aligned}\rho_1\,[\mathbf{U}_1] &= \operatorname{div}\Sigma_1+\mathbf{F}_{p1},\\ 0 &= \operatorname{div}\mathbf{H}_{p1}-B_{p1}+A_1,\ in\ \Omega_1,\end{aligned}\tag{11.29}$$

$$\begin{aligned}\rho_2\,[\mathbf{U}_2] &= \operatorname{div}\Sigma_2+\mathbf{F}_{p2},\\ 0 &= \operatorname{div}\mathbf{H}_{p2}-B_{p2}+A_2,\ in\ \Omega_2,\end{aligned}\tag{11.30}$$

$$\begin{aligned}\Sigma_1\mathbf{N}_1 &= \mathbf{R}_p+\mathbf{T}_{p1},\ \mathbf{H}_{p1}\cdot\mathbf{N}_1=a_1,\\ \Sigma_2\mathbf{N}_2 &= -\mathbf{R}_p+\mathbf{T}_{p2},\ \mathbf{H}_{p2}\cdot\mathbf{N}_2=a_2,\ on\ \partial\Omega_1\cap\partial\Omega_2,\end{aligned}\tag{11.31}$$

$$\Sigma_1\mathbf{N}_1=\mathbf{T}_{p1},\ \mathbf{H}_{p1}\cdot\mathbf{N}_1=a_1,\ on\ \partial\Omega_1\backslash(\partial\Omega_1\cap\partial\Omega_2),\tag{11.32}$$

$$\Sigma_2\mathbf{N}_2=\mathbf{T}_{p2},\ \mathbf{H}_{p2}\cdot\mathbf{N}_2=a_2,\ on\partial\Omega_2\backslash(\partial\Omega_1\cap\partial\Omega_2).\tag{11.33}$$

The thermal equations are

$$[e_1] + \mathrm{div}(2\underline{T}_1 \mathbf{Q}_{p1})$$
$$= \Sigma_1 : \mathrm{D}(\frac{\mathbf{U}_1^+ + \mathbf{U}_1^-}{2}) + B_{p1}[\beta_1] + \mathbf{H}_{p1} \cdot \mathrm{grad}[\beta_1] + \Sigma(T_1 \mathscr{B}_1), \ in\ \Omega_1, \tag{11.34}$$

$$[e_2] + \mathrm{div}(2\underline{T}_2 \mathbf{Q}_{p2})$$
$$= \Sigma_2 : \mathrm{D}(\frac{\mathbf{U}_2^+ + \mathbf{U}_2^-}{2}) + B_{p2}[\beta_2] + \mathbf{H}_{p2} \cdot \mathrm{grad}[\beta_2] + \Sigma(T_2 \mathscr{B}_2), \ in\ \Omega_2, \tag{11.35}$$

$$[e_{s1}] = \underline{T}_1 2\mathbf{Q}_{p1} \cdot \mathbf{N}_1 + \Sigma\{T_1(Q_{p1} + \mathscr{B}_{s1})\}, \ on\ \partial\Omega_1,$$

*Remark 11.6.* The last relationship and the two following ones are boundary conditions for the partial differential equations which are to be satisfied by the temperature in $\Omega_1$ and in $\Omega_2$.

$$[e_{s2}] = \underline{T}_2 2\mathbf{Q}_{p2} \cdot \mathbf{N}_2 + \Sigma\{T_2(Q_{p2} + \mathscr{B}_{s2})\}, \ on\ \partial\Omega_2,$$
$$[e_{s1}] + [e_{s2}] = \mathbf{R}_p \cdot \mathbf{D}_s(\frac{\mathbf{U}^+ + \mathbf{U}^-}{2}) + \Sigma\{T_2 \mathscr{B}_{s2}\} + \Sigma\{T_1 \mathscr{B}_{s1}\}$$
$$+ 2\underline{T}_1 \mathbf{Q}_{p1} \cdot \mathbf{N}_1 + 2\underline{T}_2 \mathbf{Q}_{p2} \cdot \mathbf{N}_2, \ on\ \partial\Omega_1 \cap \partial\Omega_2.$$

The equations are completed by the relationships

$$\Delta(\mathscr{B}_1) + S_1 = 0, \ in\ \Omega_1,$$
$$\Delta(\mathscr{B}_2) + S_2 = 0, \ in\ \Omega_2,$$
$$\Delta\{Q_{p1} + \mathscr{B}_{s1}\} + S_{s1} = 0, \ on\ \partial\Omega_1,$$
$$\Delta(Q_{p2} + \mathscr{B}_{s2}) + S_{s2} = 0, \ on\ \partial\Omega_2,$$
$$\Delta(Q_{p1}) = 0, \ \Delta(Q_{p2}) = 0, \ on\ \partial\Omega_1 \cap \partial\Omega_2.$$

Equation (11.4) with the data $\Sigma(T_1 \mathscr{B}_1)$ give the two heat impulses $\mathscr{B}_1^+$ and $\mathscr{B}_1^-$ in $\Omega_1$, depending on the temperatures $T_1^+$ and $T_1^-$. With the data $\Sigma(T_1 Q_{p1})$, $\Sigma(T_1 \mathscr{B}_{s1})$, and (11.8), the two heat impulses $\mathscr{B}_1^+$ and $\mathscr{B}_1^-$ give the heat flux $\underline{T}_1 2\mathbf{Q}_{p1} \cdot \mathbf{N}_1$ on $\partial\Omega_1 \backslash (\partial\Omega_1 \cap \partial\Omega_2)$ which is the boundary condition for (11.34). Equations (11.9), (11.13), (11.8), (11.11), (11.14), the constitutive laws (11.28) and the data $\Sigma(T_1 \mathscr{B}_{s1})$, $\Sigma(T_2 \mathscr{B}_{s2})$ give the $\mathscr{B}_s$, $Q_p$, and the two heat fluxes $\underline{T}_1 2\mathbf{Q}_{p1} \cdot \mathbf{N}_1$ and $\underline{T}_2 2\mathbf{Q}_{p2} \cdot \mathbf{N}_2$ on $\partial\Omega_1 \cap \partial\Omega_2$ depending on the temperatures and the mechanical works. The heat fluxes $\underline{T}_1 2\mathbf{Q}_{p1} \cdot \mathbf{N}_1$ and $\underline{T}_2 2\mathbf{Q}_{p2} \cdot \mathbf{N}_2$ on $\partial\Omega_1 \cap \partial\Omega_2$ are the boundary conditions on $\partial\Omega_1 \cap \partial\Omega_2$ for (11.34) and (11.35). Thus the equations provide the boundary conditions for solving the two (11.34) and (11.35) which give the two future temperatures $T_1^+$ and $T_2^+$. In the following sections examples of those mechanical and thermal equations are investigated.

### 11.5.1 The Mechanical Evolution When Decoupled from the Thermal Evolution

The equations of motion are (11.29)–(11.33). Let us recall that we have assumed that all the mechanical quantities are independent of the temperature and that the constitutive laws have been chosen simple in order to emphasize the properties of the equations

$$\Sigma_1 = 2k_M \mathsf{D}\left(\mathbf{U}_1^+ + \mathbf{U}_1^-\right), \; in \; \Omega_1,$$
$$\Sigma_2 = 2k_M \mathsf{D}\left(\mathbf{U}_2^+ + \mathbf{U}_2^-\right), \; in \; \Omega_2,$$

the positive parameter $k_M$ describes the dissipative properties of the material in a collision;

$$\mathbf{R}_p = k_{ls}\left(\mathbf{U}_2^+ - \mathbf{U}_1^+ + \mathbf{U}_2^- - \mathbf{U}_1^-\right) + P^{dr}\mathbf{N}_1, \; on \; \partial\Omega_1 \cap \partial\Omega_2,$$

the positive parameter $k_{ls}$ describes the dissipative properties of the contact surface in a collision. The properties of non-interpenetration reaction $P^{dr}$ result from its definition

$$P^{dr} \in \partial I_+(\mathbf{D}_s(\mathbf{U}^+)\cdot\mathbf{N}_1) = \partial I_-((\mathbf{U}_2^+ - \mathbf{U}_1^+)\cdot\mathbf{N}_1).$$

Introducing the constitutive laws in the equations of motion, it results by denoting $\Delta\mathbf{V}$ the laplacian of vector $\mathbf{V}$

$$\rho_1\mathbf{U}_1^+ = k_M\left(\mathrm{grad}\,\mathrm{div}\,\mathbf{U}_1^+ + \Delta\mathbf{U}_1^+\right) + \mathbf{F}_{p1} + \rho_1\mathbf{U}_1^- + k_M\left(\mathrm{grad}\,\mathrm{div}\,\mathbf{U}_1^- + \Delta\mathbf{U}_1^-\right), \; in \; \Omega_1, \quad (11.36)$$

$$\rho_2\mathbf{U}_2^+ = k_M\left(\mathrm{grad}\,\mathrm{div}\,\mathbf{U}_2^+ + \Delta\mathbf{U}_2^+\right) + \mathbf{F}_{p2} + \rho_2\mathbf{U}_2^- + k_M\left(\mathrm{grad}\,\mathrm{div}\,\mathbf{U}_2^- + \Delta\mathbf{U}_2^-\right), \; in \; \Omega_2, \quad (11.37)$$

$$\Sigma_1\mathbf{N}_1 = k_{ls}\left(\mathbf{U}_2^+ - \mathbf{U}_1^+ + \mathbf{U}_2^- - \mathbf{U}_1^-\right) + P^{dr}\mathbf{N}_1, \Sigma_2\mathbf{N}_2 = -k_{ls}\left(\mathbf{U}_2^+ - \mathbf{U}_1^+ + \mathbf{U}_2^- - \mathbf{U}_1^-\right) + P^{dr}\mathbf{N}_2, \; on \; \partial\Omega_1 \cap \partial\Omega_2, \quad (11.38)$$

$$\Sigma_1\mathbf{N}_1 = \mathbf{T}_{p1}, \; on \; \partial\Omega_1 \backslash (\partial\Omega_1 \cap \partial\Omega_2), \quad (11.39)$$

and

$$\Sigma_2\mathbf{N}_2 = \mathbf{T}_{p2}, \; on \; \partial\Omega_2 \backslash (\partial\Omega_1 \cap \partial\Omega_2). \quad (11.40)$$

The preceding equations (11.36)–(11.40) have a variational formulation. Let $C$ be the convex set of the kinematically admissible future velocities

$$C = \{(\mathbf{V}_1, \mathbf{V}_2) \in \mathscr{V}_1 \times \mathscr{V}_2 | (\mathbf{V}_2 - \mathbf{V}_1) \cdot \mathbf{N}_1 \geq 0, \text{ on } \partial\Omega_1 \cap \partial\Omega_2\}.$$

It is easy to show that a solution of (11.36)–(11.40) satisfies

$$\begin{aligned}
&(\mathbf{U}_1^+, \mathbf{U}_2^+) \in C, \forall (\mathbf{V}_1, \mathbf{V}_2) \in C, \\
&a_1(\mathbf{U}_1^+, \mathbf{V}_1 - \mathbf{U}_1^+) + a_2(\mathbf{U}_2^+, \mathbf{V}_2 - \mathbf{U}_2^+) + b(\mathbf{U}_2^+ - \mathbf{U}_1^+, \mathbf{V}_2 - \mathbf{V}_1 - (\mathbf{U}_2^+ - \mathbf{U}_1^+)) \\
&\geq L_1(\mathbf{V}_1 - \mathbf{U}_1^+) + L_2(\mathbf{V}_2 - \mathbf{U}_2^+) - b(\mathbf{U}_2^- - \mathbf{U}_1^-, \mathbf{V}_2 - \mathbf{V}_1 - (\mathbf{U}_2^+ - \mathbf{U}_1^+)),
\end{aligned} \tag{11.41}$$

with

$$\begin{aligned}
a_1(\mathbf{U}, \mathbf{V}) &= \int_{\Omega_1} 2k\mathsf{D}(\mathbf{U}) : \mathsf{D}(\mathbf{V}) d\Omega_1 + \int_{\Omega_1} \rho_1 \mathbf{U} \cdot \mathbf{V} d\Omega_1, \\
b(\mathbf{U}, \mathbf{V}) &= \int_{\partial\Omega_1 \cap \partial\Omega_2} k_{ls} \mathbf{U} \cdot \mathbf{V}) d\Gamma_1, \\
L_1(\mathbf{V}) &= \int_{\Omega_1} \mathbf{F}_{p1} \cdot \mathbf{V} d\Omega_1 + \int_{\partial\Omega_1 \setminus (\partial\Omega_1 \cap \partial\Omega_2)} \mathbf{T}_{p1} \cdot \mathbf{V} d\Gamma_1 \\
&\quad - \int_{\Omega_1} 2k\mathsf{D}(\mathbf{U}_1^-) : \mathsf{D}(\mathbf{V}) d\Omega_1 + \int_{\Omega_1} \rho_1 \mathbf{U}_1^- \cdot \mathbf{V} d\Omega_1.
\end{aligned}$$

The functions $a_2$ and $L_2$ are defined in the same way. Problem (11.41) is a classical variational inequality which has a unique solution in a convenient functional framework [105, 158, 181, 194]. Thus the future velocities $\mathbf{U}^+$ are uniquely determined by the past.

## 11.5.2 An Example: Collision of a Bar with a Rigid Support

Consider the bar shown in Fig. 11.1 falling on a rigid fixed support ($\mathbf{U}_2 = 0$), [83]. The bar is assumed to be elastic when evolving smoothly (with Young modulus $2.5\ 10^{10}$ Pa, Poisson coefficient 0.3 and density 7877 kg/m$^3$). When colliding with the rigid support, it is dissipative.

The bar equation of motion when colliding the support is

$$\rho_1 \mathbf{U}_1^+ = k_M \operatorname{grad} \operatorname{div} \mathbf{U}_1^+ + k_M \Delta \mathbf{U}_1^+ + \rho_1 \mathbf{U}_1^- + k_M \operatorname{grad} \operatorname{div} \mathbf{U}_1^- + k_M \Delta \mathbf{U}_1^-,$$

in the bar

$$\Sigma_1 \mathbf{N}_1 = -k_{ls} \left( \mathbf{U}_1^+ + \mathbf{U}_1^- \right) - P^{dr} \mathbf{N}_1,$$

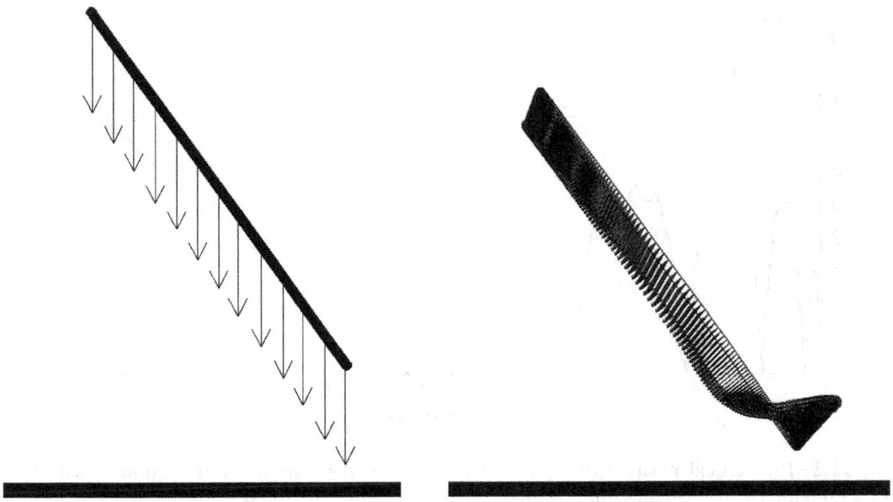

**Fig. 11.1** A bar falls with a vertical velocity of $-1\ \mathrm{m/s}$ on a rigid support with an angle $\theta = 54°$ with the horizontal. Its length is 0.6 m. Its width is 0.012 m. The two ends are rounded. The velocities after the first collisions are shown

on the contact surface, and

$$\Sigma_1 \mathbf{N}_1 = 0,$$

outside the contact surface. The dissipative parameters are chosen as: $k_M = 10^5\ \mathrm{Ns^2/m^2}$ and $k_{ls} = 8\ 10^6\ \mathrm{Ns^2/m^3}$. The velocities of the bar after the first collision where contact is not maintained, are not rigid body velocities. The bar vibrates afterwards. There is a competition between the slow rising motion after the collision and the fast vibrating motion resulting from the elasticity of the bar. The first oscillations are such that the bar which has not risen far above the support collides with it other times, [86]. The vertical position of a point of the lower part of the bar versus the time is shown in Fig. 11.2. Experiments have shown this behaviour, [201].

### 11.5.3 Thermal Evolution When the Mechanical Equations Are Decoupled from the Thermal Equations

The equations to get the future temperatures and volume fractions are,

$$\rho_1 [e_1] + \mathrm{div}(2\underline{T}_1 \mathbf{Q}_{p1})$$
$$= \Sigma_1 : \mathsf{D}\left(\frac{\mathbf{U}_1^+ + \mathbf{U}_1^-}{2}\right) + B_{p1}[\beta_1] + \mathbf{H}_{p1} \cdot \mathrm{grad}[\beta_1] + \Sigma(T_1 \mathscr{B}_1),$$

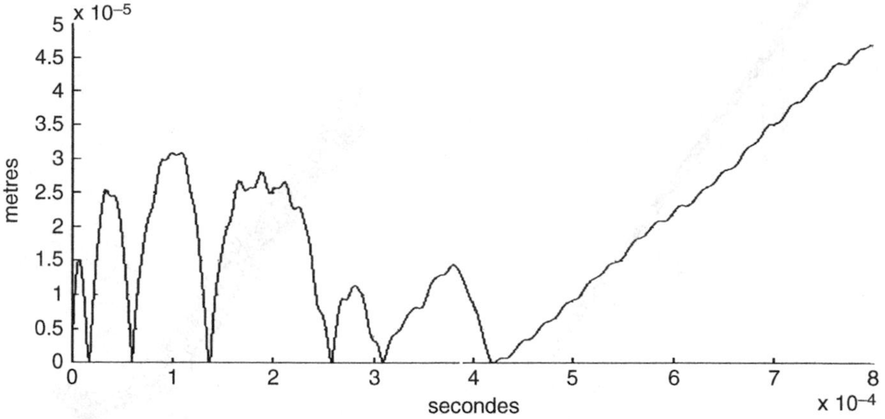


**Fig. 11.2** The vertical position of a point of the lower part of the bar which collides with the rigid support. The bar vibrates after the first collision. It results microrebounds before the bar rises sufficiently above the rigid support [83]

$$
\begin{aligned}
&0 = \operatorname{div} \mathbf{H}_{p1} - B_{p1} + A_1, \ in\ \Omega_1, \\
&\rho_2[e_2] + \operatorname{div}(2\underline{T}_2 \mathbf{Q}_{p2}) \\
&\quad = \Sigma_2 : \mathsf{D}(\frac{\mathbf{U}_2^+ + \mathbf{U}_2^-}{2}) + B_{p2}[\beta_2] + \mathbf{H}_{p2} \cdot \operatorname{grad}[\beta_2] + \Sigma(T_2 \mathscr{B}_2), \\
&0 = \operatorname{div} \mathbf{H}_{p2} - B_{p2} + A_2, \ in\ \Omega_2, \\
&[e_{s1}] = \underline{T}_1 2\mathbf{Q}_{p1} \cdot \mathbf{N}_1 + \Sigma\left\{T_1(Q_{p1} + \mathscr{B}_{s1})\right\}, \\
&\mathbf{H}_{p1} \cdot \mathbf{N}_1 = a_1, \ on\ \partial\Omega_1 \backslash (\partial\Omega_1 \cap \partial\Omega_2), \\
&[e_{s2}] = \underline{T}_2 2\mathbf{Q}_{p2} \cdot \mathbf{N}_2 + \Sigma\left\{T_2(Q_{p2} + \mathscr{B}_{s2})\right\}, \\
&\mathbf{H}_{p2} \cdot \mathbf{N}_2 = a_2, \ on\ \partial\Omega_2 \backslash (\partial\Omega_1 \cap \partial\Omega_2),
\end{aligned}
$$

by using (11.8), (11.11), and (11.14) together with the constitutive law (11.28)

$$
\begin{aligned}
[e_{s1}] &= \frac{\underline{T}_1}{\underline{T}_1 + \underline{T}_2} \mathbf{R}_p \cdot \mathbf{D}_s(\frac{\mathbf{U}^+ + \mathbf{U}^-}{2}) \\
&\quad + \underline{T}_1 2\mathbf{Q}_{p1} \cdot \mathbf{N}_1 - \frac{2\underline{T}_1 \underline{T}_2}{\underline{T}_1 + \underline{T}_2}(Q_{p2} - Q_{p1}) + \Sigma\{T_1 \mathscr{B}_{s1}\},
\end{aligned} \tag{11.42}
$$

$$
\begin{aligned}
[e_{s2}] &= \frac{\underline{T}_2}{\underline{T}_1 + \underline{T}_2} \left( \mathbf{R}_p \cdot \mathbf{D}_s(\frac{\mathbf{U}^+ + \mathbf{U}^-}{2}) \right) \\
&\quad + \underline{T}_2 2\mathbf{Q}_{p2} \cdot \mathbf{N}_2 + \frac{2\underline{T}_1 \underline{T}_2}{\underline{T}_1 + \underline{T}_2}(Q_{p2} - Q_{p1}) + \Sigma\{T_2 \mathscr{B}_{s2}\}, \ on\ \partial\Omega_1 \cap \partial\Omega_2,
\end{aligned} \tag{11.43}
$$

and

$$\mathbf{H}_{p1}\cdot\mathbf{N}_1 = a_1,\ \mathbf{H}_{p2}\cdot\mathbf{N}_2 = a_2,\ on\ \partial\Omega_1\cap\partial\Omega_2.$$

The equations are completed by relationships (11.4), (11.7), (11.9), (11.13) and, (11.15). The constitutive laws have already been chosen, (see formulas (11.28)). Let us recall those we need

$$\begin{aligned}
-\mathbf{Q}_{p1} &= \frac{k}{\underline{T}_1}\operatorname{grad}\underline{T}_1,\ -\mathbf{Q}_{p2} = \frac{k}{\underline{T}_2}\operatorname{grad}\underline{T}_2,\\
-(Q_{p2}-Q_{p1}) &= \frac{\underline{T}_1+\underline{T}_2}{\underline{T}_1\underline{T}_2}k^d_{ps}\delta\underline{T},\\
B_{p1}\in B^{fe}_{p1}+c[\beta_1] &= -\frac{L_1}{T_0}(T_1^+-T_0)+\partial I(\beta_1^+)+c[\beta_1],\\
\mathbf{H}_{p1} &= \mathbf{H}^{fe}_{p1}+k_m\operatorname{grad}[\beta_1],\\
B_{p2}\in B^{fe}_{p2}+c[\beta_2] &= -\frac{L_2}{T_0}(T_2^+-T_0)+\partial I(\beta_2^+)+c[\beta_2],\\
\mathbf{H}_{p2} &= \mathbf{H}^{fe}_{p2}+k_m\operatorname{grad}[\beta_2],
\end{aligned}$$

where thc constants $k$ and $k^d_{ps}$ denote the different thermal conductivities and $c$ and $k_m$ are dissipative parameters. We have already chosen the volume free energies

$$\Psi(T,\beta) = -C_iT\ln T-\beta\frac{L_i}{T_0}(T-T_0)+\frac{k_i}{2}(\operatorname{grad}\beta)^2+I(\beta).$$

Since we assume the mechanical properties do not depend on the temperature, the surface free energies have a unique temperature dependent term which is chosen simple

$$\Psi_s(T) = -C_sT\ln T,$$

where the $C$'s are heat capacities.

We assume that the exterior heat impulses are null: $\Sigma(T_1\mathscr{B}_1)=\Sigma(T_2\mathscr{B}_2)=0$, $\Sigma(T_1Q_{p1})=0$ on $\partial\Omega_1\backslash(\partial\Omega_1\cap\partial\Omega_2)$, $\Sigma(T_1\mathscr{B}_{s1})=0$ on $\partial\Omega_1$, $\Sigma(T_2Q_{p2})=0$ on $\partial\Omega_2\backslash(\partial\Omega_1\cap\partial\Omega_2)$, $\Sigma(T_2\mathscr{B}_{s2})=0$ on $\partial\Omega_2$ and that no external percussion work is provided $A_1=A_2=0$, and $a_1=a_2=0$.

*Remark 11.7.* It is also possible to assume that the exterior entropy impulses are given.

The energy balances (11.34), (11.35), (11.8), (11.40) and (11.42), (11.43), with the constitutive laws give the equations

$$\begin{aligned}
&C_1[T_1]+L_1[\beta_1]-2k\Delta\underline{T}_1\\
&\quad=\Sigma_1:\mathsf{D}(\frac{\mathbf{U}_1^++\mathbf{U}_1^-}{2})+B_{p1}[\beta_1]+\mathbf{H}_{p1}\cdot\operatorname{grad}[\beta_1],
\end{aligned}\tag{11.44}$$

$$c[\beta_1] - k_m\Delta[\beta_1] + \partial I(\beta_1^+) \ni \frac{L_1}{T_0}(T_1^+ - T_0),\ in\ \Omega_1, \tag{11.45}$$

$$C_2[T_2] + L_2[\beta_2] - 2k\Delta\underline{T}_2$$
$$= \Sigma_2 : \mathsf{D}(\frac{\mathbf{U}_2^+ + \mathbf{U}_2^-}{2}) + B_{p2}[\beta_2] + \mathbf{H}_{p2}\cdot \mathrm{grad}[\beta_2], \tag{11.46}$$

$$c[\beta_2] - k_m\Delta[\beta_2] + \partial I(\beta_2^+) \ni \frac{L_2}{T_0}(T_2^+ - T_0),\ in\ \Omega_2, \tag{11.47}$$

where $\Delta$ denotes the laplacian operator,

$$C_{1s}[T_1] + 2k\frac{\partial \underline{T}_1}{\partial \mathbf{N}_1} = 0,\ k_m\frac{\partial[\beta_1]}{\partial \mathbf{N}_1} = 0,\ on\ \partial\Omega_1\backslash(\partial\Omega_1\cap\partial\Omega_2), \tag{11.48}$$

$$C_{2s}[T_2] + 2k\frac{\partial \underline{T}_2}{\partial \mathbf{N}_2} = 0,\ k_m\frac{\partial[\beta_2]}{\partial \mathbf{N}_2} = 0,\ on\ \partial\Omega_2\backslash(\partial\Omega_1\cap\partial\Omega_2), \tag{11.49}$$

these relationships are the boundary conditions on $\partial\Omega_1\backslash(\partial\Omega_1\cap\partial\Omega_2)$ and $\partial\Omega_2\backslash(\partial\Omega_1\cap\partial\Omega_2)$ of (11.44) and (11.46) in $\Omega_1$ and $\Omega_2$,

$$C_{1s}[T_1] + 2k\frac{\partial \underline{T}_1}{\partial \mathbf{N}_1} = \frac{\underline{T}_1}{\underline{T}_1 + \underline{T}_2}\left(\mathbf{R}_p\cdot\mathbf{D}_s\left(\frac{\mathbf{U}^+ + \mathbf{U}^-}{2}\right)\right) + 2k_{ps}^d\delta\underline{T}, \tag{11.50}$$

$$C_{2s}[T_2] + 2k\frac{\partial \underline{T}_2}{\partial \mathbf{N}_2} = \frac{\underline{T}_2}{\underline{T}_1 + \underline{T}_2}\left(\mathbf{R}_p\cdot\mathbf{D}_s\left(\frac{\mathbf{U}^+ + \mathbf{U}^-}{2}\right)\right) - 2k_{ps}^d\delta\underline{T}, \tag{11.51}$$

$$k_m\frac{\partial[\beta_1]}{\partial \mathbf{N}_1} = 0,\ k_m\frac{\partial[\beta_2]}{\partial \mathbf{N}_2} = 0,\ on\ \partial\Omega_1\cap\partial\Omega_2.$$

As already mentioned, relationships (11.50) and (11.51) together with (11.42) and (11.43) give the boundary conditions on $\partial\Omega_1\cap\partial\Omega_2$ for the two partial differential equations (11.44) and (11.46) equations in $\Omega_1$ and $\Omega_2$ giving the future temperature $T_1^+$ and $T_2^+$. The equations giving the temperatures and the volume fractions are coupled due to the phase change latent heat.

### 11.5.4 The Temperature Variation in a Collision

Because collisions are dissipative, their thermal effect should be to warm the colliding solids. In this section, we investigate this problem. Let us assume that the temperatures before the collision are uniform (they do not depend on $\mathbf{x}$) in each solid and lower than the phase change temperature, i.e., $T_1^- < T_0$, $T_2^- < T_0$. We assume also that $\beta_1^- = \beta_2^- = 0$. The effect of $k_{ps}^d$ is to equalize the temperature on the boundary thus we assume that $k_{ps}^d = 0$ in order to avoid this effect. We assume also that $\mathbf{H}_{p1} = \mathbf{H}_{p2} = 0$. By using the relationship $\underline{T} = T^- + [T]/2$, it results the equations,

$$
\begin{aligned}
C_1[T_1]+L_1[\beta_1]-k\Delta[T_1] &= \Sigma_1 : \mathsf{D}(\frac{\mathbf{U}_1^+ + \mathbf{U}_1^-}{2}) = Y_1,\\
c[\beta_1]+\partial I(\beta_1^+) &\ni \frac{L_1}{T_0}(T_1^+ - T_0),\ in\ \Omega_1,\\
C_2[T_2]+L_2[\beta_2]-k\Delta[T_2] &= \Sigma_2 : \mathsf{D}(\frac{\mathbf{U}_2^+ + \mathbf{U}_2^-}{2}) = Y_2,\\
c[\beta_2]+\partial I(\beta_2^+) &\ni \frac{L_2}{T_0}(T_2^+ - T_0),\ in\ \Omega_2,
\end{aligned}
\tag{11.52}
$$

because we have equations of motion $B_{p1}=0$ and $B_{p2}=0$,

$$
\begin{aligned}
C_{1s}[T_1]+k\frac{\partial[T_1]}{\partial \mathbf{N}_1} &= 0 = Y_{1s},\ on\ \partial\Omega_1\backslash(\partial\Omega_1\cap\partial\Omega_2),\\
C_{2s}[T_2]+k\frac{\partial[T_2]}{\partial \mathbf{N}_2} &= 0 = Y_{2s},\ on\ \partial\Omega_2\backslash(\partial\Omega_1\cap\partial\Omega_2),
\end{aligned}
$$

and

$$
\begin{aligned}
C_{1s}[T_1]+k\frac{\partial[T_1]}{\partial \mathbf{N}_1} &= \frac{\underline{T}_1}{\underline{T}_1+\underline{T}_2}\left(\mathbf{R}_p\cdot\mathbf{D}_s(\frac{\mathbf{U}^+ + \mathbf{U}^-}{2})\right) = Y_{1s},\\
C_{2s}[T_2]+k\frac{\partial[T_2]}{\partial \mathbf{N}_2} &= \frac{\underline{T}_2}{\underline{T}_1+\underline{T}_2}\left(\mathbf{R}_p\cdot\mathbf{D}_s(\frac{\mathbf{U}^+ + \mathbf{U}^-}{2})\right) = Y_{2s},\ on\ \partial\Omega_1\cap\partial\Omega_2.
\end{aligned}
$$

Let us define $X_1 = np([T_1])$ and $X_2 = np([T_2])$, where $np(A)$ is the negative part of $A$, $np(A) = \sup\{-A, 0\}$. The equation of motion for $\beta_1$ is

$$
\begin{aligned}
0 = B_{p1} \in B_{p1}^{fe} + \partial\Phi_1([\beta_1]) &= -\frac{L_1}{T_0}(T_1^+ - T_0) + \partial I(\beta_1^+) + \partial\Phi_1([\beta_1])\\
&= -\frac{L_1}{T_0}(T_1^+ - T_0) + \partial I([\beta_1]) + \partial\Phi_1([\beta_1]),
\end{aligned}
$$

because $\beta_1^+ = [\beta_1]$. We get

$$
\frac{L_1}{T_0}(T_1^+ - T_0)[\beta_1] \in (\partial I([\beta_1]) + \partial\Phi_1([\beta_1]))[\beta_1],
$$

and

$$
\frac{L_1}{T_0}(T_1^+ - T_0)[\beta_1] \geq 0. \tag{11.53}
$$

It results that if $[\beta_1] > 0$, $(T_1^+ - T_0) \geq 0$ and $(T_1^+ - T_1^-) \geq (T_1^+ - T_0) \geq 0$: then $X_1 = np([T_1]) = 0$. Thus we have proved that

$$L_1\,[\beta_1]X_1 = 0, \tag{11.54}$$

because either $[\beta_1]$ or $X_1$ is zero. By multiplying (11.44), (11.48), (11.42) by $X_1 = np([T_1])$ and (11.46), (11.49), (11.43) by $X_2 = np([T_2])$, using relationship (11.54), and integrating over the $\Omega$'s and $\partial\Omega$'s

$$\begin{aligned}
&-\int_{\Omega_1} C_1(X_1)^2 d\Omega_1 - \int_{\Omega_1} k\,(\mathrm{grad}\,X_1)^2\, d\Omega_1 - \int_{\partial\Omega_1} C_{1s}(X_1)^2 d\Gamma_1 \\
&-\int_{\Omega_2} C_2(X_2)^2 d\Omega_2 - \int_{\Omega_2} (k\,\mathrm{grad}\,X_2)^2\, d\Omega_2 - \int_{\partial\Omega_2} C_{2s}(X_2)^2 d\Gamma_2 \\
&= \int_{\Omega_1} Y_1 X_1 d\Omega_1 + \int_{\partial\Omega_1} Y_{1s} X_1 d\Gamma_1 + \int_{\Omega_2} Y_2 X_2 d\Omega_2 + \int_{\partial\Omega_2} Y_{2s} X_2 d\Gamma_2.
\end{aligned} \tag{11.55}$$

Due to the constitutive laws and properties of the subdifferentials, the right-hand sides $Y$'s and $Y_s$'s of the previous equations (11.42)–(11.49) are non negative. Then the right-hand side of (11.55) is non negative. But the left-hand side is non positive. Thus both of them are zero and $X_1 = np([T_1])$ and $X_2 = np([T_2])$ are equal to zero. Thus $[T_1] \geq 0$ and $[T_2] \geq 0$.

As one expects, the temperatures increase when two solids collide. Phase change can occur. Relationship (11.53) shows that if it occurs at point $\mathbf{x}$, temperature $T_i^+(\mathbf{x})$ at that point is larger than the phase change temperature $T_0$.

# Chapter 12
# Phase Change Depending on a State Quantity: Liquid–Vapor Phase Change

Water may be solid, liquid and gaseous. We investigate in this chapter the liquid vapor phase change. Compared to the solid liquid phase change, there is a major difference: the phase change temperature depends on a parameter, for instance on the pressure. It is well known that water boils at the Mont Blanc summit at a lower temperature than at sea level. Moreover, if the temperature is larger than $T_c$, the critical temperature, water is only in its gaseous phase. At temperature lower than critical temperature $T_c$, water may be either liquid or gaseous. Experiments show that the phase change temperature depends on the pressure as shown in Fig. 12.1. A complete physical description may be found in [140, 147, 209].

Experiments show also that all the physical properties depend on two parameters as shown in Fig. 12.1, for instance $T$ and pressure $p$. Numerous abaci, called thermodynamical diagrams, may be built. For instance, if $P$ and specific volume $\tau$ are chosen, isovalues of temperature $T(P,\tau)$ are drawn in the $\tau, P$ frame producing the diagram shown in Fig. 12.2. The diagram $\tau, T$, the Clapeyron diagram, is shown in Fig. 12.3 with isovalues of pressure $P(\tau, T)$. These diagrams are obtained with the two phases being in contact on a surface. This is a surfacic phase change as it occurs when water boils.

But in a shower room after taking a shower, the mist is a mixture of air, of vapor which is transparent and of microscopic droplets of water which make the mixture not so transparent. The same mixture may be seen in fog and one may guess that clouds are also mixtures of air, vapor and liquid water (and even of ice which is assumed not to be present in the following predictive theories). These mixtures can be in contact with either liquid or pure vapor. These surfacic phase changes are also described in the diagrams. For instance, in Fig. 12.3, the vapor content $\beta_v$ is shown.

Such mixtures, mist, fog or clouds, evolve in volumes. When a phase change occurs, it is a volumic phase change.

In this chapter, we investigate the surfacic phase change and in the next chapter the volumic phase change. The results are similar.

M. Frémond, *Phase Change in Mechanics*, Lecture Notes of the Unione Matematica Italiana 13, DOI 10.1007/978-3-642-24609-8_12,

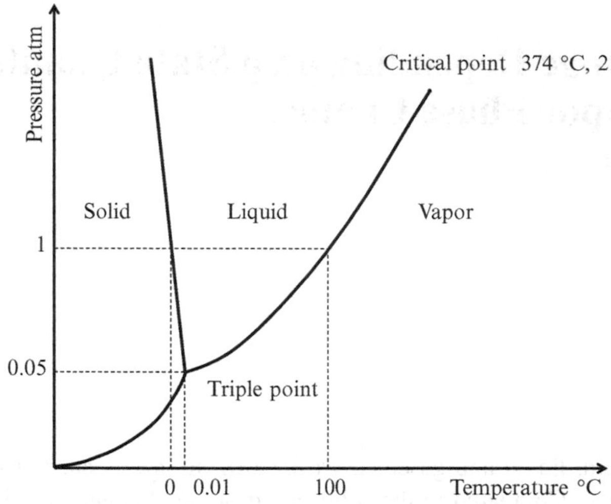


**Fig. 12.1** Schematic pressure $P$ versus temperature $T$ for water in contact with vapor

**Fig. 12.2** A typical pressure $P$ versus specific volume $\tau$ for different temperatures

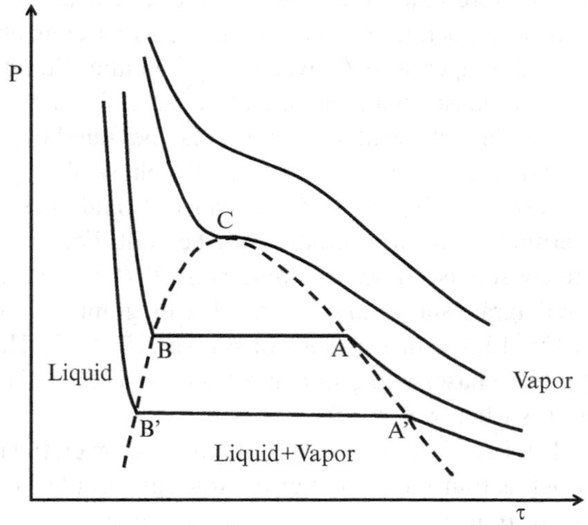


## 12.1 The Temperature Is Lower than Critical Temperature $T_c$

Let us consider liquid water in contact with vapor at an equilibrium, for instance boiling water. The contact surface, the phase change surface, is a free surface between a liquid domain with liquid volume fraction $\beta_l = 1$ and a vapor domain with vapor volume fraction $\beta_v = 1 - \beta_l = 1$. On this surface $[\beta_l] = [\beta_v] = 1$. Applying the

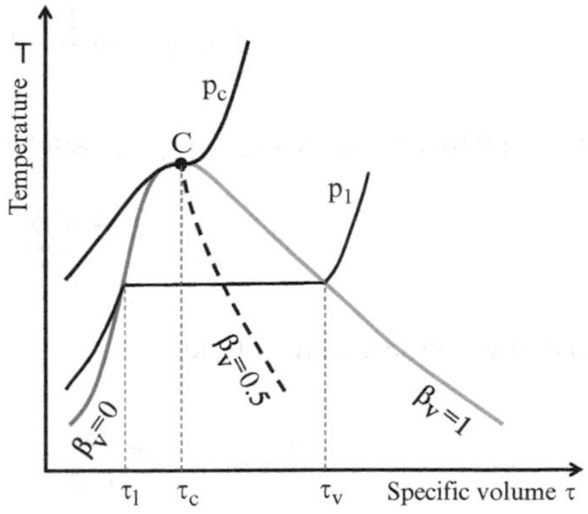


**Fig. 12.3** The thermodynamical diagram temperature $T$ versus specific volume $\tau$ for vapor in contact with liquid water. On the right of the green line there is vapor. On the left of the blue line there is liquid. In between the two lines there is a mixture of liquid and vapor. Some isopressure and isomixture curves are shown

results of Sect. 3.5, relationships (3.34), (3.35) and (3.36) give

$$-([\widehat{\Psi}] + \underline{\widehat{s}}[T] - \underline{T}_N[\frac{1}{\rho}]) = \frac{\partial \Phi}{\partial m},$$

$$\underline{Q} = \frac{\partial \Phi}{\partial [T]},$$

$$\underline{\mathbf{T}}_T = \frac{\partial \Phi}{\partial [\mathbf{U}_T]},$$

$$\mathbf{H} \cdot \mathbf{N} = \frac{\partial \Phi}{\partial [\dot{\beta}]},$$

where density is $\rho = 1/\tau$ with specific volume $\tau$, $\widehat{\Psi} = \tau\Psi$ and $\widehat{s} = \tau s$ are the specific free energy and specific entropy whereas $\Psi$ and $s$ are the volumic free energy and volumic entropy. Following experiments, we assume no dissipation and no discontinuity of temperature on the phase change surface, and we choose surface pseudopotential of dissipation

$$\Phi = I_0([T]),$$

where $I_0$ is the indicator function of the origin of $\mathbb{R}$. Because the two phases in contact are at an equilibrium, the velocity discontinuity $[\mathbf{U}]$ is null and we get from (3.28)

$$[\mathbf{T}] = 0, \tag{12.1}$$

the continuity of the stress which is the continuity of the pressure ($\mathbf{T} = \sigma\mathbf{N} = -P_{cloud}\mathbf{N}$ is the stress on the phase change surface, normal vector $\mathbf{N}$ is defined in Sect. 3.5). The definition of pressures (13.46) given below (see Sect. 13.11.2)

$$P_{cloud} = p_l \frac{1}{\tau_l} - \Psi_l = p_v \frac{1}{\tau_v} - \Psi_v, \tag{12.2}$$

with specific pressures $p_l$ and $p_v$ given by formula (13.38) in Remark 13.9

$$p_l = -\tau_l^2 \frac{\partial \Psi_l}{\partial \tau_l}, \ p_v = -\tau_v^2 \frac{\partial \Psi_v}{\partial \tau_v},$$

give the pressure in the cloud

$$P_{cloud} = -\tau_l \frac{\partial \Psi_l}{\partial \tau_l} - \Psi_l = -\tau_v \frac{\partial \Psi_v}{\partial \tau_v} - \Psi_v\,. \tag{12.3}$$

The surface constitutive laws give

$$-([\widehat{\Psi}] + P_{cloud}[\frac{1}{\rho}]) = 0, \tag{12.4}$$

$$[T] = 0, \ \underline{Q} \in \mathbb{R},$$

$$\underline{\mathbf{T}}_T = 0,$$

$$\mathbf{H} \cdot \mathbf{N} = 0\,. \tag{12.5}$$

We get from relationships (12.2) and (12.4)

$$p_l = p_v\,. \tag{12.6}$$

On the phase change surface there are three relationships, (12.3), (12.6), between the four parameters $\tau_l$, $\tau_v$, $P_{cloud}$ and $T$. Thus they depend on one parameter. If we choose the volume fraction, we get the functions giving pressure $P_{cloud}(\tau_l)$ (the $B'BC$ dotted line on the left) and $P_{cloud}(\tau_v)$ (the $CAA'$ dotted line on the right) in Fig. 12.2. In the diagram of Fig. 12.3, the functions giving the temperature are $T(\tau_l)$ (the blue curve) and $T(\tau_v)$ (the green curve). If we choose temperature $T$, we get $P_{cloud}(T)$, the curve separating liquid and vapor domains in Fig. 12.1.

The phase change depends on a parameter, for instance $P_{cloud}$ or $T$. When $P_{cloud}$ or $T$ is chosen, all the physical quantities are known. Whereas for the solid liquid phase change, the ice water phase change, the phase change does not depend on any parameter: the phase change temperature is fixed.

The last constitutive law, (12.5), prescribes that the work flux $\mathbf{H} \cdot \mathbf{N}$ is continuous. With constitutive law (13.26) for $\mathbf{H}_v$ and $\mathbf{H}_l$, $\text{grad}\,\beta_v \cdot \mathbf{N}$ and $\text{grad}\,\beta_l \cdot \mathbf{N}$ are continuous but $\beta_v$ and $\beta_l$ are discontinuous.

*Remark 12.1.* Formula (12.4) is analogous to the formula which gives the pressure in the volume as formula (13.46).

### 12.1.1 Partial Phase Change

Let us consider liquid water ($\beta_{l1} = 1$) in contact with a mixture of vapor and liquid water ($\beta_{l2} \neq 1$). Due to relationship (12.1), the pressures on each side of the phase change surface are equal. Their value given by (13.46) is

$$P_{cloud} = -\tau_l \frac{\partial \Psi_l}{\partial \tau_l}(T, \tau_l) - \Psi_l(T, \tau_l)$$
$$= -\tau_{l2} \frac{\partial \Psi_l}{\partial \tau}(T, \tau_{l2}) - \Psi_l(T, \tau_{l2}) = -\tau_{v2} \frac{\partial \Psi_v}{\partial \tau}(T, \tau_{v2}) - \Psi_v(T, \tau_{v2}). \quad (12.7)$$

It results from these equations that $\tau_{l2} = \tau_{l1} = \tau_l$, $\tau_l$ being the common value. Relationship (12.4)

$$-([\widehat{\Psi}] + P_{cloud}[\frac{1}{\rho}]) = 0,$$

gives with (12.7)

$$p_{v2} = -\tau_{v2}^2 \frac{\partial \Psi_v}{\partial \tau_{v2}}(\tau_{v2}, T) = p_l = -\tau_l^2 \frac{\partial \Psi_l}{\partial \tau}(\tau_l, T) \,. \quad (12.8)$$

The two relationships (12.7) and (12.8), which are identical to relationships (12.2) and (12.6), give $\tau_l = \tau_l(T)$ and $\tau_{v2} = \tau_{v2}(T)$. These relationships with mixture specific volume $\tau$ defined by

$$\frac{1}{\tau} = \frac{\beta_{v2}}{\tau_{v2}} + \frac{1-\beta_{v2}}{\tau_l} \,.$$

imply that $T = T(\tau, \beta_{v2})$. Such a relationship $\tau \to T(\tau, \beta_{v2})$ is shown in Fig. 12.3 for $\beta_{v2} = 0$, 0.5 and 1.

Let us note that the results are the same for the contact of liquid water and vapor and for the contact of mixtures of liquid and vapor.

## 12.2 The Temperature Is Larger than Critical Temperature $T_c$

Let us consider liquid water in contact with vapor at an equilibrium. The contact surface, the phase change surface, is a free surface between a liquid domain with liquid volume fraction $\beta_l = 1$ and a vapor domain with vapor volume fraction $\beta_v = 1 - \beta_l = 1$. On this surface $[\beta_l] = [\beta_v] = 1$. Applying the results of previous section, we have

$$P_{cloud} = \frac{p_l}{\tau_l} - \Psi_l = \frac{p_v}{\tau_v} - \Psi_v, \tag{12.9}$$

and

$$p_l = -\tau_l^2 \frac{\partial \Psi_l}{\partial \tau_l} = p_v = -\tau_v^2 \frac{\partial \Psi_v}{\partial \tau_v}. \tag{12.10}$$

When temperature $T$ is larger than critical temperature $T_c$, the free energies are such that

$$\Psi_l(\tau_l, T) = F(\tau_l, T),\ \Psi_v(\tau_v, T) = F(\tau_v, T),$$

see Sect. 13.15 below. Relationship (12.10) gives

$$-\tau_l^2 \frac{\partial F}{\partial \tau}(\tau_l, T) = -\tau_v^2 \frac{\partial F}{\partial \tau}(\tau_v, T).$$

Thus $\tau_l = \tau_v$ and $[\tau] = [1/\rho] = 0$. Relationship (12.4) gives $[\widehat{\Psi}] = 0$, which is satisfied for any temperature. Thus at temperature larger than critical temperature $T_c$, it is impossible to have a difference between liquid and vapor. Thus there is no liquid vapor phase change. Volume fractions $\beta_l$ and $\beta_v$ are no longer state quantities.

# Chapter 13
# Clouds: Mixture of Air, Vapor and Liquid Water

Clouds are mixtures of air, vapor, liquid water and even of ice. For the sake of simplicity, we assume there is no ice phase. At each point of the mixture there is air and water either gaseous or liquid. Different indices are used: $v$ for vapor, $l$ for liquid, $a$ for air, $w$ for water. The clouds we see in the sky have a structure evolving with time. We think that the clouds cohesion results from local interactions: at a point $\mathbf{x}$, the physical quantities, for instance, the velocities, the vapor content and liquid water content,.. do depend on their values in the neighbourhood of point $\mathbf{x}$. A way to take into account these interactions is to introduce space derivatives which clearly depend on the neighbourhood values.

## 13.1 The Temperature Is Lower than Critical Temperature $T_c$

When the temperature is lower than temperature $T_c$, called the critical temperature, vapor and liquid may coexist. In the sequel we assume $T \leq T_c$ up to Sect. 13.15 where the temperature is assumed to satisfy $T \geq T_c$.

## 13.2 State Quantities

The state quantities are:

- The vapor, liquid and air specific volumes

$$\tau_v,\ \tau_l,\ \tau_a.$$

M. Frémond, *Phase Change in Mechanics*, Lecture Notes of the Unione Matematica Italiana 13, DOI 10.1007/978-3-642-24609-8_13,

- Air-vapor and liquid volume fractions

$$\beta_v,\ \beta_l.$$

We introduce two volume fractions related by

$$1 = \beta_v + \beta_l,$$

assuming that air and vapor coexist in the gaseous phase. It is necessary to have the two volume fractions because they have different velocities.

- The gradients of the volume fractions

$$\operatorname{grad}\beta_v,\ \operatorname{grad}\beta_l.$$

They describe the influence of the volume fractions at a point onto their neighbourhood volume fractions. Such an influence is responsible for the cohesion of the clouds which is conspicuous when looking at them.

- The temperature

$$T,$$

assuming the phases are in thermal equilibrium and have the same temperature. The non thermal equilibrium situation in case of fast phenomena, may be investigated, [97, 143].

We denote

$$\begin{aligned}
E_v &= (\tau_v, \beta_v, \beta_l, \operatorname{grad}\beta_v, T),\\
E_l &= (\tau_l, \beta_v, \beta_l, \operatorname{grad}\beta_l, T),\\
E_a &= (\tau_a, \beta_v, T),\\
E &= E_v \cup E_l \cup E_a.
\end{aligned}$$

## 13.3 Quantities Which Describe the Evolution and the Thermal Heterogeneity

We denote

$$\mathbf{U}_v,\ \mathbf{U}_l,\ \mathbf{U}_a,$$

the velocities of the three phases. The quantities which describe the evolution are the classical deformation velocities and other velocities accounting for different physical properties. These velocities may be called generalized deformation velocities.

### 13.3.1 Classical and Generalized Deformation Velocities

They are:

- The classical deformation velocities

$$\mathsf{D}(\mathbf{U}_v),\ \mathsf{D}(\mathbf{U}_l),\ \mathsf{D}(\mathbf{U}_a),$$

  where $\mathsf{D}$ is the usual strain rate operator.
- The material derivatives of the volume fractions

$$\frac{d^v\beta_v}{dt}=\frac{\partial\beta_v}{\partial t}+\mathbf{U}_v\cdot\operatorname{grad}\beta_v,\quad \frac{d^l\beta_v}{dt}=\frac{\partial\beta_v}{\partial t}+\mathbf{U}_l\cdot\operatorname{grad}\beta_v,$$
$$\frac{d^l\beta_l}{dt}=\frac{\partial\beta_l}{\partial t}+\mathbf{U}_l\cdot\operatorname{grad}\beta_l,\quad \frac{d^v\beta_l}{dt}=\frac{\partial\beta_l}{\partial t}+\mathbf{U}_v\cdot\operatorname{grad}\beta_l,$$
$$\frac{d^a\beta_v}{dt}=\frac{\partial\beta_v}{\partial t}+\mathbf{U}_a\cdot\operatorname{grad}\beta_v.$$

- The gradients of the volume fractions velocities accounting for local interactions

$$\operatorname{grad}\frac{d^v\beta_v}{dt},\ \operatorname{grad}\frac{d^l\beta_l}{dt}.$$

- The material derivatives of the specific volumes

$$\frac{d^v\tau_v}{dt},\ \frac{d^l\tau_l}{dt},\ \frac{d^a\tau_a}{dt}.$$

- The relative velocities of the phases

$$\mathbf{U}_v-\mathbf{U}_l,\ \mathbf{U}_v-\mathbf{U}_a,\ \mathbf{U}_l-\mathbf{U}_a.$$

  These relative velocities may take into account dissipative interactions between the phases.

### 13.3.2 Thermal Heterogeneity

It is described by

$$\operatorname{grad}T.$$

We denote $\delta E$ the set of all these quantities: velocities of deformation and gradient of temperature

$$\begin{aligned}\delta E = (&\mathsf{D}(\mathbf{U}_v),\ \mathsf{D}(\mathbf{U}_l),\ \mathsf{D}(\mathbf{U}_a),\\ &\frac{\mathrm{d}^v\beta_v}{\mathrm{d}t}, \frac{\mathrm{d}^l\beta_l}{\mathrm{d}t}, \frac{\mathrm{d}^a\beta_v}{\mathrm{d}t}, \mathrm{grad}\,\frac{\mathrm{d}^v\beta_v}{\mathrm{d}t}, \mathrm{grad}\,\frac{\mathrm{d}^l\beta_l}{\mathrm{d}t},\\ &\mathbf{U}_v-\mathbf{U}_l,\ \mathbf{U}_v-\mathbf{U}_a,\ \mathbf{U}_l-\mathbf{U}_a,\\ &\frac{\mathrm{d}^v\tau_v}{\mathrm{d}t}, \frac{\mathrm{d}^l\tau_l}{\mathrm{d}t}, \frac{\mathrm{d}^a\tau_a}{\mathrm{d}t}, \frac{\mathrm{d}^l\beta_v}{\mathrm{d}t}, \frac{\mathrm{d}^v\beta_l}{\mathrm{d}t},\\ &\mathrm{grad}\,T).\end{aligned} \tag{13.1}$$

Velocities

$$\frac{\mathrm{d}^v\tau_v}{\mathrm{d}t}, \frac{\mathrm{d}^l\tau_l}{\mathrm{d}t}, \frac{\mathrm{d}^a\tau_a}{\mathrm{d}t}, \frac{\mathrm{d}^l\beta_v}{\mathrm{d}t}, \frac{\mathrm{d}^v\beta_l}{\mathrm{d}t},$$

do not appear in the power of the interior forces of the following section, they intervene like velocities $d\eta/dt$ of Sect. 3.3.1.

## 13.4 Mass Balances

Let us consider subdomains $\mathcal{O}_v(\tau)$ and $\mathcal{O}_l(\tau)$ of domain $\Omega(\tau)$ occupied by the mixture at time $\tau$. The mass of vapor which is contained in $\mathcal{O}_v(\tau)$ is

$$\int_{\mathcal{O}_v(\tau)} \frac{\beta_v(\mathbf{x},\tau)}{\tau_v(\mathbf{x},\tau)} d\Omega.$$

The mass of liquid water which is contained in $\mathcal{O}_l(\tau)$ is

$$\int_{\mathcal{O}_l(\tau)} \frac{\beta_l(\mathbf{x},\tau)}{\tau_l(\mathbf{x},\tau)} d\Omega.$$

Let us choose $\mathcal{O}_v(\tau)$, $\mathcal{O}_l(\tau)$ such that at time $t$, $\mathcal{O}_v(t) = \mathcal{O}_l(t) = \mathcal{O}$. The mass of water which is contained in subdomain $\mathcal{O}$ at time $t$ is constant. It results

$$\forall t,\ \forall \mathcal{O},\ \mathcal{O}_v(t) = \mathcal{O}_l(t) = \mathcal{O},$$

$$\frac{\mathrm{d}^v}{\mathrm{d}\tau}\left\{\int_{\mathcal{O}_v(\tau)} \frac{\beta_v(\mathbf{x}(\tau),\tau)}{\tau_v(\mathbf{x}(\tau),\tau)} d\Omega\right\}(t) + \frac{\mathrm{d}^l}{\mathrm{d}\tau}\left\{\int_{\mathcal{O}_l(\tau)} \frac{\beta_l(\mathbf{x}(\tau),\tau)}{\tau_l(\mathbf{x}(\tau),\tau)} d\Omega\right\}(t) = 0,$$

where the $d^v/dt$ and $d^l/dt$ are the material derivatives.

The air mass balance is

$$\forall t,\ \forall \mathcal{O},\ \mathcal{O}_a(t) = \mathcal{O},\ \frac{\mathrm{d}^a}{\mathrm{d}\tau}\left\{\int_{\mathcal{O}_a(\tau)} \frac{\beta_v(\mathbf{x}(\tau),\tau)}{\tau_a(\mathbf{x}(\tau),\tau)} d\Omega\right\}(t) = 0.$$

By using relationship

$$\frac{\mathrm{d}}{\mathrm{d}\tau}\left\{\int_{\Omega(\tau)} f(\mathbf{x}(\tau),\tau)d\Omega\right\}(t)=\int_{\Omega(t)}\left\{\frac{\mathrm{d}f}{\mathrm{d}t}(\mathbf{x},t)+f(\mathbf{x},t)\,\mathrm{div}\,\mathbf{U}(\mathbf{x},t)\right\}d\Omega, \quad (13.2)$$

where material points $\mathbf{x}(t)$ move with velocity $\mathbf{U}$, it results the mass balance equations:

- For the water

$$\frac{\mathrm{d}^v}{\mathrm{d}t}\left(\frac{\beta_v}{\tau_v}\right)+\frac{\mathrm{d}^l}{\mathrm{d}t}\left(\frac{\beta_l}{\tau_l}\right)+\frac{\beta_v}{\tau_v}\,\mathrm{div}\,\mathbf{U}_v+\frac{\beta_l}{\tau_l}\,\mathrm{div}\,\mathbf{U}_l=0. \quad (13.3)$$

- And for the air which occupies the same domain than the vapor

$$\frac{\mathrm{d}^a}{\mathrm{d}t}\left(\frac{\beta_v}{\tau_a}\right)+\frac{\beta_v}{\tau_a}\,\mathrm{div}\,\mathbf{U}_a=0. \quad (13.4)$$

They are relationships between elements of $\delta E$ and they are going to be taken into account by the pseudopotential of dissipation.

## 13.5 Equations of Motion

They result from the principle of virtual power. The important power is the power of the interior forces which introduces the different internal forces by their power.

### 13.5.1 The Power of the Interior Forces

It is the sum of different powers, each of them being the scalar product of a velocity of deformation by the related internal force. The power we choose is

$$\begin{aligned}\mathscr{P}_{int}(\Omega) = -\int_{\Omega} &\{\sigma_v : \mathsf{D}(\mathbf{U}_v)+\sigma_l : \mathsf{D}(\mathbf{U}_l)+\sigma_a : \mathsf{D}(\mathbf{U}_a)\\ &+B_{vv}\frac{\mathrm{d}^v\beta_v}{\mathrm{d}t}+B_{ll}\frac{\mathrm{d}^l\beta_l}{\mathrm{d}t}+B_{va}\frac{\mathrm{d}^a\beta_v}{\mathrm{d}t}\\ &+\mathbf{H}_v\cdot\mathrm{grad}\frac{\mathrm{d}^v\beta_v}{\mathrm{d}t}+\mathbf{H}_l\cdot\mathrm{grad}\frac{\mathrm{d}^l\beta_l}{\mathrm{d}t}\\ &+\mathbf{f}_{vl}\cdot(\mathbf{U}_v-\mathbf{U}_l)+\mathbf{f}_{va}\cdot(\mathbf{U}_v-\mathbf{U}_a)+\mathbf{f}_{la}\cdot(\mathbf{U}_l-\mathbf{U}_a)\}\,d\Omega,\end{aligned}$$

where $\Omega$ is the domain occupied by the mixture at time $t$. Works $B$ and works flux vectors $\mathbf{H}$ account for the microscopic motions which intervene in the vapor-liquid phase change. The novel quantity $\mathbf{H}$ describes the influence of the neighbourhood of a point onto this point for the microscopic motion (velocity $\mathrm{d}\beta/\mathrm{d}t$ represents the effect at the macroscopic level of the velocities at the microscopic level). Let us note, that in the same way, stress $\sigma$ describes the effect of the neighbourhood of a point onto this point for the macroscopic motion. Forces $\mathbf{f}_{vl}$ account for interactions, for instance frictions, between vapor and liquid with relative velocity $\mathbf{U}_v - \mathbf{U}_l$.

*Remark 13.1.* Let us recall that the choice of the power of the internal forces depends on the scientist aiming to predict the motion of a mechanical system, [112]. The choice results mostly from the choice of $\delta E$ which contains the deformation velocities which are seen as being important and useful. The choice of the power of the internal forces is not fixed once for all. There are many possibilities resulting in different models.

*Remark 13.2.* In terms of mathematics, an internal force is an element of the dual space of the linear space spanned by the related velocities of deformation. The two spaces are equipped with a bilinear form which is the power, see Appendix A. Let us recall that the quantity which is chosen by the engineer is the velocity of deformation. It is the quantity which is seen and measured in experiments. The internal forces are abstract quantities whose basic property is to be the dual quantity of the velocity of deformation in the bilinear form to give the power which is also measured in experiments.

Let us define vector

$$F^{mec} = (\sigma_v, \sigma_l, \sigma_a, B_{vv}, B_{ll}, B_{va}, \mathbf{H}_v, \mathbf{H}_l, \mathbf{f}_{vl}, \mathbf{f}_{va}, \mathbf{f}_{la}, 0, 0, 0, 0, 0, 0),$$

and get

$$\mathscr{P}_{int}(\Omega) = -\int_{\Omega} F^{mec} \cdot \delta E d\Omega.$$

The last five but one 0 of vector $F^{mec}$ are related to velocities $\mathrm{d}\eta/\mathrm{d}t$ which do not intervene in the power of the interior forces. The last 0 is related to $\mathrm{grad}\,T$ which do not either intervene in the power of the interior forces.

### 13.5.2 The Power of the Acceleration Forces

The variation of the linear momentum of the vapor

$$\frac{\mathrm{d}^v}{\mathrm{d}t}\left\{\int_{\mathscr{O}} \frac{\beta_v}{\tau_v}\mathbf{U}_v d\Omega\right\} = \int_{\Omega(t)} \left\{\frac{\mathrm{d}^v}{\mathrm{d}t}(\frac{\beta_v}{\tau_v}\mathbf{U}_v) + (\frac{\beta_v}{\tau_v}\mathbf{U}_v)\,\mathrm{div}\,\mathbf{U}_v\right\} d\Omega,$$

using formula (13.2) introduces the acceleration. Thus the actual power of the acceleration forces is

$$\mathscr{P}_{acc}(\Omega) = \int_\Omega \left\{ \frac{\mathrm{d}^v}{\mathrm{d}t}(\frac{\beta_v}{\tau_v}\mathbf{U}_v) + (\frac{\beta_v}{\tau_v}\mathbf{U}_v)\,\mathrm{div}\,\mathbf{U}_v \right\} \cdot \mathbf{U}_v$$

$$+ \left\{ \frac{\mathrm{d}^l}{\mathrm{d}t}(\frac{\beta_l}{\tau_l}\mathbf{U}_l) + (\frac{\beta_l}{\tau_l}\mathbf{U}_l)\,\mathrm{div}\,\mathbf{U}_l \right\} \cdot \mathbf{U}_l$$

$$+ \left\{ \frac{\mathrm{d}^a}{\mathrm{d}t}(\frac{\beta_a}{\tau_l}\mathbf{U}_a) + (\frac{\beta_a}{\tau_a}\mathbf{U}_a)\,\mathrm{div}\,\mathbf{U}_a \right\} \cdot \mathbf{U}_a d\Omega.$$

The variation of the linear momentum

$$\frac{\mathrm{d}^v}{\mathrm{d}t}(\frac{\beta_v}{\tau_v}\mathbf{U}_v) + (\frac{\beta_v}{\tau_v}\mathbf{U}_v)\,\mathrm{div}\,\mathbf{U}_v,$$

is the sum of the usual vapor linear momentum variation

$$\frac{\beta_v}{\tau_v}\frac{\mathrm{d}^v}{\mathrm{d}t}\mathbf{U}_v,$$

and of the linear momentum variation resulting from the production of vapor

$$\left\{ \frac{\mathrm{d}^v}{\mathrm{d}t}(\frac{\beta_v}{\tau_v}) + \frac{\beta_v}{\tau_v}\,\mathrm{div}\,\mathbf{U}_v \right\} \mathbf{U}_v,$$

due to the phase change.

Let us note that due to the air mass balance (13.4), we have

$$\left\{ \frac{\mathrm{d}^a}{\mathrm{d}t}(\frac{\beta_a}{\tau_l}\mathbf{U}_a) + (\frac{\beta_a}{\tau_a}\mathbf{U}_a)\,\mathrm{div}\,\mathbf{U}_a \right\} = \frac{\beta_a}{\tau_l}\frac{\mathrm{d}^a}{\mathrm{d}t}\mathbf{U}_a.$$

### 13.5.3 Power of the Exterior Forces

The actual power is

$$\mathscr{P}_{ext}(\Omega) = \int_\Omega \mathbf{f}_v^{ext} \cdot \mathbf{U}_v + \mathbf{f}_l^{ext} \cdot \mathbf{U}_l + \mathbf{f}_a^{ext} \cdot \mathbf{U}_a d\Omega$$

$$+ \int_{\partial\Omega} \mathbf{g}_v^{ext} \cdot \mathbf{U}_v + \mathbf{g}_l^{ext} \cdot \mathbf{U}_l + \mathbf{g}_a^{ext} \cdot \mathbf{U}_a d\Gamma,$$

where the $\mathbf{f}_v^{ext}$ are the volume exterior forces and the $\mathbf{g}_v^{ext}$ are the surface exterior tractions, $\mathbf{N}$ is the normal outward vector to domain $\Omega$.

### 13.5.4 The Equations of Motion

The equations of motion resulting from the principle of virtual power are:

- The macroscopic linear momentum equations of motion. The linear momentum equations of motion

$$\begin{aligned}
\frac{d^v}{dt}(\frac{\beta_v}{\tau_v}\mathbf{U}_v)+(\frac{\beta_v}{\tau_v}\mathbf{U}_v)\operatorname{div}\mathbf{U}_v &= \operatorname{div}\sigma_v-\mathbf{f}_{vl}-\mathbf{f}_{va}+\mathbf{f}_v^{ext},\ in\ \Omega,\\
\sigma_v\mathbf{N} &= \mathbf{g}_v^{ext},\ on\ \partial\Omega,\\
\frac{d^l}{dt}(\frac{\beta_l}{\tau_l}\mathbf{U}_l)+(\frac{\beta_l}{\tau_l}\mathbf{U}_l)\operatorname{div}\mathbf{U}_l &= \operatorname{div}\sigma_l+\mathbf{f}_{vl}-\mathbf{f}_{la}+\mathbf{f}_l^{ext},\ in\ \Omega,\\
\sigma_l\mathbf{N} &= \mathbf{g}_l^{ext},\ on\ \partial\Omega,\\
\frac{\beta_a}{\tau_l}\frac{d^a}{dt}\mathbf{U}_a &= \operatorname{div}\sigma_a+\mathbf{f}_{la}+\mathbf{f}_{va}+\mathbf{f}_a^{ext},\ in\ \Omega,\\
\sigma_a\mathbf{N} &= \mathbf{g}_a^{ext},\ on\ \partial\Omega,
\end{aligned}\tag{13.5}$$

or

$$\begin{aligned}
\frac{\beta_v}{\tau_v}\frac{d^v}{dt}\mathbf{U}_v+\mathbf{U}_v\left\{\frac{d^v}{dt}(\frac{\beta_v}{\tau_v})+\frac{\beta_v}{\tau_v}\operatorname{div}\mathbf{U}_v\right\} &= \operatorname{div}\sigma_v-\mathbf{f}_{vl}-\mathbf{f}_{va}+\mathbf{f}_v^{ext},\ in\ \Omega,\\
\sigma_v\mathbf{N} &= \mathbf{g}_v^{ext}\ on\ \partial\Omega,\\
\frac{\beta_l}{\tau_l}\frac{d^l}{dt}\mathbf{U}_l+\mathbf{U}_l\left\{\frac{d^l}{dt}(\frac{\beta_l}{\tau_l})+\frac{\beta_l}{\tau_l}\operatorname{div}\mathbf{U}_l\right\} &= \operatorname{div}\sigma_l+\mathbf{f}_{vl}-\mathbf{f}_{la}+\mathbf{f}_l^{ext},\ in\ \Omega,\\
\sigma_l\mathbf{N} &= \mathbf{g}_l^{ext},\ on\ \partial\Omega,\\
\frac{\beta_a}{\tau_l}\frac{d^a}{dt}\mathbf{U}_a &= \operatorname{div}\sigma_a+\mathbf{f}_{la}+\mathbf{f}_{va}+\mathbf{f}_a^{ext},\ in\ \Omega,\\
\sigma_a\mathbf{N} &= \mathbf{g}_a^{ext},\ on\ \partial\Omega.
\end{aligned}$$

*Remark 13.3.* Note that the sum of the two first equation is the water linear momentum equation of motion where the inertia term is

$$\begin{aligned}
&\frac{\beta_v}{\tau_v}\frac{d^v}{dt}\mathbf{U}_v+\mathbf{U}_v\left\{\frac{d^v}{dt}(\frac{\beta_v}{\tau_v})+\frac{\beta_v}{\tau_v}\operatorname{div}\mathbf{U}_v\right\}+\frac{\beta_l}{\tau_l}\frac{d^l}{dt}\mathbf{U}_l+\mathbf{U}_l\left\{\frac{d^l}{dt}(\frac{\beta_l}{\tau_l})+\frac{\beta_l}{\tau_l}\operatorname{div}\mathbf{U}_l\right\}\\
&=\frac{\beta_v}{\tau_v}\frac{d^v}{dt}\mathbf{U}_v+\frac{\beta_l}{\tau_l}\frac{d^l}{dt}\mathbf{U}_l+(\mathbf{U}_v-\mathbf{U}_l)\left\{\frac{d^v}{dt}(\frac{\beta_v}{\tau_v})+\frac{\beta_v}{\tau_v}\operatorname{div}\mathbf{U}_v\right\},
\end{aligned}$$

taking into account the water mass balance (13.3). The last quantity is the linear momentum of the material points changing phase from liquid water with velocity $\mathbf{U}_l$ to vapor with velocity $\mathbf{U}_v$. The amount of mass changing phase and becoming vapor is

$$\left\{ \frac{d^v}{dt}(\frac{\beta_v}{\tau_v}) + \frac{\beta_v}{\tau_v} \operatorname{div} \mathbf{U}_v \right\}.$$

- The macroscopic angular momentum equations of motion

$$\sigma_v - \sigma_v^T = 0, \ \sigma_l - \sigma_l^T = 0, \ \sigma_a - \sigma_a^T = 0,$$

where $\sigma^T$ is the transposed matrix of matrix $\sigma$. They are satisfied if the constitutive laws give symmetric stresses. That is going to be the case in the sequel.
- The microscopic motion equations of motion

$$\begin{aligned} -B_{vv} + \operatorname{div} \mathbf{H}_v &= 0, \ in\ \Omega, \ \mathbf{H}_v \cdot \mathbf{N} = 0, \ on\ \partial\Omega, \\ -B_{ll} + \operatorname{div} \mathbf{H}_l &= 0, \ in\ \Omega, \ \mathbf{H}_l \cdot \mathbf{N} = 0, \ on\ \partial\Omega, \\ -B_{va} &= 0, \ in\ \Omega. \end{aligned} \tag{13.6}$$

These equations of motion are new. They describe the effects at the macroscopic level of the microscopic motions responsible for the liquid vapor phase changes, [107, 112].

## 13.6 Entropy Balance

Because the specific volumes intervene, the results of Chap. 3 which assume the densities are constant are to be modified. The changes result from the derivatives of volume integrals, for instance the time derivative of the internal energy present in a volume $\Omega(t)$.

### *13.6.1 The Energy Balance*

The energy balance of the cloud system is:

*The actual work of the acceleration forces plus the variation of the internal energy is equal to the work of the exterior forces plus the quantity of heat provided to the system.*

The actual power of the acceleration forces is

$$\begin{aligned} \mathscr{P}_{acc}(\mathscr{O}) = & \int_{\mathscr{O}} \left\{ \frac{d^v}{dt}(\frac{\beta_v}{\tau_v}\mathbf{U}_v) + (\frac{\beta_v}{\tau_v}\mathbf{U}_v) \operatorname{div} \mathbf{U}_v \right\} \cdot \mathbf{U}_v d\Omega \\ & + \int_{\mathscr{O}} \left\{ \frac{d^l}{dt}(\frac{\beta_l}{\tau_l}\mathbf{U}_l) + (\frac{\beta_l}{\tau_l}\mathbf{U}_l) \operatorname{div} \mathbf{U}_l \right\} \cdot \mathbf{U}_l d\Omega \end{aligned}$$

$$
\begin{aligned}
&+\int_{\mathcal{O}} \left\{ \frac{\mathrm{d}^a}{\mathrm{d}t}(\frac{\beta_v}{\tau_a}\mathbf{U}_a) + (\frac{\beta_v}{\tau_a}\mathbf{U}_a)\operatorname{div}\mathbf{U}_a \right\} \cdot \mathbf{U}_a d\Omega \\
&= \frac{\mathrm{d}^v}{\mathrm{d}\tau}\left\{ \int_{\mathcal{O}_v(\tau)} \frac{\beta_v}{\tau_v}\frac{\mathbf{U}_v^2}{2} d\Omega \right\}(t) + \int_{\mathcal{O}} \left( \frac{\mathrm{d}^v}{\mathrm{d}t}(\frac{\beta_v}{\tau_v}) + \frac{\beta_v}{\tau_v}\operatorname{div}\mathbf{U}_v \right) \frac{\mathbf{U}_v^2}{2} d\Omega \\
&+\frac{\mathrm{d}^l}{\mathrm{d}\tau}\left\{ \int_{\mathcal{O}_v(\tau)} \frac{\beta_l}{\tau_l}\frac{\mathbf{U}_l^2}{2} d\Omega \right\}(t) + \int_{\mathcal{O}} \left( \frac{\mathrm{d}^l}{\mathrm{d}t}(\frac{\beta_l}{\tau_l}) + \frac{\beta_l}{\tau_l}\operatorname{div}\mathbf{U}_l \right) \frac{\mathbf{U}_l^2}{2} d\Omega \\
&+\frac{\mathrm{d}^a}{\mathrm{d}\tau}\left\{ \int_{\mathcal{O}_v(\tau)} \frac{\beta_v}{\tau_a}\frac{\mathbf{U}_a^2}{2} d\Omega \right\}(t).
\end{aligned}
$$

Denoting

$$
\mathcal{K} = \int_{\mathcal{O}(\tau)} \frac{\beta}{\tau}\frac{\mathbf{U}^2}{2} d\Omega,
$$

the different kinetic energies, we get with the water mass balance (13.3)

$$
\mathcal{P}_{acc}(\mathcal{O}) = \frac{\mathrm{d}^v \mathcal{K}_v}{\mathrm{d}\tau} + \frac{\mathrm{d}^l \mathcal{K}_l}{\mathrm{d}\tau} + \frac{\mathrm{d}^a \mathcal{K}_a}{\mathrm{d}\tau} + \int_{\mathcal{O}} \left( \frac{\mathrm{d}^v}{\mathrm{d}t}(\frac{\beta_v}{\tau_v}) + \frac{\beta_v}{\tau_v}\operatorname{div}\mathbf{U}_v \right) \left( \frac{\mathbf{U}_v^2}{2} - \frac{\mathbf{U}_l^2}{2} \right) d\Omega.
$$

Thus the energy balance is

$$
\begin{aligned}
&\forall t,\ \forall \mathcal{O},\ \mathcal{O}_v(t) = \mathcal{O}_l(t) = \mathcal{O}_a(t) = \mathcal{O}, \\
&\frac{\mathrm{d}^v \mathcal{K}_v}{\mathrm{d}\tau} + \frac{\mathrm{d}^l \mathcal{K}_l}{\mathrm{d}\tau} + \frac{\mathrm{d}^a \mathcal{K}_a}{\mathrm{d}\tau} + \int_{\mathcal{O}} \left( \frac{\mathrm{d}^v}{\mathrm{d}t}(\frac{\beta_v}{\tau_v}) + \frac{\beta_v}{\tau_v}\operatorname{div}\mathbf{U}_v \right) \left( \frac{\mathbf{U}_v^2}{2} - \frac{\mathbf{U}_l^2}{2} \right) d\Omega \\
&+\frac{\mathrm{d}^v}{\mathrm{d}\tau}\left\{ \int_{\mathcal{O}_v(\tau)} e_v d\Omega \right\} + \frac{\mathrm{d}^l}{\mathrm{d}\tau}\left\{ \int_{\mathcal{O}_l(\tau)} e_l d\Omega \right\} + \frac{\mathrm{d}^a}{\mathrm{d}\tau}\left\{ \int_{\mathcal{O}_a(\tau)} e_a d\Omega \right\} \\
&= \mathcal{P}_{ext}(\mathcal{O}) + \int_{\mathcal{O}} T R^{ext} d\Omega + \int_{\partial\mathcal{O}} -T\mathbf{Q}\cdot\mathbf{N} d\Gamma,
\end{aligned}
\tag{13.7}
$$

where the $e$'s are the volumic internal energies, $T\mathbf{Q}$ is the heat flux vector, $\mathbf{Q}$ being the entropy flux vector and $R^{ext}T$ is the exterior volumic source of heat, $R^{ext}$ being the exterior volumic source of entropy. Relationship (13.7) means that the heat and work provided to the cloud modify the internal energy, the kinetic energy of each phase and adapt the kinetic energy of material changing phase to the kinetic energy of the new phase.

Subtracting the principle of virtual power with the actual velocities relationship

$$
\mathcal{P}_{acc}(\mathcal{O}) = \mathcal{P}_{int}(\mathcal{O}) + \mathcal{P}_{ext}(\mathcal{O}),
$$

from the energy balance (13.7), we get the classical relationship

$$\forall t,\ \forall \mathscr{O},\ \mathscr{O}_v(t) = \mathscr{O}_l(t) = \mathscr{O}_a(t) = \mathscr{O},$$

$$\frac{\mathrm{d}^v}{\mathrm{d}\tau}\left\{\int_{\mathscr{O}_v(\tau)} e_v(\mathbf{x},\tau)d\Omega\right\} + \frac{\mathrm{d}^l}{\mathrm{d}\tau}\left\{\int_{\mathscr{O}_l(\tau)} e_l(\mathbf{x},\tau)d\Omega\right\} + \frac{\mathrm{d}^a}{\mathrm{d}\tau}\left\{\int_{\mathscr{O}_a(\tau)} e_a(\mathbf{x},\tau)d\Omega\right\}$$

$$= -\mathscr{P}_{int}(\mathscr{O}) + \int_{\mathscr{O}} TR^{ext} d\Omega + \int_{\partial\mathscr{O}} -T\mathbf{Q}\cdot\mathbf{N}d\Gamma.$$

It results the energy balance is

$$\left\{\frac{\mathrm{d}^v e_v}{\mathrm{d}t} + e_v \operatorname{div}\mathbf{U}_v\right\} + \left\{\frac{\mathrm{d}^l e_l}{\mathrm{d}t} + e_l \operatorname{div}\mathbf{U}_l\right\} + \left\{\frac{\mathrm{d}^a e_a}{\mathrm{d}t} + e_a \operatorname{div}\mathbf{U}_a\right\}$$

$$+\operatorname{div} T\mathbf{Q} = F^{mec}\cdot\delta E + R^{ext}T, \text{ in } \Omega, \tag{13.8}$$

$$-T\mathbf{Q}\cdot\mathbf{N} = T\pi^{ext}, \text{ in } \partial\Omega,$$

where $T\pi^{ext}$ is the exterior surfacic source of heat, $\pi^{ext}$ being the exterior surfacic source of entropy.

We define vector

$$F = (\sigma_v,\sigma_l,\sigma_a,B_{vv},B_{ll},B_{va},\mathbf{H}_v,\mathbf{H}_l,\mathbf{f}_{vl},\mathbf{f}_{va},\mathbf{f}_{la},0,0,0,0,0,-\mathbf{Q})$$

$$= F^{mec} + (\underbrace{16\ times\ 0}, -\mathbf{Q}).$$

Vector $F$ is equal to vector $F^{mec}$ where the last component of $F^{mec}$ which is equal to 0, is replaced by $-\mathbf{Q}$. Relationship (13.8) gives

$$\left\{\frac{\mathrm{d}^v e_v}{\mathrm{d}t} + e_v \operatorname{div}\mathbf{U}_v\right\} + \left\{\frac{\mathrm{d}^l e_l}{\mathrm{d}t} + e_l \operatorname{div}\mathbf{U}_l\right\} + \left\{\frac{\mathrm{d}^a e_a}{\mathrm{d}t} + e_a \operatorname{div}\mathbf{U}_a\right\}$$

$$+T\operatorname{div}\mathbf{Q} = F\cdot\delta E + R^{ext}T,\ in\ \Omega,$$

$$-T\mathbf{Q}\cdot\mathbf{N} = T\pi^{ext},\ on\ \partial\Omega. \tag{13.9}$$

### 13.6.2 The Second Law

It is

$$\forall t,\ \forall \mathscr{O},\ \mathscr{O}_v(t) = \mathscr{O}_l(t) = \mathscr{O}_a(t) = \mathscr{O},$$

$$\frac{\mathrm{d}^v}{\mathrm{d}\tau}\left\{\int_{\mathscr{O}_v(\tau)} s_v(\mathbf{x},\tau)d\Omega\right\} + \frac{\mathrm{d}^l}{\mathrm{d}\tau}\left\{\int_{\mathscr{O}_l(\tau)} s_l(\mathbf{x},\tau)d\Omega\right\} + \frac{\mathrm{d}^a}{\mathrm{d}\tau}\left\{\int_{\mathscr{O}_a(\tau)} s_a(\mathbf{x},\tau)d\Omega\right\}$$

$$\geq \int_{\mathscr{O}} R^{ext} d\Omega + \int_{\partial\mathscr{O}} -\mathbf{Q}\cdot\mathbf{N}d\Gamma,$$

and

$$T > 0. \tag{13.10}$$

It results

$$\left\{\frac{\mathrm{d}^v s_v}{\mathrm{d}t} + s_v \operatorname{div} \mathbf{U}_v\right\} + \left\{\frac{\mathrm{d}^l s_l}{\mathrm{d}t} + s_l \operatorname{div} \mathbf{U}_l\right\} + \left\{\frac{\mathrm{d}^a s_a}{\mathrm{d}t} + e_a \operatorname{div} \mathbf{U}_a\right\} + \operatorname{div} \mathbf{Q}$$
$$\geq R^{ext}, \text{ in } \Omega. \tag{13.11}$$

## *13.6.3 The Free Energies*

Non dissipative interior forces are defined with a potential. We choose the volumic free energy.

### 13.6.3.1 Volumic Vapor Free Energy and Its Time Derivative

It is

$$\tilde{\Psi}_v(E_v) = \beta_v \Psi_v(\tau_v, T) + I(\beta_v) + \frac{k_v}{2} (\operatorname{grad} \beta_v)^2 + I_0(\beta_v + \beta_l - 1),$$

where $\Psi_v(\tau_v, T)$ is given by physics, $I$ is the indicator function of interval $[0,1]$ and $I_0$ is the indicator function of the origin of $\mathbb{R}$. Dependence on $\operatorname{grad} \beta_v$ has been chosen quadratic for the sake of simplicity, with parameter $k_v$ non negative.

This free energy implies that $\beta_v + \beta_l - 1 = 0$, $\beta_v \in [0,1]$, $\beta_l \in [0,1]$. Water is either liquid or vapor (there is no ice). Moreover relationship $\beta_v + \beta_l - 1 = 0$ implies that there are no voids in the vapor liquid mixture and that no interpenetration occurs between liquid and vapor phases.

In energy balance equations (13.9) and (13.14) below, the internal constraints on the state quantities are satisfied because the laws of mechanics apply for actual evolutions. Thus time derivatives $\mathrm{d}^v e_v/\mathrm{d}t$ and $\mathrm{d}^v \tilde{\Psi}_v/\mathrm{d}t$ which intervene in (13.9) and (13.14) below are computed with the internal constraints on the state quantities which are satisfied. Thus in these relationships we have

$$\frac{\mathrm{d}^v \tilde{\Psi}_v}{\mathrm{d}t} = \beta_v \frac{\partial \Psi_v}{\partial \tau_v} \frac{\mathrm{d}^v \tau_v}{\mathrm{d}t} + \Psi_v \frac{\mathrm{d}^v \beta_v}{\mathrm{d}t}$$
$$+ k_v \operatorname{grad} \beta_v \cdot \operatorname{grad} \frac{\mathrm{d}^v \beta_v}{\mathrm{d}t} - k_v \operatorname{grad} \beta_v \otimes \operatorname{grad} \beta_v : \operatorname{grad} \mathbf{U}_v + \beta_v \frac{\partial \Psi_v}{\partial T} \frac{\mathrm{d}^v T}{\mathrm{d}t},$$

where we have used

$$\frac{\mathrm{d}^v \operatorname{grad} \beta_v}{\mathrm{d}t} = \operatorname{grad} \frac{\mathrm{d}^v \beta_v}{\mathrm{d}t} - \operatorname{grad} \beta_v \cdot \operatorname{grad} \mathbf{U}_v,$$

with

$$(\operatorname{grad}\beta_v \cdot \operatorname{grad}\mathbf{U}_v)_j = (\beta_v)_i(\mathbf{U}_v)_{i,j},$$

and

$$(\operatorname{grad}\beta_v \otimes \operatorname{grad}\beta_v)_{ij} = (\beta_v)_{,i}(\beta_v)_{,j},$$

which is symmetric.

The reactions to the internal constraints are defined by

$$B_{reacvv} \in \partial I(\beta_v),\ B_{reacv} \in \partial I_0(\beta_v + \beta_l - 1).$$

They satisfy, following Theorem 3.2

$$(B_{reacvv} + B_{reacv})\frac{\mathrm{d}^v\beta_v}{\mathrm{d}t} + B_{reacv}\frac{\mathrm{d}^v\beta_l}{\mathrm{d}t} \geq 0.$$

*Remark 13.4.* The reactions depend on the values of $\beta_v$ and of $\beta_v + \beta_l - 1$ but they depend also on $\mathbf{x}$ and $t$: they are $B_{reacvv}(\beta_v, \mathbf{x}, t)$ and $B_{reacv}(\beta_v + \beta_l - 1, \mathbf{x}, t)$. In the sequel we write $B_{reacvv}$ and $B_{reacv}$ instead of $B_{reacvv}(\beta_v, \mathbf{x}, t)$ and of $B_{reacv}(\beta_v + \beta_l - 1, \mathbf{x}, t)$.

Let us recall that if the evolution is smooth (see definition in Remark 3.9), the inequality is an equality and the reactions are workless. For instance, this is the case, if either functions $t \to \beta$ are differentiable or reactions $t \to B_{reacvv}$ and $t \to B_{reacv}$ are continuous.

#### 13.6.3.2 Volumic Liquid Water Free Energy and Its Time Derivative

It is

$$\tilde{\Psi}_l(E_l) = \beta_l \Psi_l(\tau_l, T) + I(\beta_l) + \frac{k_l}{2}(\operatorname{grad}\beta_l)^2 + I_0(\beta_v + \beta_l - 1),$$

where $\Psi_l(\tau_l, T)$ is given by physics.

*Remark 13.5.* The dependence of $\tilde{\Psi}_v$ and $\tilde{\Psi}_l$ on $\operatorname{grad}\beta_v$ and $\operatorname{grad}\beta_l$ has been chosen quadratic for the sake of simplicity. In a less schematic theory, we may choose a non quadratic dependence with $k_v = k_v(\tau_v, T)$ and $k_l = k_l(\tau_l, T)$. Moreover we may assume $k_v = k_l$. This assumption is made in the sequel, see formula (13.28).

In an actual evolution, the time derivative of the liquid water free energy satisfies

$$\frac{\mathrm{d}^l\tilde{\Psi}_l}{\mathrm{d}t} = \beta_l \frac{\partial \Psi_l}{\partial \tau_l}\frac{\mathrm{d}^l\tau_l}{\mathrm{d}t} + \Psi_l\frac{\mathrm{d}^l\beta_l}{\mathrm{d}t}$$
$$+k_l \operatorname{grad}\beta_l \cdot \operatorname{grad}\frac{\mathrm{d}^l\beta_l}{\mathrm{d}t} - k_l \operatorname{grad}\beta_l \otimes \operatorname{grad}\beta_l : \operatorname{grad}\mathbf{U}_l + \beta_l\frac{\partial \Psi_l}{\partial T}\frac{\mathrm{d}^l T}{\mathrm{d}t}.$$

The reactions to the internal constraints on the state quantities are defined by

$$B_{reacll} \in \partial I(\beta_v),\ B_{reacl} \in \partial I_0(\beta_v + \beta_l - 1).$$

They satisfy, following Theorem 3.2

$$(B_{reacll} + B_{reacl})\frac{d^l \beta_l}{dt} + B_{reacl}\frac{d^l \beta_v}{dt} \geq 0,$$

the inequality being an equality in case the evolution is smooth (see Remark 3.9).

#### 13.6.3.3 Volumic Air Free Energy

It is

$$\tilde{\Psi}_a(E_l) = \beta_v \Psi_a(\tau_a, T) + I(\beta_v),$$

where $\Psi_a(\tau_a, T)$ is given by physics, for instance

$$\Psi_a(\tau_a, T) = -rT\frac{\ln \tau_a}{\tau_a}, \tag{13.12}$$

where $r$ is a constant. This free energy is the ideal gas free energy. We do not add the indicator function $I_0(\beta_v + \beta_l - 1)$ because air does not change phase, ($\beta_v$ being the volume fraction occupied by the air).

In an actual evolution, the time derivative of the air free energy satisfies

$$\frac{d^a \tilde{\Psi}_a}{dt} = \beta_v \frac{\partial \Psi_a}{\partial \tau_a}\frac{d^a \tau_a}{dt} + \Psi_a \frac{d^v \beta_v}{dt} + \beta_v \frac{\partial \Psi_a}{\partial T}\frac{d^a T}{dt}.$$

The reaction to the internal constraint is defined by

$$B_{reacva} \in \partial I(\beta_v),$$

which satisfies, following Theorem 3.2

$$B_{reacva}\frac{d^v \beta_v}{dt} \geq 0,$$

the inequality being an equality in case the evolution is smooth (see Remark 3.9).

*Remark 13.6.* The three specific volumes $\tau$ are positive quantities. This physical property is to be taken into account by each free energy $\Psi$ by adding to $\Psi$ the indicator function $I_+(\tau)$ of $\mathbb{R}^+$, $\Psi = \Psi + I_+(\tau)$. For the sake of simplicity, we do not mention this property in the sequel. The free energies we are going to choose will satisfy $\lim_{\tau \to 0} \Psi(\tau) = \infty$. Thus if specific volume is non negative at time 0, it remains non negative.

Let us note that the specific free energies are $\hat{\Psi}_v = \tau_v \Psi_v$ and $\hat{\Psi}_l = \tau_l \Psi_l$.

#### 13.6.3.4 The Non Dissipative Interior Forces and the Non Dissipative Reactions Forces

We define the non dissipative interior forces vector $F^{nd}$ with the derivatives of free energies $\tilde{\Psi}_v$, $\tilde{\Psi}_l$ and $\tilde{\Psi}_a$. It is

$$
\begin{aligned}
F^{nd} &= \Big\{ \sigma_v^{nd} = -k_v \operatorname{grad} \beta_v \otimes \operatorname{grad} \beta_v + \beta_v \Psi_v 1, \\
\sigma_l^{nd} &= -k_l \operatorname{grad} \beta_l \otimes \operatorname{grad} \beta_l + \beta_l \Psi_l 1, \ \sigma_a^{nd} = \beta_v \Psi_a 1, \\
B_{vv}^{nd} &= \Psi_v, \ B_{ll}^{nd} = \Psi_l, \\
B_{va}^{nd} &= \Psi_a, \\
\mathbf{H}_v^{nd} &= k_v \operatorname{grad} \beta_v, \ \mathbf{H}_l^{nd} = k_l \operatorname{grad} \beta_l, \\
\mathbf{f}_{vl}^{nd} &= 0, \ \mathbf{f}_{va}^{nd} = 0, \\
\mathbf{f}_{la}^{nd} &= 0, \\
Z_v^{nd} &= \beta_v \frac{\partial \Psi_v}{\partial \tau_v}, \ Z_l^{nd} = \beta_l \frac{\partial \Psi_l}{\partial \tau_l}, \ Z_a^{nd} = \beta_v \frac{\partial \Psi_a}{\partial \tau_a}, \\
\hat{B}_{vl}^{nd} &= 0, \ \hat{B}_{lv}^{nd} = 0, \ -\mathbf{Q}^{nd} = 0 \Big\},
\end{aligned}
$$

where 1 is the identity matrix.

We define the non dissipative reaction forces vector $F^{reac}$ with the reactions to the internal constraints taken into account by the free energies. It is

$$
\begin{aligned}
F^{reac} &= \{ \sigma_v^{reac} = 0, \ \sigma_l^{reac} = 0, \ \sigma_a^{reac} = 0, \\
B_{vv}^{reac} &= B_{reacvv} + B_{reacv}, \ B_{ll}^{reac} = B_{reacll} + B_{reacl}, \\
B_{va}^{reac} &= B_{reacva}, \\
\mathbf{H}_v^{reac} &= 0, \ \mathbf{H}_l^{reac} = 0, \\
\mathbf{f}_{vl}^{reac} &= 0, \ \mathbf{f}_{va}^{reac} = 0, \\
\mathbf{f}_{la}^{reac} &= 0, \\
Z_v^{reac} &= 0, \ Z_l^{reac} = 0, \ Z_a^{reac} = 0, \\
\hat{B}_{vl}^{reac} &= B_{reacv}, \ \hat{B}_{lv}^{reac} = B_{reacl}, \\
& \qquad -\mathbf{Q}^{reac} = 0 \},
\end{aligned}
$$

Let us recall that the entropies are (see Sect. 3.2.3)

$$s_v = -\beta_v \frac{\partial \Psi_v}{\partial T}, \; s_l = -\beta_l \frac{\partial \Psi_l}{\partial T}, \; s_a = -\beta_v \frac{\partial \Psi_a}{\partial T}.$$

With the computations of the time derivatives of the free energies and the definition of $F^{nd}$, we get

$$\begin{aligned}
& s_v \frac{d^s T}{dt} + s_l \frac{d^l T}{dt} + s_a \frac{d^a T}{dt} \\
& + \frac{d^v \tilde{\Psi}_v}{dt} + \tilde{\Psi}_v \operatorname{div} \mathbf{U}_v + \frac{d^l \tilde{\Psi}_l}{dt} + \tilde{\Psi}_l \operatorname{div} \mathbf{U}_l + \frac{d^a \tilde{\Psi}_a}{dt} + \tilde{\Psi}_a \operatorname{div} \mathbf{U}_a \\
& = F^{nd} \cdot \delta E. \qquad (13.13)
\end{aligned}$$

We have also

$$F^{reac} \cdot \delta E \geq 0.$$

Let us recall that assuming smooth evolution, i.e., the non dissipative reactions are workless, the inequality becomes an equality (see Remark 3.9). This is why vector $F^{reac}$ is called non dissipative reaction vector.

### 13.6.4 The Entropy Balance

Following the definition of the free energy given in Sect. 3.2.3, we use relationship $e = \tilde{\Psi} + Ts$, and get from energy balance (13.9)

$$\begin{aligned}
& T \left\{ \frac{d^v s_v}{dt} + s_v \operatorname{div} \mathbf{U}_v + \frac{d^l s_l}{dt} + s_l \operatorname{div} \mathbf{U}_l + \frac{d^a s_a}{dt} + s_a \operatorname{div} \mathbf{U}_a + \operatorname{div} \mathbf{Q} \right\} \\
& + s_v \frac{d^s T}{dt} + s_l \frac{d^l T}{dt} + s_a \frac{d^a T}{dt} \\
& + \frac{d^v \tilde{\Psi}_v}{dt} + \tilde{\Psi}_v \operatorname{div} \mathbf{U}_v + \frac{d^l \tilde{\Psi}_l}{dt} + \tilde{\Psi}_l \operatorname{div} \mathbf{U}_l + \frac{d^a \tilde{\Psi}_a}{dt} + \tilde{\Psi}_a \operatorname{div} \mathbf{U}_a \\
& = F \cdot \delta E + R^{ext} T, \; in \; \Omega, \qquad (13.14)
\end{aligned}$$

By relationship (13.13), we get

$$\begin{aligned}
& T \left\{ \frac{d^v s_v}{dt} + s_v \operatorname{div} \mathbf{U}_v + \frac{d^l s_l}{dt} + s_l \operatorname{div} \mathbf{U}_l + \frac{d^a s_a}{dt} + s_a \operatorname{div} \mathbf{U}_a + \operatorname{div} \mathbf{Q} \right\} \\
& = \left( F - F^{nd} \right) \cdot \delta E + R^{ext} T, \; in \; \Omega, \qquad (13.15)
\end{aligned}$$

The interior forces and entropy flux vector $F$ is split between a dissipative part $F^d$ with index $^d$, a non dissipative part $F^{nd}$ with index $^{nd}$ and a reaction part $F^{reac}$ with index $^{reac}$

$$F = F^d + F^{nd} + F^{reac},$$

with

$$F = (\sigma_v, \sigma_l, \sigma_a, B_{vv}, B_{ll}, B_{va}, \mathbf{H}_v, \mathbf{H}_l, \ \mathbf{f}_{vl}, \mathbf{f}_{va}, \mathbf{f}_{la}, 0,0,0,0,0, -\mathbf{Q}),$$

$$F^d = (\sigma_v^d, \sigma_l^d, \sigma_a^d, B_{vv}^d, B_{ll}^d, B_{va}^d, \mathbf{H}_v^d, \mathbf{H}_l^d, \ \mathbf{f}_{vl}^d, \mathbf{f}_{va}^d, \mathbf{f}_{la}^d, Z_v^d, Z_l^d, Z_a^d, \hat{B}_{vl}^d, \hat{B}_{lv}^d, -\mathbf{Q}^d),$$

$$\begin{aligned} F^{nd} = (&\sigma_v^{nd}, \sigma_l^{nd}, \sigma_a^{nd}, B_{vv}^{nd}, B_{ll}^{nd}, B_{va}^{nd}, \mathbf{H}_v^{nd}, \mathbf{H}_l^{nd}, \\ &\mathbf{f}_{vl}^{nd}, \mathbf{f}_{va}^{nd}, \mathbf{f}_{la}^{nd}, Z_v^{nd}, Z_l^{nd}, Z_a^{nd}, \hat{B}_{vl}^{nd}, \hat{B}_{lv}^{nd}, -\mathbf{Q}^{nd}), \end{aligned}$$

and

$$\begin{aligned} F^{reac} = (&\sigma_v^{reac}, \sigma_l^{reac}, \sigma_a^{reac}, B_{vv}^{reac}, B_{ll}^{reac}, B_{va}^{reac}, \mathbf{H}_v^{reac}, \mathbf{H}_l^{reac}, \\ &\mathbf{f}_{vl}^{reac}, \mathbf{f}_{va}^{reac}, \mathbf{f}_{la}^{reac}, Z_v^{reac}, Z_l^{reac}, Z_a^{reac}, \hat{B}_{vl}^{reac}, \hat{B}_{lv}^{reac}, -\mathbf{Q}^{reac}). \end{aligned}$$

It results from energy balance (13.15) and relationship (13.10)

$$\begin{aligned} &\left\{ \frac{\mathrm{d}^v s_v}{\mathrm{d}t} + s_v \operatorname{div} \mathbf{U}_v \right\} + \left\{ \frac{\mathrm{d}^l s_l}{\mathrm{d}t} + s_l \operatorname{div} \mathbf{U}_l \right\} + \left\{ \frac{\mathrm{d}^a s_a}{\mathrm{d}t} + s_a \operatorname{div} \mathbf{U}_a \right\} + \operatorname{div} \mathbf{Q} \\ &= \mathscr{D} + \mathscr{D}^{reac} + R^{ext}, \ in\ \Omega, \\ &-\mathbf{Q} \cdot \mathbf{N} = \pi^{ext}, \ on\ \partial\Omega, \end{aligned} \tag{13.16}$$

where

$$\mathscr{D} = \frac{1}{T} \left( \delta E \cdot F^d \right),$$

the volume interior rate of entropy production is as usual the product of $\delta E$ by the dissipative forces $F^d$ and

$$\mathscr{D}^{reac} = \frac{1}{T} (\delta E \cdot F^{reac}) \geq 0, \tag{13.17}$$

is the non dissipative reaction forces entropy production which is null when the evolution is smooth. We have already assumed this property and have the entropy balance

$$\left\{\frac{\mathrm{d}^v s_v}{\mathrm{d}t} + s_v \operatorname{div} \mathbf{U}_v\right\} + \left\{\frac{\mathrm{d}^l s_l}{\mathrm{d}t} + s_l \operatorname{div} \mathbf{U}_l\right\} + \left\{\frac{\mathrm{d}^a s_a}{\mathrm{d}t} + s_a \operatorname{div} \mathbf{U}_a\right\} + \operatorname{div} \mathbf{Q}$$
$$= \mathscr{D} + R^{ext}, \ in\ \Omega,$$
$$-\mathbf{Q} \cdot \mathbf{N} = \pi^{ext}, \ on\ \partial\Omega, \tag{13.18}$$

which is equivalent to the energy balance.

## 13.7 The Second Law: An Equivalent Formulation

By comparing relationships (13.11) and (13.18), we get an equivalent formulation of the second law

$$T > 0, \ \mathscr{D} \geq 0. \tag{13.19}$$

It is satisfied if $T > 0$ (which we assume or which may be ensured by having the free energies to depend on $\ln T$) and if constitutive law for $F^d$ is defined by a pseudopotential of dissipation $\Phi(E, \delta E)$

$$F^d \in \partial\Phi(E, \delta E), \tag{13.20}$$

where the subdifferential set is computed with respect to $\delta E$. The proof is given in Theorem 3.1.

*Remark 13.7.* In case the evolution is not smooth, second law is

$$T > 0, \ \mathscr{D} + \mathscr{D}^{reac} \geq 0.$$

Let us note that constitutive law (13.20) and relationship (13.17) ensure the second law is satisfied in this situation.

## 13.8 Pseudopotential of Dissipation

The pseudopotential of dissipation describes the dissipative properties which depend on the velocities and on the thermal heterogeneity, $\delta E$. Among these properties, there are the internal constraints on the velocities.

### 13.8.1 Internal Constraints on the Velocities

We have internal constraints on velocities

$$\frac{\mathrm{d}^l\beta_v}{\mathrm{d}t} = \frac{\partial\beta_v}{\partial t} + \mathbf{U}_l\cdot\mathrm{grad}\,\beta_v = \left(\frac{\partial\beta_v}{\partial t} + \mathbf{U}_v\cdot\mathrm{grad}\,\beta_v\right) + (\mathbf{U}_l - \mathbf{U}_v)\cdot\mathrm{grad}\,\beta_v$$
$$= \frac{\mathrm{d}^v\beta_v}{\mathrm{d}t} + (\mathbf{U}_l - \mathbf{U}_v)\cdot\mathrm{grad}\,\beta_v,$$

and

$$\frac{\mathrm{d}^v\beta_l}{\mathrm{d}t} = \frac{\mathrm{d}^l\beta_l}{\mathrm{d}t} + (\mathbf{U}_v - \mathbf{U}_l)\cdot\mathrm{grad}\,\beta_l.$$

These two internal constraints with $\beta_v + \beta_l = 1$ are equivalent to

$$\frac{\mathrm{d}^l\beta_v}{\mathrm{d}t} + \frac{\mathrm{d}^v\beta_l}{\mathrm{d}t} = \frac{1}{2}(\mathbf{U}_v - \mathbf{U}_l)\cdot\mathrm{grad}(\beta_l - \beta_v), \tag{13.21}$$
$$\beta_v + \beta_l = 1.$$

We have also

$$\frac{\mathrm{d}^a\beta_v}{\mathrm{d}t} = \frac{\mathrm{d}^v\beta_v}{\mathrm{d}t} + (\mathbf{U}_a - \mathbf{U}_v)\cdot\mathrm{grad}\,\beta_v. \tag{13.22}$$

Constraints (13.21) and (13.22) involve velocities. Thus they are taken into account by the pseudopotential of dissipation. In the same way, mass balances (13.3) and (13.4) are constraints on the velocities which are taken into account by the pseudopotential of dissipation.

### 13.8.2 The Pseudopotential of Dissipation

We choose

$$\Phi(E,\delta E) = I_0\left(\frac{\mathrm{d}^v}{\mathrm{d}t}\left(\frac{\beta_v}{\tau_v}\right) + \frac{\mathrm{d}^l}{\mathrm{d}t}\left(\frac{\beta_l}{\tau_l}\right) + \frac{\beta_v}{\tau_v}\mathrm{div}\,\mathbf{U}_v + \frac{\beta_l}{\tau_l}\mathrm{div}\,\mathbf{U}_l\right)$$
$$+I_0\left(\frac{\mathrm{d}^a}{\mathrm{d}t}\left(\frac{\beta_v}{\tau_a}\right) + \frac{\beta_v}{\tau_a}\mathrm{div}\,\mathbf{U}_a\right)$$
$$+I_0\left(\frac{\mathrm{d}^l\beta_v}{\mathrm{d}t} + \frac{\mathrm{d}^v\beta_l}{\mathrm{d}t} - \frac{1}{2}(\mathbf{U}_v - \mathbf{U}_l)\cdot\mathrm{grad}(\beta_l - \beta_v)\right)$$
$$+I_0\left(\frac{\mathrm{d}^a\beta_v}{\mathrm{d}t} - \frac{\mathrm{d}^v\beta_v}{\mathrm{d}t} - (\mathbf{U}_a - \mathbf{U}_v)\cdot\mathrm{grad}\,\beta_v\right)$$

$$
\begin{aligned}
&+\frac{\beta_v\beta_l k_{vl}}{2}(\mathbf{U}_v-\mathbf{U}_l)^2+\frac{\beta_v k_{va}}{2}(\mathbf{U}_v-\mathbf{U}_a)^2+\frac{\beta_v\beta_l k_{la}}{2}(\mathbf{U}_l-\mathbf{U}_a)^2\\
&+\frac{\beta_l}{2}\left\{\lambda_l(\operatorname{div}\mathbf{U}_l)^2+2\mu_l\mathsf{D}(\mathbf{U}_l):\mathsf{D}(\mathbf{U}_l)\right\}\\
&+\frac{\beta_v}{2}\left\{\lambda_v(\operatorname{div}\mathbf{U}_v)^2+2\mu_v\mathsf{D}(\mathbf{U}_v):\mathsf{D}(\mathbf{U}_v)\right\}\\
&+\frac{\beta_v}{2}\left\{\lambda_a(\operatorname{div}\mathbf{U}_a)^2+2\mu_a\mathsf{D}(\mathbf{U}_a):\mathsf{D}(\mathbf{U}_a)\right\}\\
&+\frac{\lambda}{2T}(\operatorname{grad}T)^2.
\end{aligned}
$$

We assume air, vapor and liquid water to be viscous. The $\lambda_i$ and $\mu_i$ are the viscosity parameters. The quadratic terms as $(\mathbf{U}_v-\mathbf{U}_l)^2$ account for frictions between the phases, the intensity of which is proportional to the velocity difference $(\mathbf{U}_v-\mathbf{U}_l)$. We have assume thermal conductivity $\lambda$ to be the same in any phase, it is possible to have $\lambda=\lambda(\beta_l,\beta_v)$ depending on the volume fractions.

## 13.9 Constitutive Laws

As usual and already done, the internal forces $F$ are split between non dissipative forces $F^{nd}$, non dissipative reaction forces $F^{reac}$ and dissipative forces $F^d$. Forces $F^{nd}$ and $F^{reac}$ have already been defined.

### *13.9.1 The Dissipative Internal Forces*

They are defined by relationship (13.20)

$$F^d\in\partial\Phi(E,\delta E).$$

We define the specific pressures $p_w$, $p_a$ and reaction works $p$ and $q$

$$
\begin{aligned}
&-p_w\in\partial I_0\left(\frac{d^v}{dt}\left(\frac{\beta_v}{\tau_v}\right)+\frac{d^l}{dt}\left(\frac{\beta_l}{\tau_l}\right)+\frac{\beta_v}{\tau_v}\operatorname{div}\mathbf{U}_v+\frac{\beta_l}{\tau_l}\operatorname{div}\mathbf{U}_l\right),\\
&-p_a\in\partial I_0\left(\frac{d^a}{dt}\left(\frac{\beta_v}{\tau_a}\right)+\frac{\beta_v}{\tau_a}\operatorname{div}\mathbf{U}_a\right),
\end{aligned}
$$

$$-p \in \partial I_0\left(\frac{\mathrm{d}^l\beta_v}{\mathrm{d}t}+\frac{\mathrm{d}^v\beta_l}{\mathrm{d}t}-\frac{1}{2}(\mathbf{U}_v-\mathbf{U}_l)\cdot \operatorname{grad}(\beta_l-\beta_v)\right),$$

$$-q \in \partial I_0\left(\frac{\mathrm{d}^a\beta_v}{\mathrm{d}t}-\frac{\mathrm{d}^v\beta_v}{\mathrm{d}t}-(\mathbf{U}_a-\mathbf{U}_v)\cdot \operatorname{grad}\beta_v\right).$$

It results vector $F^d$ is

$$F^d = \left\{\sigma_v^d = -p_w\frac{\beta_v}{\tau_v}1+\beta_v\left\{\lambda_v \operatorname{div}\mathbf{U}_v 1+2\mu_v \mathsf{D}(\mathbf{U}_v)\right\},\right.$$

$$\sigma_l^d = -p_w\frac{\beta_l}{\tau_l}1+\beta_l\left\{\lambda_l \operatorname{div}\mathbf{U}_l 1+2\mu_l \mathsf{D}(\mathbf{U}_l)\right\},$$

$$\sigma_a^d = -p_a\frac{\beta_v}{\tau_a}1+\beta_v\left\{\lambda_a \operatorname{div}\mathbf{U}_a 1+2\mu_a \mathsf{D}(\mathbf{U}_a)\right\},$$

$$B_{vv}^d = -p_w\frac{1}{\tau_v}+q,\ B_{ll}^d = -p_w\frac{1}{\tau_l},$$

$$B_{va}^d = -p_a\frac{1}{\tau_a}-q,$$

$$\mathbf{H}_v^d - 0,\ \mathbf{H}_l^{nd} = 0,$$

$$\mathbf{f}_{vl}^d = \beta_l\beta_v k_{vl}(\mathbf{U}_v-\mathbf{U}_l)-\frac{p}{2}\operatorname{grad}(\beta_l-\beta_v),\ \mathbf{f}_{va}^d = \beta_v k_{va}(\mathbf{U}_v-\mathbf{U}_a)-q\operatorname{grad}\beta_v,$$

$$\mathbf{f}_{la}^d = \beta_l\beta_v k_{la}(\mathbf{U}_l-\mathbf{U}_a),$$

$$Z_v^d = p_w\frac{\beta_v}{\tau_v^2},\ Z_l^d = p_w\frac{\beta_l}{\tau_l^2},\ Z_a^d = p_a\frac{\beta_v}{\tau_a^2},$$

$$\hat{B}_{vl}^d = -p,\ \hat{B}_{lv}^d = -p,$$

$$\left.-\mathbf{Q}^d = \frac{\lambda}{T}\operatorname{grad}T\right\}.$$

### 13.9.2 The Constitutive Laws

They are given by vector $F$

$$F = F^d + F^{nd} + F^{reac}, \tag{13.23}$$

the components of which are

$$\sigma_v = -p_w \frac{\beta_v}{\tau_v} 1 + \beta_v \Psi_v(\tau_v, T) 1 + \beta_v \{\lambda_v \operatorname{div} \mathbf{U}_v 1 + 2\mu_v \mathrm{D}(\mathbf{U}_v)\}$$
$$-k_v \operatorname{grad} \beta_v \otimes \operatorname{grad} \beta_v,$$
$$\sigma_l = -p_w \frac{\beta_l}{\tau_l} 1 + \beta_l \Psi_l(\tau_l, T) 1 + \beta_l \{\lambda_l \operatorname{div} \mathbf{U}_l 1 + 2\mu_l \mathrm{D}(\mathbf{U}_l)\} - k_l \operatorname{grad} \beta_l \otimes \operatorname{grad} \beta_l,$$
$$\sigma_a = -p_a \frac{\beta_v}{\tau_a} 1 + \beta_v \Psi_a(\tau_a, T) 1 + \beta_v \{\lambda_a \operatorname{div} \mathbf{U}_a 1 + 2\mu_a \mathrm{D}(\mathbf{U}_a)\},$$
$$B_{vv} = -p_w \frac{1}{\tau_v} + q + \Psi_v(\tau_v, T) + B_{reacvv} + B_{reacv},$$
$$B_{ll} = -p_w \frac{1}{\tau_l} + \Psi_l(\tau_l, T) + B_{reacll} + B_{reacl},$$
$$B_{va} = -p_a \frac{1}{\tau_a} - q + \Psi_a(\tau_a, T) + B_{reacva},$$
$$\mathbf{H}_v = k_v \operatorname{grad} \beta_v, \ \mathbf{H}_l = k_l \operatorname{grad} \beta_l,$$
$$\mathbf{f}_{vl} = \beta_l \beta_v k_{vl} (\mathbf{U}_v - \mathbf{U}_l) - \frac{p}{2} \operatorname{grad}(\beta_l - \beta_v), \mathbf{f}_{va} = \beta_v k_{va} (\mathbf{U}_v - \mathbf{U}_a) - q \operatorname{grad} \beta_v,$$
$$\mathbf{f}_{la} = \beta_l \beta_v k_{la} (\mathbf{U}_l - \mathbf{U}_a),$$
$$0 = p_w \frac{\beta_v}{\tau_v^2} + \beta_v \frac{\partial \Psi_v}{\partial \tau_v}, \ 0 = p_w \frac{\beta_l}{\tau_l^2} + \beta_l \frac{\partial \Psi_l}{\partial \tau_l}, \ 0 = p_a \frac{\beta_v}{\tau_a^2} + \beta_v \frac{\partial \Psi_a}{\partial \tau_a}, \tag{13.24}$$
$$0 = -p + B_{reacv}, \ 0 = -p + B_{reacl}, \tag{13.25}$$
$$-\mathbf{Q} = \frac{\lambda}{T} \operatorname{grad} T. \tag{13.26}$$

### 13.9.3 Consequences of the Constitutive Laws

We have

$$\beta_v \in [0,1], \ \beta_l \in [0,1], \ \beta_v + \beta_l = 1,$$

because there exist reaction forces which imply that the different subdifferential sets are not empty, [112], see also Theorem A.2 of Appendix A. The constitutive laws imply also that mass balances (13.3), (13.4) and internal constraints (13.21), (13.22) on the velocities are satisfied. This is also a consequence of the existence of the dissipative reactions forces, the specific pressures, $p_w$ and $p_a$, and reaction works, $p$ and $q$.

## 13.10 The Equations of the Predictive Theory

They result from the equations of motion, (13.5) and (13.6), the entropy balance, (13.18) and the constitutive laws, (13.23). Note again that the constitutive laws imply mass balances (13.3), (13.4) and internal constraints (13.21), (13.22) on the velocities are satisfied. They imply also the constraints on the volume fractions are satisfied.

With relationships (13.25), it is possible to eliminate $B_{reacv}$ and $B_{reacl}$ from the equations for the microscopic motion

$$-\left\{\Psi_v(\tau_v,T)-p_w\frac{1}{\tau_v}\right\}-B_{reacvv}-p-q+k_v\Delta\beta_v=0,$$

$$-\left\{\Psi_l(\tau_l,T)-p_w\frac{1}{\tau_l}\right\}-B_{reacll}-p+k_l\Delta\beta_l=0,$$

$$-\left\{\Psi_a(\tau_a,T)-p_a\frac{1}{\tau_a}\right\}-B_{reacva}+q=0.$$

By adding and subtracting the two first equations and eliminating $q$, we get

$$-\left\{\Psi_a(\tau_a,T)-p_a\frac{1}{\tau_a}\right\}-\left\{\Psi_v(\tau_v,T)-p_w\frac{1}{\tau_l}\right\}+\left\{\Psi_l(\tau_l,T)-p_w\frac{1}{\tau_v}\right\}$$
$$-B_{reacva}-B_{reacvv}+B_{reacll}+k_v\Delta\beta_v-k_l\Delta\beta_l=0,$$

$$-\left\{\Psi_v(\tau_v,T)-p_w\frac{1}{\tau_l}\right\}-\left\{\Psi_l(\tau_l,T)-p_w\frac{1}{\tau_v}\right\}-\left\{\Psi_a(\tau_a,T)-p_a\frac{1}{\tau_a}\right\}$$
$$-B_{reacvv}-B_{reacll}-2p+k_v\Delta\beta_v+k_l\Delta\beta_l=0,$$

$$-\left\{\Psi_a(\tau_a,T)-p_a\frac{1}{\tau_a}\right\}-B_{reacva}+q=0.$$

Using relationship $\beta_v+\beta_l=1$, we get

$$-\left\{\Psi_a(\tau_a,T)-p_a\frac{1}{\tau_a}\right\}-\left\{\Psi_v(\tau_v,T)-p_w\frac{1}{\tau_v}\right\}+\left\{\Psi_l(\tau_l,T)-p_w\frac{1}{\tau_l}\right\}$$
$$-B_{reacvv}+B_{reacll}+(k_v+k_l)\Delta\beta_v=0,$$

$$-\left\{\Psi_v(\tau_v,T)-p_w\frac{1}{\tau_l}\right\}-\left\{\Psi_l(\tau_l,T)-p_w\frac{1}{\tau_v}\right\}-\left\{\Psi_a(\tau_a,T)-p_a\frac{1}{\tau_a}\right\}$$
$$-B_{reacvv}-B_{reacll}-2p+(k_v-k_l)\Delta\beta_v=0,$$

$$-\left\{\Psi_a(\tau_a,T)-p_a\frac{1}{\tau_a}\right\}-B_{reacva}+q=0.$$

For the sake of simplicity we choose $k_v = k_l = k_\beta$, (see Remark 13.5). By noting that

$$B_{reacvv} - B_{reacll} \in \partial I(\beta_v) - \partial I(1-\beta_v) = \partial I(\beta_v),$$

we get

$$\left\{\Psi_a(\tau_a,T) - p_a\frac{1}{\tau_a}\right\} + \left\{\Psi_v(\tau_v,T) - p_w\frac{1}{\tau_v}\right\} - \left\{\Psi_l(\tau_l,T) - p_w\frac{1}{\tau_l}\right\}$$

$$+B^{reac} - 2k_\beta \Delta\beta_v = 0, \tag{13.27}$$

$$-\left\{\Psi_v(\tau_v,T) - p_w\frac{1}{\tau_v}\right\} - \left\{\Psi_l(\tau_l,T) - p_w\frac{1}{\tau_l}\right\} - \left\{\Psi_a(\tau_a,T) - p_a\frac{1}{\tau_a}\right\}$$

$$-B_{reacvv} - B_{reacll} = 2p, \tag{13.28}$$

$$-\left\{\Psi_a(\tau_a,T) - p_a\frac{1}{\tau_a}\right\} - B_{reacva} + q = 0. \tag{13.29}$$

$$B^{reac} \in \partial I(\beta_v),\ B_{reacvv} \in \partial I(\beta_v),\ B_{reacll} \in \partial I(1-\beta_v),\ B_{reacva} \in \partial I(\beta_v).$$

The other equations are

$$\frac{\beta_v}{\tau_v}\frac{\mathrm{d}^v}{\mathrm{d}t}\mathbf{U}_v + \mathbf{U}_v\left\{\frac{\mathrm{d}^v}{\mathrm{d}t}\left(\frac{\beta_v}{\tau_v}\right) + \frac{\beta_v}{\tau_v}\operatorname{div}\mathbf{U}_v\right\}$$

$$= -\operatorname{div}(k\operatorname{grad}\beta_v \otimes \operatorname{grad}\beta_v) + \operatorname{grad}\left(\beta_v\left(\Psi_v - p_w\frac{1}{\tau_v}\right)\right)$$

$$+\beta_v\left\{(\lambda_v+\mu_v)\operatorname{grad}\operatorname{div}\mathbf{U}_v + \mu_v\Delta\mathbf{U}_v\right\}$$

$$-\beta_l\beta_v k_{vl}(\mathbf{U}_v - \mathbf{U}_l) - \frac{p}{2}\operatorname{grad}(1-2\beta_v) - \beta_v k_{va}(\mathbf{U}_v - \mathbf{U}_a) + q\operatorname{grad}\beta_v + \mathbf{f}_v^{ext},$$

$$\frac{1-\beta_v}{\tau_l}\frac{\mathrm{d}^l}{\mathrm{d}t}\mathbf{U}_l + \mathbf{U}_l\left\{\frac{\mathrm{d}^l}{\mathrm{d}t}\left(\frac{1-\beta_v}{\tau_l}\right) + \frac{1-\beta_v}{\tau_l}\operatorname{div}\mathbf{U}_l\right\}$$

$$= -\operatorname{div}(k\operatorname{grad}\beta_v \otimes \operatorname{grad}\beta_v) + \operatorname{grad}\left((1-\beta_v)\left(\Psi_l - p_w\frac{1}{\tau_l}\right)\right)$$

$$+(1-\beta_v)\left\{(\lambda_l+\mu_l)\operatorname{grad}\operatorname{div}\mathbf{U}_l + \mu_l\Delta\mathbf{U}_l\right\}$$

$$+\beta_v\beta_l k_{vl}(\mathbf{U}_v - \mathbf{U}_l) + \frac{p}{2}\operatorname{grad}(1-2\beta_v) - \beta_v\beta_l k_{la}(\mathbf{U}_l - \mathbf{U}_a) + \mathbf{f}_l^{ext}, \tag{13.30}$$

$$\frac{\beta_v}{\tau_a}\frac{\mathrm{d}^a}{\mathrm{d}t}\mathbf{U}_a = \operatorname{grad}\left(\beta_v\left(\Psi_a - p_a\frac{1}{\tau_a}\right)\right)$$

$$+\beta_v\left\{(\lambda_a+\mu_a)\operatorname{grad}\operatorname{div}\mathbf{U}_a + \mu_a\Delta\mathbf{U}_a\right\}$$

$$+\beta_v\beta_l k_{la}(\mathbf{U}_l - \mathbf{U}_a) + \beta_v k_{va}(\mathbf{U}_v - \mathbf{U}_a) - q\operatorname{grad}\beta_v + \mathbf{f}_a^{ext}, \tag{13.31}$$

$$0 = \beta_v \left( \frac{\partial \Psi_v}{\partial \tau_v} + p_w \frac{1}{\tau_v^2} \right) = (1-\beta_v) \left( \frac{\partial \Psi_l}{\partial \tau_l} + p_w \frac{1}{\tau_l^2} \right)$$

$$= \beta_v \left( \frac{\partial \Psi_a}{\partial \tau_a} + p_a \frac{1}{\tau_a^2} \right), \tag{13.32}$$

$$\frac{\mathrm{d}^v s_v}{\mathrm{d}t} + s_v \operatorname{div} \mathbf{U}_v + \frac{\mathrm{d}^l s_l}{\mathrm{d}t} + s_l \operatorname{div} \mathbf{U}_l + \frac{\mathrm{d}^a s_a}{\mathrm{d}t} + s_a \operatorname{div} \mathbf{U}_a$$

$$-\lambda \Delta \ln T = \mathscr{D} + R^{ext}, \tag{13.33}$$

$$\frac{\mathrm{d}^v}{\mathrm{d}t}\left(\frac{\beta_v}{\tau_v}\right) + \frac{\mathrm{d}^l}{\mathrm{d}t}\left(\frac{1-\beta_v}{\tau_l}\right) + \frac{\beta_v}{\tau_v}\operatorname{div}\mathbf{U}_v + \frac{1-\beta_v}{\tau_l}\operatorname{div}\mathbf{U}_l = 0, \tag{13.34}$$

$$\frac{\mathrm{d}^a}{\mathrm{d}t}\left(\frac{\beta_v}{\tau_a}\right) + \frac{\beta_v}{\tau_a}\operatorname{div}\mathbf{U}_a = 0. \tag{13.35}$$

External actions $\mathbf{f}_v^{ext}$, $\mathbf{f}_l^{ext}$, $\mathbf{f}_a^{ext}$ and $R^{ext}$ are given, for instance

$$\mathbf{f}_v^{ext} = \frac{\beta_v}{\tau_v}\mathbf{g},\ \mathbf{f}_l^{ext} = \frac{1-\beta_v}{\tau_l}\mathbf{g},\ \mathbf{f}_a^{ext} = \frac{\beta_v}{\tau_v}\mathbf{g},\ R^{ext} = 0,$$

where $\mathbf{g}$ is the gravity acceleration vector.

There are 18 equations, (13.27)–(13.35), for 9 velocities $\mathbf{U}$, 3 specific volumes $\tau$, 1 volume fraction $\beta_v$, 1 temperature $T$, 2 pressures $p_w$, $p_a$, and 2 reaction works $p$, $q$, unknowns, i.e., for 18 unknowns. Boundary and initial conditions are to be added.

The mechanical novelty is the local interaction in the liquid water-vapor phase change: the volume fractions at a material point influence the neighbourhood volume fractions. This interaction introduces the gradients of the $\beta_i$'s which result in the laplacian in (13.27) for $\beta_v$.

*Remark 13.8.* Equations (13.28)–(13.29) give reaction works $p$ and $q$ with value:

- If $0 < \beta_v < 1$,

$$2p = -\left\{\Psi_v(\tau_v, T) - p_w \frac{1}{\tau_v}\right\}$$

$$-\left\{\Psi_l(\tau_l, T) - p_w \frac{1}{\tau_l}\right\} - \left\{\Psi_a(\tau_a, T) - p_a \frac{1}{\tau_a}\right\}, \tag{13.36}$$

$$q = \left\{\Psi_a(\tau_a, T) - p_a \frac{1}{\tau_a}\right\}. \tag{13.37}$$

In the liquid vapor air mixture, it is possible that there is no vapor by having $\tau_v = \infty$. Note that we have

$$\lim_{\tau_v \to \infty} \left\{\Psi_v(\tau_v, T) - p_w \frac{1}{\tau_v}\right\} = 0,$$

because vapor with large $\tau_v$ behaves like an ideal gas. Equation (13.27) becomes

$$\left\{\Psi_a(\tau_a,T)-p_a\frac{1}{\tau_a}\right\}-\left\{\Psi_l(\tau_l,T)-p_w\frac{1}{\tau_l}\right\}+B^{reac}-2k_\beta\Delta\beta_v=0.$$

It is also possible that there is no air by having $\tau_a=\infty$ with

$$\lim_{\tau_a\to\infty}\left\{\Psi_a(\tau_a,T)-p_a\frac{1}{\tau_a}\right\}=0.$$

Equation (13.27) becomes

$$\left\{\Psi_v(\tau_v,T)-p_w\frac{1}{\tau_v}\right\}-\left\{\Psi_l(\tau_l,T)-p_w\frac{1}{\tau_l}\right\}+B^{reac}-2k_\beta\Delta\beta_v=0.$$

- If $\beta_v=1$, there is only air and vapor. The reaction works are

$$p\in\mathbb{R},\; q\geq\left\{\Psi_a(\tau_a,T)-p_a\frac{1}{\tau_a}\right\}.$$

It is possible that the air vapor mixture contains either only vapor with $\tau_a=\infty$ or only air with $\tau_v=\infty$ (see Sect. 13.11.4 down below).

- If $\beta_v=0$, there is no air which can only appear by diffusion with $\tau_a=\infty$, thus we have

$$p\in\mathbb{R},\; q\leq\lim_{\tau_a\to\infty}\left\{\Psi_a(\tau_a,T)-p_a\frac{1}{\tau_a}\right\}=0,$$

because air with large $\tau_a$ behaves like an ideal gas.

In case $\beta_v$ is 0 (there is only liquid water), or in case $\beta_v$ is 1 (there is only vapor and air), reaction work $p$ does not intervene in the equations (because $\text{grad}\,(1-2\beta_v)=0$). Reaction $q$ does not intervene either because $\text{grad}\,\beta_v=0$.

*Remark 13.9.* Equation (13.32) may by replaced by their sum and their difference, giving

$$-p_w=\beta_v\tau_v^2\frac{\partial\Psi_v}{\partial\tau_v}+(1-\beta_v)\tau_l^2\frac{\partial\Psi_l}{\partial\tau_l},\tag{13.38}$$

$$-p_w(2\beta_v-1)=\beta_v\tau_v^2\frac{\partial\Psi_v}{\partial\tau_v}-(1-\beta_v)\tau_l^2\frac{\partial\Psi_l}{\partial\tau_l}.$$

First relationship gives water specific pressure $p_w$ which can be eliminated and reducing the number of unknowns to 17.

### 13.10.1 A Potential and the Free Enthalpy

Free energies $\Psi$ are assumed to be convex functions of densities $1/\tau$. Because quantity $p/\tau - \Psi$ is involved in numerous formulas, it may useful to introduce new potentials defined by

$$-G_v(p_v,T) = \Psi_v^*(p_v,T) = \sup_\tau \left\{ p_v \frac{1}{\tau} - \Psi_v(\tau,T) \right\},$$

$$-G_l(p_v,T) = \Psi_l^*(p_v,T) = \sup_\tau \left\{ p_l \frac{1}{\tau} - \Psi_l(\tau,T) \right\}, \tag{13.39}$$

(see Sect. A.5 of Appendix A), and have constitutive laws

$$\frac{1}{\tau_v} = \frac{\partial \Psi_v^*}{\partial (p_v)}(p_v,T) = -\frac{\partial G_v}{\partial p_v}(p_v,T),$$

$$\frac{1}{\tau_l} = -\frac{\partial G_l}{\partial p_l}(p_l,T),$$

together with

$$p_v = \frac{\partial \Psi_v}{\partial (1/\tau_v)}(\tau_v,T) = -\tau_v^2 \frac{\partial \Psi_v}{\partial \tau_v}(\tau_v,T),\ p_l = -\tau_l^2 \frac{\partial \Psi_l}{\partial \tau_l}(\tau_l,T),$$

giving

$$p_w = \beta_v p_v + (1-\beta_v) p_l.$$

*Remark 13.10.* The schematic free energies chosen below are convex functions of the densities $1/\tau$. Potential $G_a$ with free energy given by (13.12) is

$$G_a(p_a,T) = -kT \exp\left(\frac{p_a - kT}{kT}\right),$$

giving

$$\frac{1}{\tau_a} = \exp\left(\frac{p_a - kT}{kT}\right),\ p_a = kT - kT \ln \tau_a.$$

Let us note that if $1/\tau \to \Psi(\tau,T)$ is convex and if $\tau \geq 0$ (see Remark 13.6), specific free energy $\tau \to \hat{\Psi}(\tau,T) = \tau\Psi(\tau,T)$ is also convex

**Theorem 13.1.** *If $\tau \geq 0$ convexity of function $1/\tau \to \Psi(\tau,T)$ is equivalent to convexity of function $\tau \to \hat{\Psi}(\tau,T) = \tau\Psi(\tau,T)$.*

*Proof.* It results from relationship

$$\frac{\partial^2 \Psi}{\partial (1/\tau)^2} = \tau^3 \left(2\frac{\partial \Psi}{\partial \tau} + \tau \frac{\partial^2 \Psi}{\partial \tau^2}\right) = \tau^3 \frac{\partial^2 \hat{\Psi}}{\partial \tau^2}.$$

Let us recall that condition $\tau \geq 0$ has been included in the definition of the free energies $\Psi$, see Remark 13.6. □

Within this convexity assumption, free enthalpy $\hat{G}(P,T)$ is defined by

$$-\hat{G}(P,T) = \sup_x \left\{-Px - \hat{\Psi}(x,T)\right\},$$

with

$$\hat{G}(P,T) = P\tau + \hat{\Psi}(\tau,T),$$
$$P = -\frac{\partial \hat{\Psi}}{\partial \tau}(\tau,T),\ \tau = \frac{\partial \hat{G}}{\partial P}(P,T).$$

Let us note that

$$\Psi(\tau,T) - p\frac{1}{\tau} = G(p,T) \Leftrightarrow p = -\tau^2\frac{\partial \Psi}{\partial \tau}(\tau,T) \Leftrightarrow \frac{1}{\tau} = -\frac{\partial G}{\partial p}(p,T),$$
$$P\tau + \hat{\Psi}(\tau,T) = \hat{G}(P,T) \Leftrightarrow P = -\frac{\partial \hat{\Psi}}{\partial \tau}(\tau,T) \Leftrightarrow \tau = \frac{\partial \hat{G}}{\partial P}(P,T). \tag{13.40}$$

These relationships imply

$$P = -\frac{\partial\,(\tau\Psi)}{\partial \tau}(\tau,T) = p\frac{1}{\tau} - \Psi(\tau,T) = -G(p,T),$$
$$p = -\tau^2\frac{\partial \Psi}{\partial \tau}(\tau,T) = \hat{\Psi}(\tau,T) + \tau P = \hat{G}(P,T). \tag{13.41}$$

*Remark 13.11.* We have

$$\frac{\partial p}{\partial \tau} = -\frac{\partial}{\partial \tau}\left(\tau^2\frac{\partial \Psi}{\partial \tau}\right) = -\tau(2\frac{\partial \Psi}{\partial \tau} + \tau\frac{\partial^2 \Psi}{\partial \tau^2}) = -\tau\frac{\partial^2 \hat{\Psi}}{\partial \tau^2},$$
$$\frac{\partial P}{\partial \tau} = -\frac{\partial^2 \hat{\Psi}}{\partial \tau^2}.$$

It results that pressures $\tau \to p(\tau,T)$ and $\tau \to P(\tau,T)$ are decreasing functions of $\tau$.

*Remark 13.12.* Specific free energy $\hat{\Psi}$ of a material is defined up to a linear function of temperature $AT+B$. It results volume free energy $\Psi = \hat{\Psi}/\tau$ is defined up to function $(AT+B)/\tau$. These functions are the same for the free energies of all the phases, $\hat{\Psi}_l$, $\Psi_l$ and $\hat{\Psi}_v$, $\Psi_v$.

*Remark 13.13.* The air free enthalpy is

$$\hat{G}_a(P_a,T) = rT(1 - \ln\frac{rT}{P_a}),$$

with specific free energy $\hat{\Psi}(\tau_a,T) = -rT\ln\tau_a$.

## 13.11 Air, Liquid Water and Vapor in an Homogeneous Cloud

In an homogeneous cloud, (13.27) governing the evolution of the mixture is

$$\left\{\Psi_a(\tau_a,T)-p_a\frac{1}{\tau_a}\right\}+\left\{\Psi_v(\tau_v,T)-p_w\frac{1}{\tau_v})\right\}-\left\{\Psi_l(\tau_l,T)-p_w\frac{1}{\tau_l}\right\}+B^{reac}=0,$$

$$B^{reac}\in\partial I(\beta_v).$$

### 13.11.1 The Air, Liquid Water, Vapor Mixture

In this case, we have $0<\beta_v<1$ and

$$\left\{\Psi_a(\tau_a,T)-p_a\frac{1}{\tau_a}\right\}+\left\{\Psi_v(\tau_v,T)-p_w\frac{1}{\tau_v})\right\}-\left\{\Psi_l(\tau_l,T)-p_w\frac{1}{\tau_l}\right\}=0, \quad (13.42)$$

and, from relationships (13.32)

$$0=\frac{\partial\Psi_v}{\partial\tau_v}+p_w\frac{1}{\tau_v^2},\; 0=\frac{\partial\Psi_l}{\partial\tau_l}+p_w\frac{1}{\tau_l^2}.$$

By eliminating $p_w$ and $p_a$ we get

$$\frac{\partial(\tau_a\Psi_a)}{\partial\tau_a}+\frac{\partial(\tau_v\Psi_v)}{\partial\tau_v}-\frac{\partial(\tau_l\Psi_l)}{\partial\tau_l}=0, \quad (13.43)$$

$$-\tau_v^2\frac{\partial\Psi_v}{\partial\tau_v}=-\tau_l^2\frac{\partial\Psi_l}{\partial\tau_l}. \quad (13.44)$$

There are two relationships and four parameters $(\tau_a,\tau_v,\tau_l,T)$. The two relationships (13.43) and (13.44) give

$$\tau_v=\hat{\tau}_v(T,\tau_a),\tau_l=\hat{\tau}_l(T,\tau_a),$$

and the specific pressure

$$\begin{aligned}\hat{p}_w(T,\tau_a)&=-\hat{\tau}_v^2(T,\tau_a)\frac{\partial\Psi_v}{\partial\tau_v}(\hat{\tau}_v(T,\tau_a),T)\\&=-\hat{\tau}_l^2(T,\tau_a)\frac{\partial\Psi_l}{\partial\tau_l}(\hat{\tau}_l(T,\tau_a),T).\end{aligned} \quad (13.45)$$

The curves $T\to\hat{\tau}_v(T,\tau_a)$, $T\to\hat{\tau}_l(T,\tau_a)$ for $\tau_a$ fixed, are drawn in the thermodynamical diagrams. An example is given in Fig. 12.3 for the case there is no air

($\tau_a = \infty$, $\{-\Psi_a(\tau_a, T) + p_a/\tau_a\} = 0$) where function $T = \hat{\tau}_v^{-1}(\tau)$ is the green line and function $T = \hat{\tau}_l^{-1}(\tau)$ is the blue line.

### 13.11.2 The Cloud Pressure

When the cloud is homogeneous, pressure $P_{cloud}$ in the cloud results from the stresses $\sigma_v = -\beta_v (p_w 1/\tau_v - \Psi_v) 1$, $\sigma_l = -\beta_l (p_w/\tau_l - \Psi_l) 1$ and $\sigma_a = -\beta_v (p_a/\tau_a - \Psi_a) 1$. It is

$$\begin{aligned} P_{cloud} &= \beta_v \left\{ p_w \frac{1}{\tau_v} - \Psi_v \right\} + \beta_l \left\{ p_w \frac{1}{\tau_l} - \Psi_l \right\} + \beta_v \left\{ p_a \frac{1}{\tau_a} - \Psi_a \right\} \\ &= -\beta_v \left( \tau_v \frac{\partial \Psi_v}{\partial \tau_v} + \Psi_v \right) - \beta_l \left( \tau_l \frac{\partial \Psi_l}{\partial \tau_l} + \Psi_l \right) - \beta_v \left( \tau_a \frac{\partial \Psi_a}{\partial \tau_a} + \Psi_a \right) \\ &= -\beta_v \frac{\partial \hat{\Psi}_v}{\partial \tau_v} - \beta_l \frac{\partial \hat{\Psi}_l}{\partial \tau_l} - \beta_v \frac{\partial \hat{\Psi}_a}{\partial \tau_a}, \end{aligned} \tag{13.46}$$

where $\hat{\Psi} = \tau \Psi$ is the specific free energy and relationship (13.45) has been used. We define

$$P_v = \left\{ p_w \frac{1}{\tau_v} - \Psi_v \right\} = -\frac{\partial \hat{\Psi}_v}{\partial \tau_v}, \quad P_l = \left\{ p_w \frac{1}{\tau_l} - \Psi_l \right\} = -\frac{\partial \hat{\Psi}_l}{\partial \tau_l},$$

$$P_a = \left\{ p_a \frac{1}{\tau_a} - \Psi_a \right\} = -\frac{\partial \hat{\Psi}_a}{\partial \tau_a},$$

the pressures in each material. We get with relationship (13.42):

- In case the mixture contains liquid, vapor and air

$$\begin{aligned} P_{cloud} &= \left\{ p_w \frac{1}{\tau_v} - \Psi_v \right\} + \left\{ p_a \frac{1}{\tau_a} - \Psi_a \right\} = \left\{ p_w \frac{1}{\tau_l} - \Psi_l \right\} \\ &= P_v + P_a = P_l. \end{aligned}$$

- In case the mixture contains only liquid

$$P_{cloud} = \left\{ p_w \frac{1}{\tau_l} - \Psi_l \right\} = P_l.$$

- In case the mixture contains vapor and air

$$P_{cloud} = \left\{ p_w \frac{1}{\tau_v} - \Psi_v \right\} + \left\{ p_a \frac{1}{\tau_a} - \Psi_a \right\} = P_v + P_a.$$

If there is no air,

$$P_{cloud} = \left\{ p_w \frac{1}{\tau_v} - \Psi_v \right\} = P_v.$$

The vapor liquid mixture specific pressure $p_w$ given by formulas (13.38), depend on $(T, \tau_a)$. Pressure $P_{cloud}$ depends also on $(T, \tau_a)$ because $\tau_v = \hat{\tau}_v(T, \tau_a)$ and $\tau_l = \hat{\tau}_l(T, \tau_a)$ depend also on $(T, \tau_a)$. Thus $p_w$ and $P_{cloud}$ are constant when $(T, \tau_a)$ are constant. Pressure $P_{cloud}$ is shown on the Clapeyron water diagram of Fig. 12.3 in case there is no air ($\tau_a = \infty$, $(p_a 1/\tau_a - \Psi_a) = 0$).

Let us note, that pressure $P_{cloud}$ and $p_w$ are also constant on a free boundary where phase change occurs as seen in Chap. 12.

*Remark 13.14.* Relationships (13.43) and (13.44) become

$$\begin{aligned} &P_l - P_v - P_a = 0, \\ &\hat{\Psi}_v + P_v \tau_v = \hat{\Psi}_l + P_l \tau_l = p_w. \end{aligned} \tag{13.47}$$

These formulas are similar to formulas (12.2) and (12.4) of the surfacic phase change.

Relationships (13.41) and (13.47) give

$$\begin{aligned} P_v &= \hat{P}_v(T, \tau_a) = -G_v(p_w(T, \tau_a), T), \\ P_l &= \hat{P}_l(T, \tau_a) = -G_l(p_w(T, \tau_a), T). \end{aligned} \tag{13.48}$$

and

$$p_w(T, \tau_a) = \hat{G}_v(\hat{P}_v(T, \tau_a), T) = \hat{G}_l(\hat{P}_l(T, \tau_a), T). \tag{13.49}$$

The free enthalpies are equal in the mixture. The specific pressures $p_l$ and $p_v$ are equal to $p_w$ whereas pressures $P_v$ and $P_l$ are not equal because of the air. They are equal when there is no air.

### 13.11.3 The Liquid Water

In this case, we have $\beta_v = 0$ and

$$\left\{ \Psi_a(\tau_a, T) - p_a \frac{1}{\tau_a} \right\} + \left\{ \Psi_v(\tau_v, T) - p_w \frac{1}{\tau_v}) \right\} - \left\{ \Psi_l(\tau_l, T) - p_w \frac{1}{\tau_l} \right\} > 0. \tag{13.50}$$

If some vapor is to appear at temperature $T$ it may have any specific volume $\tau_v$. If some air is to appear, it appears by diffusion. Thus when appearing its density is 0, i.e., its specific volume $\tau_a$ is infinite. Thus condition (13.50) has to be satisfied for any specific volume $\tau_v$ and $\tau_a = \infty$. It results

$$\inf_{\tau_v}\left\{\Psi_v(\tau_v,T)-p_l(\tau_l,T)\frac{1}{\tau_v}\right\}>\left\{\Psi_l(\tau_l,T)-p_l(\tau_l,T)\frac{1}{\tau_l}\right\}$$
$$-\lim_{\tau_a\to 0}\left\{\Psi_a(\tau_a,T)-p_a\frac{1}{\tau_a}\right\}.$$

with

$$p_l(\tau_l,T)=-\tau_l^2\frac{\partial\Psi_l}{\partial\tau_l}(\tau_l,T).$$

Note that air being an ideal gas for large $\tau_a$, we have

$$\lim_{\tau_a\to 0}\left\{\Psi_a(\tau_a,T)-p_a\frac{1}{\tau_a}\right\}=0,$$

and the air does not intervene in the liquid vapor phase change. We have due to definition (13.39) of potentials $G$

$$\inf_{\tau_v}\left\{\Psi_v(\tau_v,T)-p_l(\tau_l,T)\frac{1}{\tau_v}\right\}=G_v(p_l(\tau_l,T),T),$$
$$\left\{\Psi_l(\tau_l,T)-p_l(\tau_l,T)\frac{1}{\tau_l}\right\}=G_l(p_l(\tau_l,T),T).$$

Thus condition (13.50) is

$$G_v(p_l(\tau_l,T),T)>G_l(p_l(\tau_l,T),T).$$

Function $p\to G_v(p,T)-G_l(p,T)$ is an increasing function because

$$\frac{\partial G_v}{\partial p}(p,T)-\frac{\partial G_l}{\partial p}(p,T)=-\frac{1}{\tau_v(p,T)}+\frac{1}{\tau_l(p,T)}>0.$$

Note that we have

$$G_v(\hat{p}_w(T),T)-G_l(\hat{p}_w(T),T)=0,$$

where specific pressure $\hat{p}_w$ is the liquid vapor mixture specific pressure $\hat{p}_w(T)=\hat{p}_w(T,\tau_a=\infty)$. Thus condition (13.50) is equivalent to

$$G_v(p_l(\tau_l,T),T)-G_l(p_l(\tau_l,T),T)>G_v(\hat{p}_w(T),T)-G_l(\hat{p}_w(T),T)=0,$$

or to

$$p_l(\tau_l,T)>\hat{p}_w(T). \tag{13.51}$$

Because pressure $p_l$ is a decreasing function of specific volume (see Remark 13.11), condition (13.50) is equivalent to

$$\tau_l < \hat{\tau}_l(T, \tau_a = \infty) \Leftrightarrow p_l(\tau_l, T) > \hat{p}_w(T).$$

Due to relationships (13.41) and (13.47), relationship (13.51) is equivalent to

$$\hat{G}_l(P_{cloud}, T) > \hat{G}_l(\hat{P}_l(T), T),$$

where $P_{cloud} = P_l$ and $\hat{P}_l(T)$ is the liquid pressure in the vapor, liquid water mixture at temperature $T$. Because free enthalpy $\hat{G}$ is an increasing function of $P$ (relationship (13.40) gives $\partial\hat{G}/\partial P = \tau > 0$), the previous inequality is equivalent to

$$P_{cloud} > \hat{P}_l(T),$$

see diagram of Fig. 12.1.

### 13.11.4 The Air, Vapor Mixture

In this case, we have $\beta_v = 1$ and

$$\left\{\Psi_a(\tau_a, T) - p_a\frac{1}{\tau_a}\right\} + \left\{\Psi_v(\tau_v, T) - p_w\frac{1}{\tau_v})\right\} - \left\{\Psi_l(\tau_l, T) - p_w\frac{1}{\tau_l}\right\} < 0. \quad (13.52)$$

This condition is

$$\left\{\Psi_a(\tau_a, T) - p_a\frac{1}{\tau_a}\right\} + \left\{\Psi_v(\tau_v, T) - p_v(\tau_v, T)\frac{1}{\tau_v}\right\} < \inf_{\tau_l}\left\{\Psi_l(\tau_l, T) - p_v(\tau_v, T)\frac{1}{\tau_l}\right\},$$

or

$$G_v(p_v(\tau_v, T), T) + \left\{\Psi_a(\tau_a, T) - p_a\frac{1}{\tau_a}\right\} < G_l(p_v(\tau_v, T), T).$$

With relationship

$$G_v(\hat{p}_w(T, \tau_a), T) + \left\{\Psi_a(\tau_a, T) - p_a\frac{1}{\tau_a}\right\} = G_l(\hat{p}_w(T, \tau_a), T),$$

and relationships (13.40), (13.47), condition (13.52) is equivalent to

$$\begin{aligned} p_v(\tau_v, T) &< \hat{p}_w(T, \tau_a) \Leftrightarrow \tau_v > \hat{\tau}_v(T, \tau_a) \\ &\Leftrightarrow P_v < \hat{P}_v(T, \tau_a) \Leftrightarrow P_{cloud} < \hat{P}_v(T, \tau_a) + P_a(T, \tau_a) \\ \Leftrightarrow \hat{G}_v(P_v, T) &< \hat{G}_v(\hat{P}_v(T, \tau_a), T), \end{aligned}$$

where specific pressure $\hat{p}_w(T,\tau_a)$ is the air liquid vapor mixture specific pressure and $\hat{P}_v(T,\tau_a)$ is the vapor pressure in the air vapor liquid water mixture at temperature $T$ with air specific volume $\tau_a$.

It is possible that there is no air in the air vapor mixture with $\tau_a=\infty$, $\Psi_a(\tau_a,T)-p_a/\tau_a=0$ and $P_a(T,\tau_a)=0$, see diagram of Fig. 12.1. The effect of the air is to favor the presence of vapor because air pressure $P_a(T,\tau_a)$ is large when there is much air, i.e., $\tau_a$ is small.

In case there is only air and no vapor, $\tau_v=\infty$ and $G_v(p_v(\tau_v=\infty),T)=0$, with ideal gas specific pressure $\lim_{\tau_v=\infty} p_v(\tau_v)=-\infty$ and $P_v=0$. The previous relationships are satisfied. In this situation vapor can only appear by diffusion because there is neither liquid nor vapor. Of course, it is impossible to have both air and vapor missing in the mixture (we have assumed that it is impossible to have voids, i.e., $\beta_v+\beta_l=1$).

## 13.12 A Schematic Example in Case There Is No Air

For $T\le T_c$ the two phases may coexist. Temperature $T_c$ is the critical temperature. We choose the following schematic free energies

$$\Psi_l(\tau_l,T)=-kT\frac{\ln\tau_l}{\tau_l}+d\frac{\tau_c-\tau_l}{\tau_l}(T-T_c),$$

$$\Psi_v(\tau_v,T)=-kT\frac{\ln\tau_v}{\tau_v},$$

where $k>0$, $\tau_c>0$ and $d>0$ are physical constants. In diagrams 13.1 and 13.2, their values are $T_c=650$ K, $\tau_c=3\times10^{-3}\,\mathrm{m}^3/\mathrm{kg}$, $k=1.001\times10^5\,\mathrm{J/(Km^3)}$ and $d=5.05\times10^7\,\mathrm{J/(Km^3)}$.

*Remark 13.15.* Note that the free energies are convex functions of $1/\tau$, giving potentials $G$ concave functions of pressures. The specific free energies $\hat{\Psi}=\tau\Psi$ are convex functions of $\tau$, giving free enthalpies $\hat{G}$ concave functions of pressures $P$. Specific free energy of the liquid phase $\hat{\Psi}_l=\tau_l\Psi_l$ involves a linear function of temperature $-d\tau_c(T_c-T)$ which has a physical significance because the specific free energy of the vapor phase does not involve the same linear function, see Remark 13.12.

In a cloud, relationships (13.43) and (13.44) give

$$-kT\Big(\frac{1}{\tau_v}-\frac{1}{\tau_l}\Big)+d(T-T_c)=0,$$

$$kT\ln\frac{\tau_v}{\tau_l}+d\tau_c(T-T_c)=0.$$

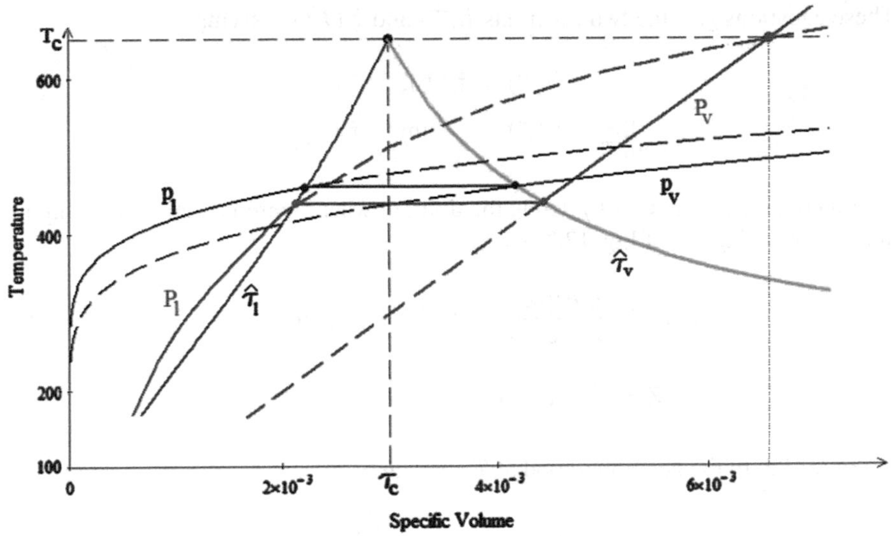


**Fig. 13.1** The schematic diagram. The isopressure $P$ curve in red and the isospecific pressure $p$ in black. They are the liquid $P_l$, $p_l$ pressure and specific pressure when specific volume and temperature are low and the vapor $P_v$, $p_v$ pressure and specific pressure when specific volume and temperature are large. At phase change temperature pressure and specific pressure are constant and specific volume is between $\hat{\tau}_l(T)$, the blue curve, and $\hat{\tau}_v(T)$ the green curve. These two schematic curves have a non smooth behaviour in the neighbourhood of critical temperature $T_c$.. They may be smoothened assuming more sophisticated free energies (see Remark 13.16 and Fig. 13.2)

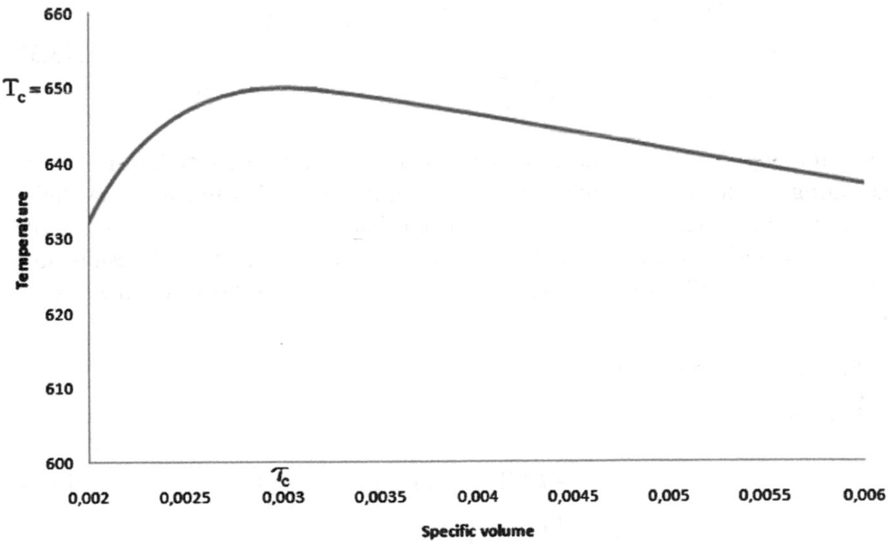


**Fig. 13.2** Functions $\hat{\tau}_l(T)$ and $\hat{\tau}_v(T)$ with smooth evolution in the neighbourhood of $\tau_c$ for liquid free energy (13.53)

These equations give the two functions $\hat{\tau}_l(T)$ and $\hat{\tau}_v(T)$ satisfying

$$\hat{\tau}_l(T) < \hat{\tau}_v(T),$$
$$\lim_{T\to T_c, T\leq T_c} \hat{\tau}_l(T) = \lim_{T\to T_c, T\leq T_c} \hat{\tau}_v(T) = \tau_c.$$

Functions $\hat{\tau}_l(T)$ and $\hat{\tau}_v(T)$ define the domain where there is a mixture of liquid and vapor in diagram of Fig. 12.3. We get

$$\hat{\tau}_l(T) = \frac{\tau_c}{Z}\frac{\exp Z - 1}{\exp Z},\ \hat{\tau}_v(T) = \frac{\tau_c}{Z}(\exp Z - 1),$$
$$Z = \frac{d\tau_c}{kT}(T_c - T),$$

see Fig. 12.3. In the neighbourhood of $T_c$, we have

$$\hat{\tau}_l(T) \simeq \tau_c(1 - \frac{Z}{2}) = \tau_c - \tau_c\frac{d}{2kT_c}(T_c - T),$$
$$\hat{\tau}_v(T) \simeq \tau_c(1 + \frac{Z}{2}) = \tau_c + \tau_c\frac{d}{2kT_c}(T_c - T).$$

*Remark 13.16.* The two functions $\hat{\tau}_l(T)$ and $\hat{\tau}_v(T)$ are not tangent for $T = T_c$ as in diagram of Fig. 12.3. To have this property free energy $\Psi_l(\tau_l, T)$ has to be upgraded with a more sophisticated formula, for instance

$$\Psi_l(\tau_l, T) = \begin{cases} -kT\frac{\ln\tau_l}{\tau_l} + d\frac{(\tau_c-\tau_l)^{1/2}}{\tau_l}(T - T_c),\ if\ \tau_c - \tau_l \geq 0,\ T - T_c \leq 0, \\ I_+(\tau_c - \tau_l),\ if\ \tau_c - \tau_l < 0, \\ -kT\frac{\ln\tau_l}{\tau_l},\ if\ \tau_c - \tau_l \geq 0,\ T - T_c \geq 0, \end{cases} \tag{13.53}$$

valid for any $T$ and $\tau_l$. Indicator function $I_+(\tau_c - \tau_l)$ keeps $\tau_l$ lower than $\tau_c$. Resulting functions $\hat{\tau}_l(T)$ and $\hat{\tau}_v(T)$ are shown in Fig. 13.2 with smooth evolution in the neighbourhood of $\tau_c$. It is to be noted that function $\tau_l \to \hat{\Psi}_l(\tau_l, T) = \tau_l\Psi_l$ is convex and function $T \to \Psi_l(\tau_l, T)$ is concave. Experimental results may be found in [147], on the behaviour in the neighbourhood of the critical point.

Pressures $p_w$ and $P$ are:

- In the liquid

$$p_l = -kT(\ln\tau_l - 1) + \tau_c d(T - T_c), \tag{13.54}$$
$$P_l = \frac{kT}{\tau_l} + d(T - T_c).$$

- In the vapor

$$p_v = -kT(\ln \tau_v - 1), \tag{13.55}$$

$$P_v = \frac{kT}{\tau_v}.$$

- They are equal in the mixture due to relationship (13.44) or (13.47), ($P_a = 0$ because there is no air).

  The isopressure and isospecific pressure curves $T$ versus specific volume $\tau$

$$T = \frac{\tau_l(P_l + dT_c)}{k + \tau_l d},\ T = \frac{p_l + \tau_c dT_c}{k(1 - \ln \tau_l) + \tau_c d},$$

$$T = \frac{P_v}{k}\tau_v,\ T = \frac{p_v}{k(1 - \ln \tau_v)},$$

have the behaviour shown in diagram of Fig. 12.3, see Fig. 13.1. In a liquid vapor mixture, the pressure and specific pressure depend only on temperature whereas in liquid phase and in vapor phase they depend on both temperature and specific volume, see Fig. 13.1.

*Remark 13.17.* The schematic liquid and vapor pressures and specific pressures are equal at critical temperature $T_c$.

### *13.12.1 The Specific Entropies and the Specific Phase Change Latent Heat*

The specific entropies are

$$\hat{s}_l(T, \tau_l) = -\frac{\partial \tau_l \Psi_l}{\partial T} = k \ln \tau_l - d(\tau_c - \tau_l),$$

$$\hat{s}_v(T, \tau_v) = -\frac{\partial \tau_v \Psi_v}{\partial T} = k \ln \tau_v.$$

The specific latent heat liquid water vapor phase change latent heat at temperature $T$ is

$$T(\hat{s}_v(T, \tau_v) - \hat{s}_l(T, \tau_l)) = T\left\{k \ln \frac{\tau_v}{\tau_l} + d(\tau_c - \tau_l)\right\} \geq 0. \tag{13.56}$$

It is non negative as required, because $\tau_v > \tau_c > \tau_l$.The specific latent heat, in agreement with experiments, is null for $T = T_c$ because $\hat{\tau}_v(T_c) = \hat{\tau}_l(T_c) = \tau_c$.

Let us note that the predictive theory depends on four parameters, $T_c$, $\tau_c$, $k$ and $d$ but they are sufficient for the theory to account for the basic phenomena.

## 13.13 Case Where Vapor, Liquid Water and Air Velocities Are Equal

We assume

$$\mathbf{U}_v = \mathbf{U}_l = \mathbf{U}_a = \mathbf{U}.$$

We add the three linear momentum macroscopic equations and get

$$\begin{aligned}
&\left(\frac{\beta_v}{\tau_a} + \frac{\beta_v}{\tau_v} + \frac{1-\beta_v}{\tau_l}\right)\frac{d\mathbf{U}}{dt} = \\
&-2\,\mathrm{div}(k\,\mathrm{grad}\,\beta_v \otimes \mathrm{grad}\,\beta_v) \\
&+\mathrm{grad}\left(\beta_v \frac{\partial(\tau_v \Psi_v)}{\partial \tau_v}(\tau_v, T) + (1-\beta_v)\frac{\partial(\tau_l \Psi_l)}{\partial \tau_l}(\tau_l, T) + \beta_v \frac{\partial(\tau_a \Psi_a)}{\partial \tau_a}(\tau_a, T)\right) \\
&+(1-\beta_v)\left\{(\lambda_l + \mu_l)\,\mathrm{grad}\,\mathrm{div}\,\mathbf{U} + \mu_l \Delta \mathbf{U}\right\} \\
&+\beta_v\left\{(\lambda_a + \lambda_v + \mu_a + \mu_v)\,\mathrm{grad}\,\mathrm{div}\,\mathbf{U} + (\mu_a + \mu_v)\Delta \mathbf{U}\right\} \\
&+\mathbf{f}_a^{ext} + \mathbf{f}_l^{ext} + \mathbf{f}_v^{ext}, \\
&\frac{\mathrm{d}}{\mathrm{d}t}(s_v + s_l + s_a) + (s_v + s_l + s_a)\,\mathrm{div}\,\mathbf{U} - \lambda \Delta \ln T = \mathscr{D} + R^{ext}, \\
&\left\{\Psi_a(\tau_a, T) - p_a \frac{1}{\tau_a}\right\} + \left\{\Psi_v(\tau_v, T) - p_w \frac{1}{\tau_v}\right\} - \left\{\Psi_l(\tau_l, T) - p_w \frac{1}{\tau_l}\right\} \\
&+B^{reac} - 2k_\beta \Delta \beta_v = 0, \\
&B^{reac} \in \partial I(\beta_v), \\
&-p_w = \beta_v \tau_v^2 \frac{\partial \Psi_v}{\partial \tau_v}(\tau_v, T) + (1-\beta_v)\tau_l^2 \frac{\partial \Psi_l}{\partial \tau_l}(\tau_l, T), \\
&-p_w(2\beta_v - 1) = \beta_v \tau_v^2 \frac{\partial \Psi_v}{\partial \tau_v}(\tau_v, T) - (1-\beta_v)\tau_l^2 \frac{\partial \Psi_l}{\partial \tau_l}(\tau_l, T), \\
&0 = \beta_v\left(\frac{\partial \Psi_a}{\partial \tau_a}(\tau_v, T) + p_a \frac{1}{\tau_a^2}\right), \\
&\frac{\mathrm{d}}{\mathrm{d}t}\left(\frac{\beta_v}{\tau_v} + \frac{1-\beta_v}{\tau_l}\right) + \left(\frac{\beta_v}{\tau_v} + \frac{1-\beta_v}{\tau_l}\right)\mathrm{div}\,\mathbf{U} = 0, \\
&\frac{\mathrm{d}}{\mathrm{d}t}\left(\frac{\beta_v}{\tau_a}\right) + \frac{\beta_v}{\tau_a}\,\mathrm{div}\,\mathbf{U} = 0.
\end{aligned}$$

There are 10 equations and 10 unknowns $\mathbf{U}$, $\beta_v$, $\tau_v$, $\tau_l$, $\tau_a$, $p_w$, $p_a$ and $T$.

## 13.14 Case Where Air Velocity Is Null, Vapor and Liquid Velocities Are Equal and Small

We have

$$\mathbf{U}_v = \mathbf{U}_l = \mathbf{U},\ \mathbf{U}_a = 0.$$

The equations of macroscopic motion of vapor and liquid of (13.30) give

$$\frac{1}{\tau}\frac{\mathrm{d}\mathbf{U}}{\mathrm{d}t} = -2\,\mathrm{div}(k\,\mathrm{grad}\,\beta_v \otimes \mathrm{grad}\,\beta_v) - \mathrm{grad}\,\tilde{P}$$
$$-(\beta_v k_{va} + \beta_l k_{la})\,\mathbf{U}$$
$$+\beta_v\{(\lambda_v+\mu_v)\,\mathrm{grad}\,\mathrm{div}\,\mathbf{U} + \mu_v\Delta\mathbf{U}\} + (1-\beta_v)\{(\lambda_l+\mu_l)\,\mathrm{grad}\,\mathrm{div}\,\mathbf{U} + \mu_l\Delta\mathbf{U}\}$$
$$+\mathbf{f}_v^{ext} + \mathbf{f}_l^{ext},$$

with

$$\frac{1}{\tau} = \frac{\beta_v}{\tau_v} + \frac{1-\beta_v}{\tau_l},$$

and

$$\tilde{P} = -\beta_v\left\{\Psi_v(\tau_v,T) - p_w\frac{1}{\tau_v}\right\} - (1-\beta_v)\left\{\Psi_l(\tau_l,T) - p_w\frac{1}{\tau_l}\right\}, \tag{13.57}$$

where $\tilde{P}$ is the vapor liquid partial pressure in the air vapor liquid mixture. It is the total pressure $P_{cloud}$ in case there is no air. We assume the pressures and the friction forces between the air and the water are the leading terms of this equation of motion. The other forces, the acceleration forces, the exterior forces and the stresses which are not pressures, are negligible compared to them. By keeping only the leading terms in this equation of motion, we get

$$(\beta_v k_{va} + \beta_l k_{la})\,\mathbf{U} = -\,\mathrm{grad}\,\tilde{P}.$$

For the sake of simplicity, we assume also $k_{va} = k_{la} = k_p$, and we have

$$\mathbf{U} = -\frac{1}{k_p}\,\mathrm{grad}\,\tilde{P}. \tag{13.58}$$

This relationship means that vapor and liquid water flow toward low pressure. Note that this type of constitutive law is rather common (see for instance, relationship (4.29)).

The mass balance (13.34) gives

$$\frac{d}{dt}\left(\frac{1}{\tau}\right) + \frac{1}{\tau}\,\mathrm{div}\,\mathbf{U} = 0. \tag{13.59}$$

The air mass balance (13.4) becomes

$$\frac{\partial}{\partial t}\left(\frac{\beta_v}{\tau_a}\right) = 0. \tag{13.60}$$

The other equations are

$$0 = \beta_v\left(\frac{\partial \Psi_a}{\partial \tau_a} + p_a\frac{1}{\tau_a^2}\right) = \beta_v\left(\frac{\partial \Psi_v}{\partial \tau_v} + p_w\frac{1}{\tau_a^2}\right)$$
$$= (1-\beta_v)\left(\frac{\partial \Psi_l}{\partial \tau_l} + p_w\frac{1}{\tau_l^2}\right), \tag{13.61}$$

$$\left\{\Psi_a - p_a\frac{1}{\tau_a}\right\} + \left\{\Psi_v - p_w\frac{1}{\tau_v}\right\} - \left\{\Psi_l - p_w\frac{1}{\tau_l}\right\}$$
$$+B^{reac} - 2k_\beta \Delta\beta_v = 0,\ B^{reac} \in \partial I(\beta_v), \tag{13.62}$$

$$\frac{\mathrm{d}s_v}{\mathrm{d}t} + \frac{\mathrm{d}s_l}{\mathrm{d}t} + \frac{\partial s_a}{\partial t} + (s_v + s_l)\,\mathrm{div}\,\mathbf{U} - \lambda\Delta\ln T = \mathscr{D} + R^{ext}. \tag{13.63}$$

The 9 unknowns are $\tilde{P}$, $\mathbf{U}$, $p_a$, $p_w$, $\tau_v$, $\tau_l$, $\tau_a$, $\beta_v$ and $T$. There are 9 equations, (13.57)–(13.63) to find them.

### 13.14.1 The Example

In the schematic example we have

$$\delta = \left\{\Psi_a - p_a\frac{1}{\tau_a}\right\} + \left\{\Psi_v - p_w\frac{1}{\tau_v}\right\} - \left\{\Psi_l - p_w\frac{1}{\tau_l}\right\}$$
$$= -\frac{rT}{\tau_a} - \frac{kT}{\tau_v} + \left\{\frac{kT}{\tau_l} + d(T_c - T)\right\}.$$

Quantity $\delta$ governs the phase change. In an homogeneous mixture, i.e., $\Delta\beta_v = 0$, we have:

- If $\delta > 0$, $\beta_v = 0$, there is only liquid water. Conditions of Sect. 13.11 apply and we have

  $$p_l(\tau_l, T) > \hat{p}_w(T) \Leftrightarrow \tau_l < \hat{\tau}_l(T, \tau_a = \infty) \Leftrightarrow P_{cloud} > \hat{P}_l(T),$$

  where specific pressure $\hat{p}_w$ is the liquid–vapor mixture specific pressure $\hat{p}_w(T) = \hat{p}_w(T, \tau_a = \infty)$, $P_{cloud} = P_l$ and $\hat{P}_l(T)$ is the liquid pressure in the vapor liquid water mixture at temperature $T$.

- If $\delta < 0$, $\beta_v = 1$, there is only vapor and air. Conditions of Sect. 13.11 apply and we have

$$p_v(\tau_v, T) < \hat{p}_w(T, \tau_a) \Leftrightarrow \tau_v > \hat{\tau}_v(T, \tau_a)$$
$$\Leftrightarrow P_v < \hat{P}_v(T, \tau_a) \Leftrightarrow P_{cloud} < \hat{P}_v(T, \tau_a) + P_a(T, \tau_a),$$

where specific pressure $\hat{p}_w$ is the air liquid vapor mixture specific pressure $\hat{p}_w(T, \tau_a)$ and $\hat{P}_v(T, \tau_a)$ is the vapor pressure in the air vapor liquid water mixture at temperature $T$ with air specific volume $\tau_a$.
- If $\delta = 0$, $0 \leq \beta_v \leq 1$, there is a mixture of air, liquid water and vapor with conditions given in Sect. 13.11

$$\begin{aligned} \tau_l &= \hat{\tau}_l(T, \tau_a),\ \tau_v = \hat{\tau}_v(T, \tau_a),\ p_v(\hat{\tau}_v, T) \\ &= p_l(\hat{\tau}_l, T) = \hat{p}_w(T, \tau_a), \\ P_{cloud} &= \hat{P}_v(T, \tau_a) + P_a(T, \tau_a) + \hat{P}_l(T, \tau_a), \\ \hat{P}_v(T, \tau_a) + P_a(T, \tau_a) &= \hat{P}_l(T, \tau_a). \end{aligned}$$

*Remark 13.18.* The preceding results based on the potentials $G$ may be obtained with the free energies. For instance, condition $\delta > 0$, is

$$\inf_{\tau_v} \left\{ \Psi_v(\tau_v, T) - p_l(\tau_l, T)\frac{1}{\tau_v} \right\} > \left\{ \Psi_l(\tau_l, T) - p_l(\tau_l, T)\frac{1}{\tau_l} \right\},$$

or

$$\left\{ \Psi_v(\bar{\tau}(\tau_l), T) - p_l(\tau_l, T)\frac{1}{\bar{\tau}} \right\} > \left\{ \Psi_l(\tau_l, T) - p_l(\tau_l, T)\frac{1}{\tau_l} \right\},$$

with

$$p_l(\tau_l, T) = -\tau_l^2 \frac{\partial \Psi_l}{\partial \tau_l}(\tau_l, T),$$

and $\bar{\tau}_v(\tau_l, T)$ defined by

$$p_l(\tau_l, T) = -\bar{\tau}_v^2 \frac{\partial \Psi_v}{\partial \tau_v}(\bar{\tau}_v, T).$$

Function

$$\begin{aligned} \tau_l \to & \left\{ \Psi_v(\bar{\tau}_v(\tau_l, T), T) - p_l(\tau_l, T)\frac{1}{\bar{\tau}_v(\tau_l, T)} \right\} \\ & - \left\{ \Psi_l(\tau_l, T) - p_l(\tau_l, T)\frac{1}{\tau_l} \right\} = \bar{\delta}(\tau_l, T), \end{aligned}$$

is decreasing because (see Remark 13.11)

$$\frac{\partial \bar{\delta}}{\partial \tau_l} = -\frac{\partial p_l}{\partial \tau_l}\left(\frac{1}{\bar{\tau}_v(\tau_l,T)} - \frac{1}{\tau_l}\right) < 0.$$

It is equal to 0 for $\tau_l = \hat{\tau}_l(T) = \hat{\tau}_l(T, \tau_a = \infty)$. Thus $\delta > 0$ is equivalent to $\tau_l < \hat{\tau}_l(T)$.

### 13.14.2 The Equations When the Temperature Is Known

Assuming temperature is known, the equations are

$$\tilde{P} = -\beta_v \left\{ \Psi_v(\tau_v,T) - p_w \frac{1}{\tau_v} \right\} - (1-\beta_v) \left\{ \Psi_l(\tau_l,T) - p_w \frac{1}{\tau_l} \right\},$$

$$\frac{\partial}{\partial t}\left(\frac{\beta_v}{\tau_a}\right) = 0,$$

$$\frac{d}{dt}\left(\frac{1}{\tau}\right) - \frac{1}{k_p \tau} \Delta \tilde{P} = 0,$$

$$\left\{ \Psi_v(\tau_v,T) - p_w \frac{1}{\tau_v} \right\} + \left\{ \Psi_a(\tau_a,T) - p_a \frac{1}{\tau_a} \right\} - \left\{ \Psi_l(\tau_l,T) - p_w \frac{1}{\tau_l} \right\}$$

$$+ B^{reac} - 2k_\beta \Delta \beta_v = 0, \; B^{reac} \in \partial I(\beta_v),$$

$$0 = \beta_v \left( \frac{\partial \Psi_v}{\partial \tau_v} + p_w \frac{1}{\tau_v^2} \right) = (1-\beta_v) \left( \frac{\partial \Psi_l}{\partial \tau_l} + p_w \frac{1}{\tau_l^2} \right) = \beta_v \left( \frac{\partial \Psi_a}{\partial \tau_a} + p_a \frac{1}{\tau_a^2} \right), \tag{13.64}$$

$$\frac{1}{\tau} = \frac{\beta_v}{\tau_v} + \frac{1-\beta_v}{\tau_l}.$$

The 8 unknowns are $\tilde{P}$, $p_w$, $p_a$, $\tau_v$, $\tau_l$, $\tau_a$, $\tau$ and $\beta_v$. The two first relationships of (13.64) are equivalent to

$$-p_w = \beta_v \tau_v^2 \frac{\partial \Psi_v}{\partial \tau_v}(\tau_v,T) + (1-\beta_v)\tau_l^2 \frac{\partial \Psi_l}{\partial \tau_l}(\tau_l,T),$$

$$2\beta_v(1-\beta_v)\tau_v^2 \frac{\partial \Psi_v}{\partial \tau_v}(\tau_v,T) = 2\beta_v(1-\beta_v)\tau_l^2 \frac{\partial \Psi_l}{\partial \tau_l}(\tau_l,T).$$

#### 13.14.2.1 An Other Formulation of the Equations

For the sake of simplicity, let us assume there is no air. We may compute $\tilde{P}$ with relationships (13.64)

$$\tilde{P} = -\beta_v \left\{ \Psi_v(\tau_v, T) - p_w \frac{1}{\tau_v} \right\} - (1-\beta_v) \left\{ \Psi_l(\tau_l, T) - p_w \frac{1}{\tau_l} \right\}$$
$$= -\left\{ \beta_v \frac{\partial (\tau_v \Psi_v)}{\partial \tau_v}(\tau_v, T) + \beta_l \frac{\partial (\tau_l \Psi_l)}{\partial \tau_l}(\tau_l, T) \right\}.$$

We get the system of partial differential equations

$$2\beta_v(1-\beta_v)\tau_v^2 \frac{\partial \Psi_v}{\partial \tau_v}(\tau_v, T) = 2\beta_v(1-\beta_v)\tau_l^2 \frac{\partial \Psi_l}{\partial \tau_l}(\tau_l, T),$$

$$\frac{d}{dt}\left(\frac{1}{\tau}\right) + \frac{1}{k_p \tau} \Delta \left\{ \beta_v \frac{\partial (\tau_v \Psi_v)}{\partial \tau_v}(\tau_v, T) + \beta_l \frac{\partial (\tau_l \Psi_l)}{\partial \tau_l}(\tau_l, T) \right\} = 0, \quad (13.65)$$

$$\frac{1}{\tau} = \frac{\beta_v}{\tau_v} + \frac{1-\beta_v}{\tau_l},$$

$$\frac{\partial (\tau_v \Psi_v)}{\partial \tau_v}(\tau_v, T) - \frac{\partial (\tau_l \Psi_l)}{\partial \tau_l}(\tau_l, T) + B^{reac} - 2k_\beta \Delta \beta_v = 0, \quad (13.66)$$

$$B^{reac} \in \partial I(\beta_v), \quad (13.67)$$

for the 4 unknowns $\tau_v$, $\tau_l$, $\tau$ and $\beta_v$.

## 13.15 The Temperature Is Larger than Critical Temperature $T_c$

When temperature $T$ tends towards $T_c$ with $T < T_c$, the physical properties of the liquid phase and of the vapor phase become similar. When temperature is larger or equal to $T_c$ they are the same. Let us note that the gas resulting from the liquid phase can mix with the air. Thus in case we assume a mixture of the three phases, we have to verify that there is no non miscible phase with the air phase, i.e., the liquid phase tends to have null volume fraction. In order to have the same mechanical properties when $T \geq T_c$, we assume

$$\lim_{T \to T_c, T < T_c} \Psi_v(\tau, T) = \lim_{T \to T_c, T < T_c} \Psi_l(\tau, T) = \mathcal{F}(\tau, T),$$

and denote $\Psi_v(\tau, T) = \mathcal{F}(\tau, T)$ and $\Psi_l(\tau, T) = \mathcal{F}(\tau, T)$ the free energies for $T > T_c$.

It we assume that the three phases coexist for $T > T_c$, constitutive laws (13.38) give

$$-p_w = \tau_v^2 \frac{\partial \mathcal{F}}{\partial \tau_v}(\tau_v, T) = \tau_l^2 \frac{\partial \mathcal{F}}{\partial \tau_l}(\tau_l, T).$$

It results

$$\tau_v = \tau_l = \tau.$$

Thus the two water phases have same density and temperature. But they have different volume fractions $\beta_v$ and $\beta_l$. Assuming air is an ideal gas, we have

$$\left\{\Psi_a(\tau_a,T) - p_a\frac{1}{\tau_a}\right\} + \left\{\Psi_v(\tau_v,T) - p_w\frac{1}{\tau_v}\right\} - \left\{\Psi_l(\tau_l,T) - p_w\frac{1}{\tau_l}\right\}$$
$$= -\frac{rT}{\tau_a} < 0. \tag{13.68}$$

Equation (13.27) which governs the evolution of the mixture

$$-\frac{rT}{\tau_a} + B^{reac} - 2k_\beta \Delta\beta_v = 0,$$

tends to impose $\beta_v = 1$ because $-rT/\tau_a$ is negative. It appears the presence of air favors the vapor phase by having $rT/\tau_a = P_a$ large. In case there is no air, the volume fractions do not intervene on the evolution. The results are the same whatever their values.

### 13.15.1 An Homogeneous Cloud

In an homogeneous domain, relationships (13.52) and (13.68) give $\beta_v = 1$. There is only vapor and air with equations of motion and entropy balance

$$\frac{1}{\tau}\frac{\mathrm{d}^v}{\mathrm{d}t}\mathbf{U}_v = -k_{va}(\mathbf{U}_v - \mathbf{U}_a) + \mathbf{f}_v^{ext}, \tag{13.69}$$
$$\frac{1}{\tau_a}\frac{\mathrm{d}^a}{\mathrm{d}t}\mathbf{U}_a = k_{va}(\mathbf{U}_v - \mathbf{U}_a) + \mathbf{f}_a^{ext},$$

$$p_w = -\tau^2\frac{\partial F}{\partial \tau}(\tau,T),\ p_a = -\tau_a^2\frac{\partial \Psi}{\partial \tau_a}(\tau_v,T), \tag{13.70}$$

$$\frac{\mathrm{d}^v s_v}{\mathrm{d}t} + \frac{\mathrm{d}^a s_a}{\mathrm{d}t} = \mathscr{D} + R^{ext}. \tag{13.71}$$

Note that the dissipative terms $k_{va}(\mathbf{U}_v - \mathbf{U}_a)$ in (13.69) and (13.70) tend to equalize the air and gas velocity in the cloud.

### 13.15.2 The Example: The Different Temperature-Specific Volume Domains

The free energies defined for any temperature and specific volume are

$$\Psi_l(\tau_l,T) = -kT\frac{\ln \tau_l}{\tau_l} - d\frac{\tau_c - \tau_l}{\tau_l}(pn(T - T_c)),$$

$$\Psi_v(\tau_v, T) = -kT\frac{\ln \tau_v}{\tau_v},$$

where function $pn\{x\} = \sup(0, -x)$ is the negative part function. It results

$$\mathscr{F}(\tau, T) = -kT\frac{\ln \tau}{\tau}. \tag{13.72}$$

#### 13.15.2.1 The Schematic Liquid–Vapor Domains

Depending on the temperature and on vapor specific volume, we have four schematic domains:

- When temperature is lower than critical temperature $T_c$, vapor and liquid may coexist. The two schematic free energies are

$$\Psi_l(\tau_l, T) = -kT\frac{\ln \tau_l}{\tau_l} + d\frac{\tau_c - \tau_l}{\tau_l}(T - T_c),$$
$$\Psi_v(\tau_v, T) = -kT\frac{\ln \tau_v}{\tau_v}. \tag{13.73}$$

  This domain, $T \le T_c$, is divided into three domains depending on the specific volume:

  1. If the specific volume is low, $\tau < \hat{\tau}_l(T)$, there is only liquid.
  2. If the specific volume is in between $\hat{\tau}_l(T)$ and $\hat{\tau}_v(T)$ there is a mixture of liquid and vapor.
  3. If the specific domain is large, $\tau > \hat{\tau}_v(T)$, there is only vapor.

- When temperature is larger than critical temperature $T_c$, there is only vapor if air is present. It is no longer possible to discriminate vapor and liquid when there is no air. They have the same free energy

$$\mathscr{F}(\tau, T) = -kT\frac{\ln \tau}{\tau},$$

  which is the free energy of an ideal gas. The only difference between vapor and liquid is that air is not miscible with liquid.

*Remark 13.19.* Functions $T \to \Psi(\tau, T)$ are concave functions for $T = T_c$ because $T \to -(\tau_c - \tau_l)\, pn(T - T_c)$ is concave because $\tau_c - \tau_l > 0$.

*Remark 13.20.* Mechanics requires that function $T \to \Psi(\tau, T)$ is concave but does not require function $1/\tau \to \Psi(\tau, T)$ is convex. This property is satisfied by the free energies we have chosen. It results the analysis of the evolution is rather easy and simple. In case function $1/\tau \to \Psi(\tau, T)$ is not convex, definition of potential $G$ is questionable and the analysis may be difficult.

*Remark 13.21.* A part of physics is not described: the case ice is present in the cloud. Ice appears when temperature is low and specific volume is small. In a predictive theory introducing the possibility of ice, a solid volume fraction intervene, $\beta_s$, with

$$\beta_s + \beta_l + \beta_v = 1,$$

assuming there are no voids and no interpenetration of the phases. The same rules provide the equations of the predictive theory. There is the possibility to have the three phases present at the triple point (see diagram of Fig. 12.1).

## 13.16 Solid–Solid and Solid–Gas Phase Changes

A gas, for instance hydrogen, circulating within a porous metal may be adsorbed by the metal acting as a gasholder. The gas-metal mixture is a mixture of two solid solutions, [154]. The solid solutions mixture composition varies involving solid–solid and solid–gas phase changes. Experiments show that the temperature and the gas pressure are the important quantities to describe this solid–solid phase change which is important when focusing on the amount of gas which is adsorbed, [154]. Assuming the macroscopic approximation that the two phases can coexist at each point together with gas and metal, a predictive theory together with analytical results are described in [45]. The state quantities are $E = (T, \tau, \beta, \operatorname{grad}\beta)$ for the metal and the solid solutions and, $E_g = (T, \tau_g)$ for the gas, $T$ is the temperature, $\tau$ the solid solution specific volume, $\beta$ the volume fraction of a solid solution, $1 - \beta$ being the volume fraction of the other solid solution and $\tau_g$ is the specific gas volume.

# Chapter 14
# Conclusion

Continuum and discontinuum mechanics give predictive theories to investigate at the engineering level, i.e., the macroscopic level, numerous phase change phenomena occurring during smooth or violent evolutions. The important elements of the theories are the choice of:

1. The work of the interior forces
2. The free energy
3. The pseudopotential of dissipation

The choice of the work of the internal forces gives the sophistication and the scope of the theories. The choices of the free energy and of the pseudopotential of dissipation which gather all the physical properties, is guided both by theory and experiment. Because experiments are difficult to perform, expensive end time consuming, the imagination and skill of the engineer and scientist is important at this step of the construction of the predictive model. The theory has to have few parameters to be useful. Note that this is the case for the cloud theory, Chap. 13, involving only four parameters.

Other elements of the modelling work have not been mentioned:

1. The mathematical analysis: it is important to know that the equations are coherent in terms of mathematics. A computer program gives always numerical answers even to problems which have no solution. It is difficult to trust them.
2. The numerical approximations.
3. The measurements of the physical parameters.

The theories may be upgraded or downgraded to apply to macroscopic engineering problems coupling mechanical and thermal actions.

M. Frémond, *Phase Change in Mechanics*, Lecture Notes of the Unione Matematica Italiana 13, DOI 10.1007/978-3-642-24609-8_14,

# Appendix A
# Some Elements of Convex Analysis

## A.1 Convex Sets

Let $C$ be a set of linear space $V$. This set is convex if

**Definition A.1.** $\forall x \in C, \forall y \in C,\ \forall \theta \in ]0,1[$, point $\theta x + (1-\theta)y \in C$.

Segment $[0,1]$ of space $V = \mathbb{R}$ is convex. The interior of a circle is a convex set of $V = \mathbb{R}^2$. On the contrary the exterior of a circle is not convex.

Convex sets are useful in mechanics to describe numerous and various properties. For instance, the possible positions of a soccer ball above the football field is a convex set: it is

$$C = \{\mathbf{x} = (x_i),\ i = 1,3\,|x_i \in \mathbb{R},\ x_3 \geq 0\}.$$

## A.2 Convex Functions

Let $\overline{\mathbb{R}} = \mathbb{R} \cup \{+\infty\}$ where the regular addition is completed by the rules

$$\forall a \in \mathbb{R},\ a + (+\infty) = +\infty,$$
$$+\infty + (+\infty) = +\infty.$$

Multiplication by positive numbers is completed by

$$\forall a \in \mathbb{R},\ a > 0,\ a \times (+\infty) = +\infty.$$

In this context, it is forbidden to multiply either by 0 or by negative numbers.

Let $f$ an application from linear space $V$ into $\overline{\mathbb{R}}$. This function is convex if

M. Frémond, *Phase Change in Mechanics*, Lecture Notes of the Unione Matematica Italiana 13, DOI 10.1007/978-3-642-24609-8,

**Definition A.2.** $\forall x \in V, \forall y \in V, \forall \theta \in ]0,1[$,

$$f(\theta x + (1-\theta)y) \leq \theta f(x) + (1-\theta) f(y).$$

*Remark A.1.* Function $f$ is actually multiplied by positive numbers because $\theta \in ]0,1[$.

### *A.2.1 Examples of Convex Functions*

It is easy to prove that functions $f_p$ where $V = \mathbb{R}$

$$x \in \mathbb{R} \to f_p(x) = \frac{1}{p} |x|^p, \; with \; p \in \mathbb{R}, \; p \geq 1,$$

are convex.

Let function $I$ from $V = \mathbb{R}$ into $\overline{\mathbb{R}}$ be defined by

$$I(x) = 0, \; if \; x \in [0,1],$$
$$I(x) = +\infty, \; if \; x \notin [0,1].$$

This function is convex. It is called the indicator function of segment $[0,1]$ (Fig. A.2). More generally, we denote indicator function of set $C \subset V$, function $I_C$ defined by

$$I_C(x) = 0, \; if \; x \in C,$$
$$I_C(x) = +\infty, \; if \; x \notin C.$$

It is easy to connect the convex function and convex set notions. It is shown that

**Theorem A.1.** *A convex $C \subset V$ is convex if and only if its indicator function $I_C$ is convex.*

Indicator functions may seem a little bit strange. In the sequel it is to be seen in the examples that they are productive tools in mechanics for dealing with internal constraints relating mechanical quantities. Three other indicator functions are useful for $V = \mathbb{R}$. They are the indicator functions $I_+$, $I_-$ and $I_0$ of the sets of the non negative and non positive numbers and of the origin:

$$I_+(x) = 0, \; if \; x \geq 0,$$
$$I_+(x) = +\infty, \; if \; x < 0,$$
$$I_-(x) = 0, \; if \; x \leq 0,$$
$$I_-(x) = +\infty, \; if \; x > 0,$$

and

$$I_0(0) = 0,$$
$$I_0(x) = +\infty, \; if \; x \neq 0.$$

## A.3 Linear Spaces in Duality

**Definition A.3.** Two linear spaces $V$ and $V^*$are in duality if there exist a bilinear form $\langle \cdot, \cdot \rangle$ defined on $V \times V^*$ such that

$$for\ any\ x \in V,\ x \neq 0,\ there\ exists\ y^* \in V^*,\ such\ that\ \langle x, y^* \rangle \neq 0;$$
$$for\ any\ y^* \in V^*,\ y^* \neq 0,\ there\ exists\ x \in V,\ such\ that\ \langle x, y^* \rangle \neq 0.$$

### *A.3.1 Examples of Linear Spaces in Duality*

Spaces
$V = \mathbb{R}$, $V^* = \mathbb{R}$ are in duality with the bilinear form which is the usual product

$$\langle x, y \rangle = x \cdot y\,; \tag{A.1}$$

$V = \mathbb{R}^n$, $V^* = \mathbb{R}^n$ are in duality with the bilinear form which is the usual scalar product

$$\langle x, y \rangle = \mathbf{x} \cdot \mathbf{y} = \sum_{i=1}^{i=n} x_i y_i, \tag{A.2}$$

where $\mathbf{x} = (x_i)$ and $\mathbf{y} = (y_i)$ are vectors of $\mathbb{R}^n$, the coordinates of which are $x_i$ and $y_i$;
$V = S$, $V^* = S$ where $S$ is the linear space of symmetric matrices $3 \times 3$, are in duality with the bilinear form

$$\begin{aligned} \mathsf{e} \in V, \mathsf{s} \in V^*, \langle \mathsf{e}, \mathsf{s} \rangle &= \mathsf{e} : \mathsf{s} = \sum_{i,j=1}^{i,j=3} e_{i,j} s_{i,j} = e_{i,j} s_{i,j} \\ &= e_{11}s_{11} + e_{22}s_{22} + e_{33}s_{33} + 2e_{12}s_{12} + 2e_{13}s_{13} + 2e_{23}s_{23}, \end{aligned} \tag{A.3}$$

where we use the Einstein summation rule. Be careful, linear spaces $V = S$ and $V^* = S$ are also in duality with the bilinear form

$$\langle\langle \mathsf{e}, \mathsf{s} \rangle\rangle = e_{11}s_{11} + e_{22}s_{22} + e_{33}s_{33} + e_{12}s_{12} + e_{13}s_{13} + e_{23}s_{23},$$

which is different from the preceding one. Let us give two more examples. Linear space of the velocities $\mathbf{U}$ in a domain $\Omega$ of $\mathbb{R}^3$

$$V = \left\{ \mathbf{U}(\mathbf{x}) \,\middle|\, \mathbf{U} \in L^2(\Omega) \right\},$$

is in duality with the linear space of the forces $\mathbf{f}$ applied to the points of the domain

$$V^* = \left\{ \mathbf{f}(\mathbf{x}) \,\middle|\, \mathbf{f} \in L^2(\Omega) \right\},$$

with bilinear form

$$\langle \mathbf{U}, \mathbf{f} \rangle = \int_\Omega \mathbf{U}(\mathbf{x}) \cdot \mathbf{f}(\mathbf{x}) d\Omega,$$

which is the power of the force applied to the domain. The linear space of the strain rates

$$V = \left\{ \mathsf{D} = (D_{ij}(\mathbf{x})) \,\middle|\, D_{ij} = D_{ji},\ D_{ij} \in L^2(\Omega) \right\},$$

is in duality with the stresses linear space

$$V^* = \left\{ \sigma = (\sigma_{ij}(\mathbf{x})) \,\middle|\, \sigma_{ij} = \sigma_{ji},\ \sigma_{ij} \in L^2(\Omega) \right\},$$

with bilinear form

$$\langle \mathsf{D}, \sigma \rangle = \int_\Omega \sigma(\mathbf{x}) : \mathsf{D}(\mathbf{x}) d\Omega = \int_\Omega \sigma_{ij}(\mathbf{x}) D_{ij}(\mathbf{x}) d\Omega,$$

which is the power of the stresses. This bilinear form is used in the definition of the power of the interior forces, see Sect. 3.1.

## A.4 Subgradients and Subdifferential Set of Convex Functions

Convex function $f_p$ given above, is differentiable for $p > 1$ but it is not for $p = 1$ where it is equal to the absolute value function $x \to |x|$, which has no derivative at the origin. In the same way, indicator function $I$ is not differentiable because its value is $+\infty$ at some points and it has no derivative at points $x = 0$ and $x = 1$. Let us recall that for smooth convex functions of one variable, the derivative is an increasing function and this property yields

$$\left( \frac{df}{dx}(y) - \frac{df}{dx}(z) \right) (y - z) \geq 0. \qquad \text{(A.4)}$$

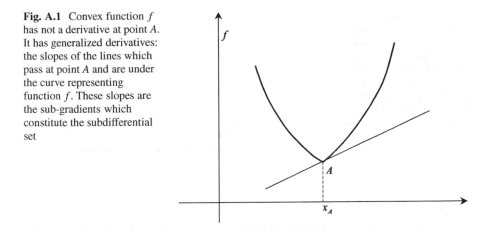

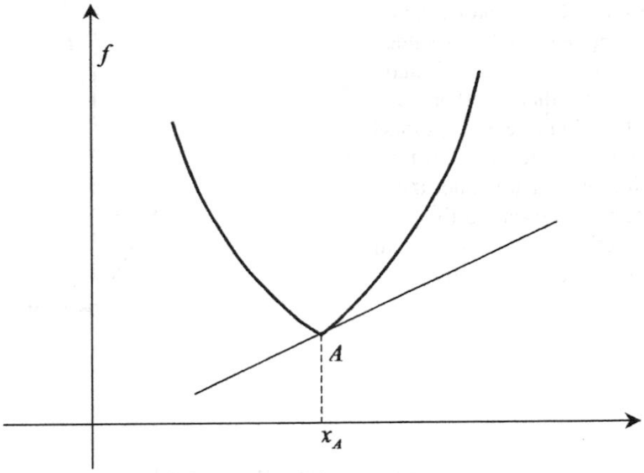


**Fig. A.1** Convex function $f$ has not a derivative at point $A$. It has generalized derivatives: the slopes of the lines which pass at point $A$ and are under the curve representing function $f$. These slopes are the sub-gradients which constitute the subdifferential set

Thus it seems that we have to loose all the calculus properties related to derivatives. Fortunately, this is not the case. Indeed, it is possible to define generalized derivatives and keep a large amount of properties related to derivatives. Consider function $f$ shown in Fig. A.1. It is convex but it is not differentiable because it has no derivative at point $A$. At a point where the function has a derivative, the curve is everywhere above the tangent. At point $A$, there exist several lines which have this properly.

The slopes of these lines are the subgradients which generalize the derivative:

**Definition A.4.** Let convex function $f$ defined on $V$ in duality with $V^*$. A subgradient of $f$ at point $x \in V$ is an element $x^* \in V^*$ which satisfies

$$\forall z \in V,\ \langle z - x, x^* \rangle + f(x) \leq f(z). \tag{A.5}$$

The set of the $x^*$which satisfy (A.5) is the subdifferential set $f$ at point $x$, denoted $\partial f(x)$.

*Remark A.2.* The subgradient depends on $f$ but depends also on the bilinear form $\langle \cdot, \cdot \rangle$.

## *A.4.1 Two Properties of the Subdifferential Set*

The subdifferential set keeps usual properties of the derivative: a function is not differentiable where its value is $+\infty$ and a differentiable convex function satisfies relationship (A.4). For a convex function first property becomes:

**Theorem A.2.** *Let convex function $f \neq +\infty$. If this function is subdifferentiable at point $x$, i.e., if $\partial f(x) \neq \emptyset$, then it is finite at that point: $f(x) < +\infty$.*

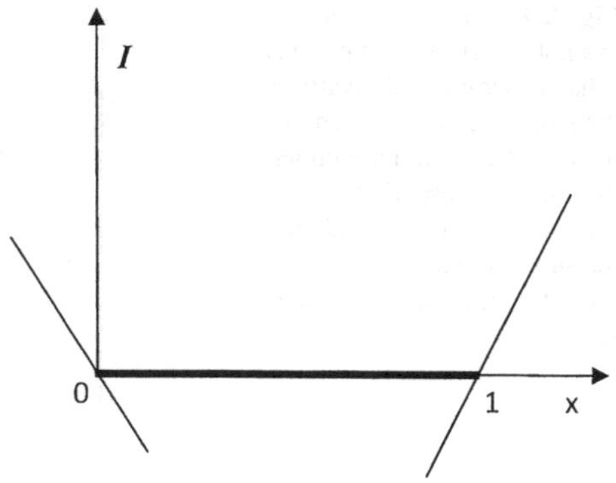


**Fig. A.2** Indicator function of segment $[0,1]$. Its value $I(x)$ is 0, if $0 \leq x \leq 1$, and $+\infty$, if either $x < 0$ or $x > 1$. The subgradients at points 0 and 1 are the slopes of the lines which are under the curve representing the function and in contact with the curve at points 0 or 1

*Proof.* Let us assume $f$ is subdifferentiable at point $x$: let $y^* \in \partial f(x)$. Let us reason ab absurdo and assume that $f$ is not finite at that point: $f(x) = +\infty$. Let us write relationship (A.5) at a point $z$ where $f(z) < +\infty$. Such a point exists because $f \neq +\infty$. Then we have

$$+\infty = \langle z - x, y^* \rangle + f(x) \leq f(z).$$

We deduce that $+\infty = f(z)$. Which is contradictory with the assumption. Let us note that if the only point $z$ where $f(z) < +\infty$ is $x$ itself, then $f(x) < +\infty$ for $f$ not to be identical to $+\infty$. □

This theorem applies in numerous constitutive laws where we have the relationship $B \in \partial I(\beta)$: because $\partial I(\beta)$ is not empty, we have $I(\beta) < +\infty$ which implies that $I(\beta) = 0$ and $0 \leq \beta \leq 1$, see Fig. A.2. Thus relationship $B \in \partial I(\beta)$ implies that quantity $\beta$ which may be a phase volume fraction is actually in between 0 and 1.

Now let us prove that relationship (A.4) is satisfied in some sense by the subdifferential set :

**Theorem A.3.** *Let $f$ a function convex. We have*

$$\forall y^* \in \partial f(y),\ \forall z^* \in \partial f(z),\ \langle y - z, y^* - z^* \rangle \geq 0.$$

*Proof.* It is sufficient to write relationship (A.5) at points $x$ and $y$. □

*Remark A.3.* It is said that the subdifferentiation operator is a monotone operator.

## A.4.2 Examples of Subdifferential Sets

Let us begin by the subdifferential set of indicator function $I$ of interval $[0,1]$. It is easy to see that the subdifferential set is (Figs. A.2 and A.3)

$$\partial I(0) = \mathbb{R}^-;\ \partial I(1) = \mathbb{R}^+;\ if\ x \in ]0,1[,\ \partial I(x) = \{0\};$$
$$if\ x \notin [0,1],\ \partial I(x) = \emptyset.$$

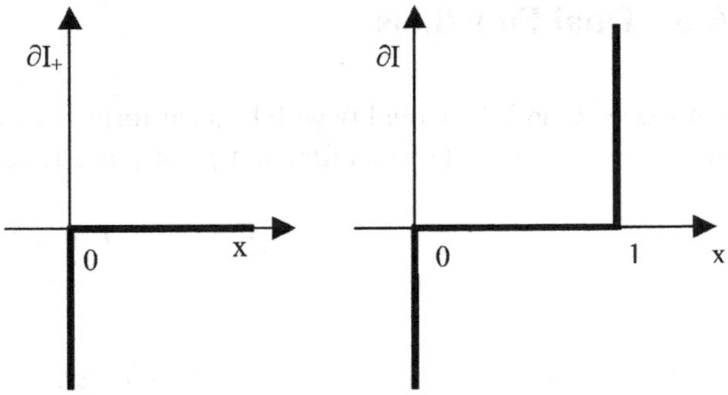


**Fig. A.3** On the left, subdifferential set $\partial I_+$ of indicator function of the set of the positive numbers $\mathbb{R}^+$: $\partial I_+(0) = \mathbb{R}^-$, $\partial I_+(x) = \{0\}$ for $x > 0$ and $\partial I_+(x) = \emptyset$ for $x < 0$. On the right, subdifferential set $\partial I$ of indicator function $I$ of segment $[0, 1]$

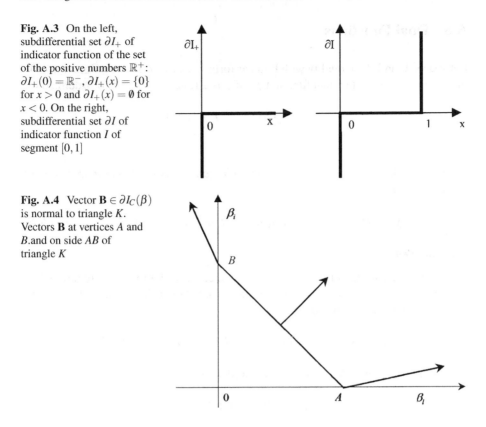

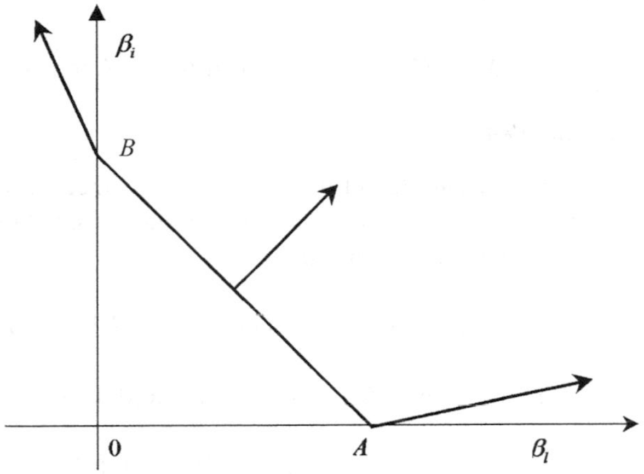


**Fig. A.4** Vector $\mathbf{B} \in \partial I_C(\beta)$ is normal to triangle $K$. Vectors $\mathbf{B}$ at vertices $A$ and $B$.and on side $AB$ of triangle $K$

The subdifferential of indicator function $I_+$ is also easily computed. It is shown in Fig. A.3.

Let us consider triangle $K$ with vertices $O$, $A$ and $B$. It is a convex set of $\mathbb{R}^2$ (Fig. A.4). Let $I_K$ be its indicator function. Let us compute its subdifferential set when the bilinear form is the usual scalar product (A.2) of $\mathbb{R}^2$. Then a subgradient of indicator function $I_K$ at point $\mathbf{x}$ is a vector $\mathbf{B}$ which satisfies (A.5)

$$\forall \mathbf{y} \in \mathbb{R}^2,\ I_K(\mathbf{y}) \geq I_K(\mathbf{x}) + (\mathbf{y} - \mathbf{x}) \cdot \mathbf{B}.$$

We deduce that:

If $\mathbf{x} \notin K$, there exists no vector $\mathbf{B}$ which satisfies the previous relationship (apply Theorem A.2).

If $\mathbf{x} \in K$, the previous relationship gives,

$$\forall \mathbf{y} \in K,\ 0 \geq (\mathbf{y} - \mathbf{x}) \cdot \mathbf{B}.$$

This relationship proves that vector $\mathbf{B}$ is normal to convex set $K$ (Fig. A.4).

## A.5 Dual Functions

Let spaces $V$ and $V^*$ in duality with bilinear form $\langle .,. \rangle$. Let $f$ a convex function of $V$ into $\overline{\mathbb{R}} = \mathbb{R} \cup \{+\infty\}$, the dual function $f^*$ of $f$ is a function of $V^*$ into $\overline{\mathbb{R}}$ defined by

$$f^*(y^*) = sup\{\langle x, y^* \rangle - f(x) \,|x \in V\}.$$

It is possible to prove

**Theorem A.4.** *Dual function $f^*$ is convex. If function $f$ is subdifferentiable at point $x$, the properties*

$$y^* \in \partial f(x),\ x \in \partial f^*(y^*),\ and\ f(x) + f^*(y^*) = < x, y^* >, \tag{A.6}$$

*are equivalent.*

As for an example, let us compute dual function of the indicator function $I$ of segment $[0,1]$ with $V^* = \mathbb{R}$ and the bilinear form being the usual multiplication $\langle x, y \rangle = xy$. Dual function is defined by

$$I^*(y) = \sup\{xy - I(x) \,|x \in V = \mathbb{R}\} = \sup\{xy \,|x \in [0,1]\}.$$

It is called the support function of segment $[0,1]$. It easy to get

$$I^*(y) = y,\ \ if\ y \geq 0, \text{and } I^*(y) = 0,\ if\ y \leq 0.$$

Function $I^*$ is the positive part function defined by

$$pp(y) = \sup\{y, 0\}.$$

### *A.5.1 The Internal Energy, Potential G and Free Enthalpy*

Free energy depends on temperature $T$ and on other state quantities $\eta$, $\Psi(T,\eta)$. It is a concave function of temperature. Dual function of its opposite function, which is convex

$$(-\Psi)^*(y,\eta) = \sup\{xy - (-\Psi)(x,\eta) \,|x \in V = \mathbb{R}\},$$

defines the internal energy depending on entropy $s$ and on $\eta$ by relationship

$$e(s,\eta) = (-\Psi)^*(s,\eta).$$

It results also the equivalent relationships

$$T \in \partial e(s,\eta),\ s \in -\partial\Psi(T,\eta),\ \text{and}\ e(s,\eta) = sT + \Psi(T,\eta),$$

where the subdifferential sets are with respect to $s$ and $T$, $\eta$ being a parameter, see also Sect. 3.2.3. Note that

$$\forall x, \forall y,\ e(y,\eta) - \Psi(x,\eta) \geq xy.$$

An other example of dual function is given in Sect. 13.10.1, where a new potential $G$ is defined with dual function of volumic free energy $\Psi$ with respect to density $1/\tau$. We define function $\Psi^*(p,T)$ by

$$\Psi^*(p,T) = \sup_x \left\{ p\frac{1}{x} - \Psi(x,T)\,|x \in V = \mathbb{R} \right\},$$

where function $1/\tau \to \Psi(\tau,T)$ is convex and potential $G(p,T)$ by

$$G(p,T) = -\Psi^*(p,T),$$

with

$$G(p,T) = -p\frac{1}{\tau} + \Psi(\tau,T) \Leftrightarrow \frac{1}{\tau} = \frac{\partial \Psi^*}{\partial p} = -\frac{\partial G}{\partial p} \Leftrightarrow p = \frac{\partial \Psi}{\partial(1/\tau)} = -\tau^2 \frac{\partial \Psi}{\partial \tau}.$$

*Remark A.4.* Note that function $p \to \Psi^*(p,T)$ is the dual function of function $y \to \Psi(1/y,T)$. To avoid too many notations, we do not denote this function.

Free enthalpy is defined with specific free energy $\hat{\Psi} = \tau\Psi$

$$-\hat{G}(P,T) = \hat{\Psi}^*(-P,T) = \sup_x \left\{ -Px - \hat{\Psi}(x,T) \right\},$$

with

$$\hat{G}(P,T) = P\tau + \hat{\Psi}(\tau,T) \Leftrightarrow \tau = \frac{\partial \hat{G}}{\partial P} \Leftrightarrow P = -\frac{\partial \hat{\Psi}}{\partial \tau}.$$

Precise presentation of convex analysis with applications is given either in books by Jean Jacques Moreau, [172] and by Ivar Ekeland and Roger Temam, [90] or by Bernard Nayroles, [174].

## A.6 Concave Functions

A function $f$ is concave if its opposite $-f$ is convex. For concave function, uppergradient, instead of subgradient, are defined. They satisfy

$$\forall z \in V,\ \langle z - x, x^* \rangle + f(x) \geq f(z).$$

We denote $\hat{\partial} f(x)$, upperdifferential set, the set of the uppergradient at point $x$. It is easy to prove that

$$\hat{\partial} f(x) = -\partial(-f)(x).$$

## A.6.1 Example of Concave Functions

The free energy $\Psi(T,\varepsilon,\beta,\operatorname{grad}\beta)$ is a concave function of temperature $T$, because it is the dual function of the opposite of internal energy $e(s,\varepsilon,\beta,\operatorname{grad}\beta)$ which is a convex function of entropy $s$, [132]. It results the heat capacity

$$-\frac{\partial^2 \Psi}{\partial T^2},$$

is non negative.

*Remark A.5.* If the internal constraint (3.21) is taken into account by

$$\Psi(T,\eta) = \Psi(T,\eta) - I_+(T),$$

the free energy $\Psi(T,\eta)$ is still a concave function of $T$, [112].

Potential $\hat{G}(P,T)$ depending on pressure $P$ and temperature $T$, is concave with respect to $T$ and concave with respect to $P$.

# Appendix B
# The Small Perturbation Assumption

Let us consider the position $\mathbf{x}$ at time $t$ of a material point whose position at time 0 is $\mathbf{a}$. We let

$$\mathbf{x}(\mathbf{a},t) = \mathbf{a} + \mathbf{u}(\mathbf{a},t), \tag{B.1}$$

where vector $\mathbf{u}(\mathbf{a},t)$ is the displacement vector. The set occupied by the material points at time 0 is the reference configuration, $\Omega_a$. The basic idea of the small perturbation assumption is to assume that the displacement

$$\mathbf{u}(\mathbf{a},t) = \mathbf{x}(\mathbf{a},t) - \mathbf{a},$$

is small compared to some reference displacement.

## B.1 Assumptions on the Displacement and on Mechanical Quantities

We assume

*ASSUMPTION 1. The function* $\mathbf{u}(\mathbf{a},t)$ *and its derivatives are of order* $\eta$

$$\left\| \frac{\partial^{i+j}\mathbf{u}(\mathbf{a},t)}{\partial t^i \partial \mathbf{a}^j} \right\| \leq c\eta,$$

*for any i,j. The parameter* $\eta$ *is small with respect to 1 and c is a constant independent of i,j.*

Let us recall that function $O(\mathbf{a},t,z)$ is such that

$$\lim_{z \to 0} \frac{\|O(\mathbf{a},t,z)\|}{z} = A,$$

where $A \in \mathbb{R}$.

M. Frémond, *Phase Change in Mechanics*, Lecture Notes of the Unione Matematica Italiana 13, DOI 10.1007/978-3-642-24609-8,

Assumption 1 allows to compare derivatives of the displacement either in the reference configuration, $\Omega_a$ or in the actual configuration, $\Omega(t)$

**Proposition B.1.** *Let* $\mathbf{u}_a(\mathbf{a},t) = \mathbf{u}(\mathbf{a}+\mathbf{u}(\mathbf{a},t),t) = \mathbf{u}(\mathbf{x},t)$, *then if assumption* 1 *is satisfied*

$$\mathbf{u}(\mathbf{x},t) = \mathbf{u}_a(\mathbf{a},t) = \mathbf{u}(\mathbf{a},t) + O(\mathbf{a},t,\eta^2,$$

$$\frac{\partial \mathbf{u}}{\partial \mathbf{x}}(\mathbf{x},t) = \frac{\partial \mathbf{u}}{\partial \mathbf{x}}(\mathbf{a},t) + O(\mathbf{a},t,\eta^2) = \frac{\partial \mathbf{u}}{\partial \mathbf{a}}(\mathbf{x},t) + O(\mathbf{a},t,\eta^2) = \frac{\partial \mathbf{u}}{\partial \mathbf{a}}(\mathbf{a},t) + O(\mathbf{a},t,\eta^2),$$

$$\frac{\mathrm{d}\mathbf{u}}{\mathrm{d}t}(\mathbf{x},t) = \frac{\partial \mathbf{u}_a}{\partial t}(\mathbf{a},t) = \frac{\partial \mathbf{u}}{\partial t}(\mathbf{x},t) + O(\mathbf{x},t,\eta^2) = \frac{\partial \mathbf{u}}{\partial t}(\mathbf{a},t) + O(\mathbf{a},t,\eta^2).$$

*Proof.* Let us prove for instance, the last relationships related to the velocity. We have

$$\begin{aligned}\frac{\mathrm{d}\mathbf{u}}{\mathrm{d}t}(\mathbf{x},t) &= \frac{\partial \mathbf{u}_a}{\partial t}(\mathbf{a},t) = \frac{\partial \mathbf{u}}{\partial t}(\mathbf{x},t) + \frac{\partial \mathbf{u}}{\partial \mathbf{x}}(\mathbf{x},t)\frac{\partial \mathbf{u}}{\partial t}(\mathbf{a},t)\\ &= \frac{\partial \mathbf{u}}{\partial t}(\mathbf{x},t) + O(\mathbf{x},t,\eta^2) = \frac{\partial \mathbf{u}}{\partial t}(\mathbf{a},t) + O(\mathbf{a},t,\eta^2),\end{aligned}$$

because

$$\begin{aligned}\frac{\partial \mathbf{u}}{\partial \mathbf{x}}(\mathbf{x},t) &= \frac{\partial \mathbf{u}}{\partial \mathbf{x}}(\mathbf{a},t) + O(\mathbf{a},t,\eta^2),\\ \frac{\partial \mathbf{u}}{\partial t}(\mathbf{x},t) &= \frac{\partial \mathbf{u}}{\partial t}(\mathbf{a},t) + O(\mathbf{a},t,\eta^2),\end{aligned}$$

due to assumption 1. □

*Remark B.1.* We may prove that

$$\frac{\partial^{i+j}\mathbf{u}(\mathbf{a},t)}{\partial t^i \partial \mathbf{a}^j} = \frac{\partial^{i+j}\mathbf{u}_a(\mathbf{a},t)}{\partial t^i \partial \mathbf{a}^j} + O(\mathbf{a},t,\eta^2).$$

It results from Proposition B.1 that the displacement and its derivatives are given, up to an $O(\eta^2)$ quantity, by the same function either in the reference configuration with position $\mathbf{a}$ or in the actual configuration with position $\mathbf{x}$ (the positions being related by relationship (B.1)). This property is important for the velocity

$$\mathbf{U}(\mathbf{x},t) = \mathbf{U}(\mathbf{a},t) + O(\mathbf{a},t,\eta^2).$$

It is also important for mechanical quantities related to displacement, for instance the deformation, computed with the displacement $\mathbf{u}(\mathbf{a},t)$ may be estimated with assumption 1. Consider gradient matrix $\mathsf{F}$

$$\mathsf{F}(\mathbf{a},t)=\frac{\partial \mathbf{x}}{\partial \mathbf{a}}(\mathbf{a},t) = \frac{\partial \mathbf{u}}{\partial \mathbf{a}}(\mathbf{a},t)+1=\mathsf{H}+1,$$

where $\mathsf{H}$ is of order 1, $\|\mathsf{H}\| \leq c\eta$. We have

$$\begin{aligned}
\mathsf{G} = \mathsf{F}^{-1} &= -\mathsf{H}+1+O(\eta^2),\\
\mathsf{C} = \mathsf{F}^T\mathsf{F} &= 1+\mathsf{H}+\mathsf{H}^T+O(\eta^2)=1+2\varepsilon+O(\eta^2),\\
\frac{\partial \mathbf{a}}{\partial \mathbf{x}}(\mathbf{x},t) &= \mathsf{F}^{-1}(\mathbf{x}^{-1}(\mathbf{x},t)) = -\mathsf{H}+1+O(\eta^2),
\end{aligned}$$

where matrix

$$\varepsilon = \frac{\mathsf{H}+\mathsf{H}^T}{2},$$

is the small deformation matrix. Proposition B.1 shows that $\mathsf{F}$, $\mathsf{G}$, $\mathsf{H}$, $\varepsilon$ have the same value up to an $O(\eta^2)$ quantity, either in the reference configuration with position $\mathbf{a}$ or in the actual configuration with position $\mathbf{x}$ (the positions being related by relationship (B.1)).

Assumption 1 and Proposition B.1 concerning the displacements are extended to other mechanical quantities with two slightly different assumptions, either

*ASSUMPTION 2.1 Let a mechanical quantity be* $g(\mathbf{a},t)$ *in the reference framework,* $\Omega_a$, *and* $f(\mathbf{x},t)$ *in the actual framework,* $\Omega(t)$. *They are related by* $g(\mathbf{a},t) = f(\mathbf{a}+\mathbf{u}(\mathbf{a},t),t)$. *It is assumed that functions* $g$, $f$ *and their first and second derivatives are of order 1 with respect to* $\eta$,

or

*ASSUMPTION 2.2 Let a mechanical quantity be* $g(\mathbf{a},t)$ *in the reference framework,* $\Omega_a$, *and* $f(\mathbf{x},t)$ *in the actual framework,* $\Omega(t)$. *They are related by* $g(\mathbf{a},t) = f(\mathbf{a}+\mathbf{u}(\mathbf{a},t),t)$. *It is assumed that functions* $g$, $f$ *are of order 0 and their first and second derivatives are of order 1 with respect to* $\eta$.

The values of the mechanical quantities at points $\mathbf{a}$ and $\mathbf{x}$ related by (B.1) may be compared

**Proposition B.2.** *If assumptions* 1 *and either* 2.1 *or* 2.2 *are satisfied, functions* $g$ *and* $f$ *satisfy*

$$\begin{aligned}
f(\mathbf{x},t) &= g(\mathbf{x},t)+O(\mathbf{x},t,\eta^2) = g(\mathbf{a},t),\ g(\mathbf{a},t) = f(\mathbf{a},t)+O(\mathbf{a},t,\eta^2),\\
\frac{\partial f}{\partial \mathbf{x}}(\mathbf{x},t) &= \frac{\partial g}{\partial \mathbf{a}}(\mathbf{a},t)+O(\mathbf{a},t,\eta^2) = \frac{\partial g}{\partial \mathbf{a}}(\mathbf{x},t)+O(\mathbf{x},t,\eta^2) = \frac{\partial f}{\partial \mathbf{x}}(\mathbf{a},t)+O(\mathbf{a},t,\eta^2),\\
\frac{\mathrm{d}f}{\mathrm{d}t}(\mathbf{x},t) &= \frac{\partial g}{\partial t}(\mathbf{a},t) = \frac{\partial f}{\partial t}(\mathbf{x},t)+O(\mathbf{x},t,\eta^2)\\
&= \frac{\partial f}{\partial t}(\mathbf{a},t)+O(\mathbf{a},t,\eta^2) = \frac{\partial g}{\partial t}(\mathbf{a},t)+O(\mathbf{a},t,\eta^2).
\end{aligned}$$

*Proof.* We have

$$\frac{\partial g}{\partial \mathbf{a}}(\mathbf{a},t) = \left(\frac{\partial f}{\partial \mathbf{x}}(\mathbf{x},t)\right)\left(\frac{\partial \mathbf{x}}{\partial \mathbf{a}}(\mathbf{a},t)\right) = \frac{\partial f}{\partial \mathbf{x}}(\mathbf{x},t) + O(\eta^2),$$

$$\frac{\partial g}{\partial \mathbf{a}}(\mathbf{a},t) = \frac{\partial g}{\partial \mathbf{a}}(\mathbf{x},t) + \left(\frac{\partial^2 g}{\partial \mathbf{a}^2}(\mathbf{x},t)\right)(-\mathbf{u}(\mathbf{a},t)) + O(\eta^2) = \frac{\partial g}{\partial \mathbf{a}}(\mathbf{x},t) + O(\eta^2),$$

because $\|\mathsf{H}\| \leq c\eta$ due to assumption 1. We have also

$$\frac{\mathrm{d}f}{\mathrm{d}t}(\mathbf{x},t) = \frac{\partial g}{\partial t}(\mathbf{a},t) = \frac{\partial f}{\partial t}(\mathbf{x},t) + \frac{\partial \mathbf{u}}{\partial t}\cdot\frac{\partial f}{\partial \mathbf{x}},$$

which gives

$$\frac{\mathrm{d}f}{\mathrm{d}t}(\mathbf{x},t) = \frac{\partial g}{\partial t}(\mathbf{a},t) = \frac{\partial f}{\partial t}(\mathbf{x},t) + O(\eta^2),$$

due to assumption 1. We have also

$$\frac{\partial f}{\partial t}(\mathbf{x},t) = \frac{\partial f}{\partial t}(\mathbf{a},t) + \left(\frac{\partial^2 f}{\partial t \partial \mathbf{a}}(\mathbf{a},t)\right)\cdot\mathbf{u}(\mathbf{a},t) + O(\eta^2),$$

$$\frac{\partial g}{\partial t}(\mathbf{a},t) = \frac{\partial g}{\partial t}(\mathbf{x},t) + \left(\frac{\partial^2 g}{\partial t \partial \mathbf{x}}(\mathbf{x},t)\right)\cdot(-\mathbf{u}(\mathbf{a},t)) + O(\eta^2),$$

which gives

$$\frac{\partial f}{\partial t}(\mathbf{x},t) + O(\eta^2) = \frac{\partial f}{\partial t}(\mathbf{a},t) + O(\eta^2) = \frac{\partial g}{\partial t}(\mathbf{a},t) + O(\eta^2) = \frac{\partial g}{\partial t}(\mathbf{x},t) + O(\eta^2).$$

The consequence of the propositions is that any property, up to an $O(\eta^2)$ quantity, is described by the same function either in the reference configuration with position **a** or in the actual configuration with position **x**, the positions being related by relationship (B.1).

### *B.1.1 The Surface Normal Vector*

Consider the surface vector

$$\mathbf{N}_a d\Gamma_a = \frac{1}{J}\mathsf{F}^T\mathbf{N}_x d\Gamma_x,$$

where $J = \det\mathsf{F} = 1 + tr\varepsilon + O(\eta^2) = 1 + O(\eta)$ and $\mathbf{a} = \mathbf{x}^{-1}(\mathbf{x},t)$ are related by (B.1). It results

$$\mathbf{N}_a d\Gamma_a = \mathbf{N}_x d\Gamma_x + O(\eta).$$

Note that the surface vector when multiplied by a mechanical quantity of order $\eta$ is given by the same formula either in the reference configuration or in the actual configuration up to an $O(\eta^2)$ quantity. Thus the boundary relationships have the same properties than the volume relationships.

## B.2 Assumptions on the Exterior Forces: The Choice of the Small Perturbation Equations

The small perturbation assumption involves two more assumptions on the exterior forces and on the choice of the equations to solve

*ASSUMPTION 3 The exterior forces and their first derivatives with respect to* $\mathbf{x}$ *and* $\mathbf{a}$ *are of order* $\eta$.

It results that the exterior forces are given by the same functions either of $\mathbf{a},t$ or of $\mathbf{x},t$, up to an $O(\eta^2)$ quantity.

Let us define $\hat{\mathbf{u}}$, $\hat{\beta}$, $\mathrm{grad}\hat{\beta}$ and $\hat{\theta}$ the solutions in reference configuration $\Omega_a$ of the equations resulting from the balance laws and constitutive laws resulting from assumptions 1, 2 and 3 where the $O(\eta^2)$ quantities are neglected. The following assumption compares the solutions of the different equations.

*ASSUMPTION 4 Functions* $\hat{\mathbf{u}}$, $\hat{\beta}$, $\mathrm{grad}\hat{\beta}$ *and* $\hat{\theta}$ *are approximations of* $\mathbf{u}$, $\beta$, $\mathrm{grad}\beta$ *and* $\theta$ *the solutions of the equations in* $\Omega(t)$ *where the* $O(\eta^2)$ *quantities are not neglected, i.e. the regular equations.*

*Remark B.2.* Let us stress the subjective status of the small perturbation assumption. It is, like the choice of the state quantities, a choice of the engineer.

*Remark B.3.* It is to have a mathematical comparison of the approximations $\hat{\mathbf{u}}$, $\hat{\beta}$, $\mathrm{grad}\hat{\beta}$ and $\hat{\theta}$ and the solutions $\mathbf{u}$, $\beta$, $\mathrm{grad}\beta$ and $\theta$. In most cases this comparison is completely out of reach due to mathematical difficulties and mostly to the fact that the constitutive laws are not clearly known outside the small perturbation assumption. Thus the difficulty is both mechanical and mathematical. A few results are available for rods, [8, 66, 192, 193].

## B.3 The Mass Balance

It is in the reference configuration

$$\rho(\mathbf{x}(\mathbf{a},t),t) = \frac{1}{J}\rho_a(\mathbf{a}) = \rho_a(\mathbf{a})(1 - tr\varepsilon(\mathbf{a},t)) + O(\eta^2),$$

where we assume functions $\mathbf{u}(\mathbf{x},t)$ satisfies Assumption 1 and $\rho(\mathbf{x},t)$, $\rho_a(\mathbf{a})$ satisfy Assumption 2.2. The mass balance is in the actual configuration

$$\frac{\mathrm{d}\rho(\mathbf{x},t)}{\mathrm{d}t}+\rho(\mathbf{x},t)\,\mathrm{div}\,\mathbf{U}(\mathbf{x},t)=0,\ in\ \Omega(t),$$

or with Assumption 1 and Proposition B.2,

$$\frac{\partial\rho}{\partial t}(\mathbf{x},t)+\rho_a(\mathbf{x}^{-1}(\mathbf{x},t))\,\mathrm{div}\,\mathbf{U}(\mathbf{x},t)=0+O(\eta^2).$$

This relationship is satisfied either in $\Omega(t)$ or in $\Omega_a$. The boundary condition

$$-\rho(\mathbf{x},t)\mathbf{U}(\mathbf{x},t)\cdot\mathbf{N}_x(\mathbf{x},t)d\Gamma_x(\mathbf{x},t)=m(\mathbf{x},t)d\Gamma_x(\mathbf{x},t),\ on\ \partial\Omega(t),$$

where $m(\mathbf{x},t)$ is the mass intake which is assumed to satisfy Assumption 3. We have

$$\begin{aligned}-\rho(\mathbf{x}(\mathbf{a},t),t)\mathbf{U}(\mathbf{x}(\mathbf{a},t),t)\cdot\mathbf{N}_x(\mathbf{x}(\mathbf{a},t),t)d\Gamma_x(\mathbf{x}(\mathbf{a},t),t)&=m(\mathbf{x}(\mathbf{a},t),t)d\Gamma_x(\mathbf{x}(\mathbf{a},t),t)\\ =-\rho(\mathbf{x}(\mathbf{a},t),t)\mathbf{U}(\mathbf{x}(\mathbf{a},t),t)\cdot(\mathbf{N}_a(\mathbf{a})d\Gamma_a(\mathbf{a})+O(\eta)),\ on\ \partial\Omega_a,\end{aligned}$$

or with Propositions B.1 and B.2

$$-\rho_a\mathbf{U}\cdot\mathbf{N}=m+O(\eta^2),$$

either in $\partial\Omega(t)$ or in $\partial\Omega_a$. From Assumption 4, the small perturbation mass balance is

$$\frac{\partial\hat{\rho}}{\partial t}+\rho_a\,\mathrm{div}\,\hat{\mathbf{U}}=0,\ in\ \Omega_a, \tag{B.2}$$

$$-\rho_a\hat{\mathbf{U}}\cdot\mathbf{N}=m,\ on\ \partial\Omega_a, \tag{B.3}$$

or

$$\rho_a(1-tr\varepsilon(\hat{\mathbf{u}}))=\hat{\rho},\ in\ \Omega_a, \tag{B.4}$$

$$-\rho_a\hat{\mathbf{u}}\cdot\mathbf{N}=\int_0^t m\,d\tau,\ on\ \partial\Omega_a. \tag{B.5}$$

In case the small displacement $\hat{\mathbf{u}}$ is known, (B.2) or (B.4) give density $\hat{\rho}$, and (B.3) and (B.5) give the mass flux. In case the density depends on other quantities, relationships (B.2) or (B.4) are internal constraints between the state quantities or their velocities.

## B.4 The Equations of Motion

### *B.4.1 The Equation for the Macroscopic Motion*

The volume and surface forces, **f** and **g**, satisfy Assumption 3. Density $\rho$ satisfies Assumption 2.2. We assume also the Cauchy stress $\Sigma$, the Piola-Kirchoff stress S and Boussinesq or Piola-Lagrange stress $\Pi$ related by

$$JG\Sigma G^T = S = G\Pi,$$

satisfy Assumption 2.1. This assumption is to result from the constitutive laws. The equation of motion in the actual configuration are

$$\rho\frac{d\mathbf{U}}{dt} = \operatorname{div}\Sigma + \mathbf{f},\ in\ \Omega(t),$$

$$\Sigma\mathbf{N}_x = \mathbf{g},\ on\ \partial\Omega(t).$$

They are in the reference configuration

$$\rho_a\frac{\partial\mathbf{U}}{\partial t} = \operatorname{div}\Pi + J\mathbf{f},\ in\ \Omega_a,$$

$$\Pi\mathbf{N_a} = \frac{J}{\|F\|}\mathbf{g},\ on\ \partial\Omega_a.$$

It results the equations of motion

$$\rho_a\frac{\partial\mathbf{U}}{\partial t} = \operatorname{div}\Sigma + \mathbf{f} + O(\eta^2),\ either\ in\ \Omega_a\ or\ in\ \Omega(t),$$

$$\Sigma\mathbf{N} = \mathbf{g} + O(\eta^2),\ either\ on\ \partial\Omega_a\ or\ on\ \partial\Omega(t),$$

because each quantity of the equations is given by the same function either in the actual or in the reference configuration up to an $O(\eta^2)$ function. With Assumption 4, the small perturbation equation for the macroscopic motion is

$$\rho_a\frac{\partial\hat{\mathbf{U}}}{\partial t} = \operatorname{div}\hat{\Sigma} + \mathbf{f},\ in\ \Omega_a,$$

$$\hat{\Sigma}\mathbf{N} = \mathbf{g},\ on\ \partial\Omega_a.$$

### *B.4.2 The Equation for the Microscopic Motion: The Classical Case*

The equations for the microscopic motion in the actual configuration are

$$-B + \operatorname{div}\mathbf{H} = A,\ in\ \Omega(t), \tag{B.6}$$

$$\mathbf{H}\cdot\mathbf{N}_x = a,\ on\ \partial\Omega(t).$$

They are in the reference configuration

$$-JB + \mathrm{div} J\mathbf{H}\mathrm{G}^T = JA, \; in \; \Omega_a,$$
$$J\mathbf{H}\mathrm{G}^T \cdot \mathbf{N}_a = \frac{J}{\|\mathbf{F}\|} a, \; on \; \partial\Omega_a.$$

We assume that exterior works $A$ and $a$ satisfy Assumption 3 and that internal works $B$ and $\mathbf{H}$ satisfy Assumption 2.1. It results from Proposition B.2 that the small perturbation equation for the microscopic motion is

$$-\hat{B} + \mathrm{div}\hat{\mathbf{H}} = A, \; in \; \Omega_a,$$
$$\hat{\mathbf{H}} \cdot \mathbf{N} = a, \; on \; \partial\Omega_a. \tag{B.7}$$

The quantities which appear in this relationship are of order $\eta$.

### B.4.3 The Equation for the Microscopic Motion: The Non Classical Case

In this particular case, the assumptions are modified: the power of the order with respect to $\eta$ is increased by 1 for the equation for the microscopic motion (thus it is $\eta^2$). For the other equations the order remains equal to $\eta$. The new assumptions are

*ASSUMPTION 2.1bis. Let a mechanical quantity which intervenes in the equation for the microscopic motion, to be $g(\mathbf{a},t)$ in the reference framework and $f(\mathbf{x},t)$ in the actual framework. They are related by $g(\mathbf{a},t) = f(\mathbf{a}+\mathbf{u}(\mathbf{a},t),t)$. It is assumed that functions $g$, $f$ and their first and second derivatives are of order $\eta^2$.*

*ASSUMPTION 3bis The external actions and their first derivatives with respect to $\mathbf{x}$ and $\mathbf{a}$ are of order $\eta^2$.*

We assume that exterior works $A$ and $a$ satisfy Assumption 3bis and that internal works $B$ and $\mathbf{H}$ satisfy Assumption 2.1bis.

Let us define $\hat{\mathbf{u}}$, $\hat{\beta}$, $\mathrm{grad}\hat{\beta}$ and $\hat{\theta}$ the solutions in reference configuration $\Omega_a$ of the equations resulting from the balance laws and constitutive laws resulting from Assumptions 1, *2bis* and *3bis* where the $O(\eta^3)$ quantities are neglected in the equation for the microscopic motion and the $O(\eta^2)$ quantities are neglected in the other equations. The following assumption compares the solutions of the different equations.

*ASSUMPTION 4bis Functions $\hat{\mathbf{u}}$, $\hat{\beta}$, $\mathrm{grad}\hat{\beta}$ and $\hat{\theta}$ are approximations of $\mathbf{u}$, $\beta$, $\mathrm{grad}\beta$ and $\theta$ the solutions of the equations in $\Omega(t)$ where the $O(\eta^3)$ quantities in the equation for the microscopic motion equation and the $O(\eta^2)$ quantities are not neglected, i.e. the regular equations.*

The small perturbation equation for the microscopic motion are the same but the quantities are of order $\eta^2$. This non classical situation is used in the damage theory where the elastic work

$$\frac{1}{2}\{\lambda_e(tr\varepsilon(\hat{\mathbf{u}}))^2+2\mu_e\varepsilon(\hat{\mathbf{u}}):\varepsilon(\hat{\mathbf{u}})\},$$

is a damage source. This quantity is of order 2 when the small deformations $\varepsilon(\hat{\mathbf{u}})$ is of order $\eta$.

## B.5 The Entropy Balance

It is in the actual configuration

$$\rho\frac{ds}{dt}+\operatorname{div}\mathbf{Q}=R+\frac{1}{T}(\Sigma^d:\mathsf{D}(\mathbf{U})+\frac{d\beta}{dt}B^d+\mathbf{H}^d\cdot\operatorname{grad}\frac{d\beta}{dt}-\frac{\mathbf{Q}\cdot\operatorname{grad}T}{T}),\ in\ \Omega(t),$$
$$-\mathbf{Q}\cdot\mathbf{N}_x=\pi,\ on\ \partial\Omega(t).$$

We assume $T=T_0+\theta$, where $\theta$ satisfies Assumption 2.1 and $T_0$ is of order 0, $\eta^0=1$, with respect to $\eta$, (T satisfies Assumption 2.2). We assume the external thermal actions $R$ and $\pi$ satisfy Assumption 3. We assume the interior forces satisfy Assumption 2.1 or 2.1bis, that the entropy flux vector satisfies Assumption 2.1 and that density $\rho$ satisfies Assumption 2.2. It results

$$\frac{1}{T}(\Sigma^d:\mathsf{D}(\mathbf{U})+\frac{d\beta}{dt}B^d+\mathbf{H}^d\cdot\operatorname{grad}\frac{d\beta}{dt}-\frac{\mathbf{Q}\cdot\operatorname{grad}T}{T})=O(\eta^2).$$

The small perturbation entropy balance is

$$\rho\frac{\partial\hat{s}}{\partial t}+\operatorname{div}\hat{\mathbf{Q}}=R,\ in\ \Omega_a,$$
$$-\hat{\mathbf{Q}}\cdot\mathbf{N}=\pi,\ on\ \partial\Omega_a.$$

## B.6 The Constitutive Laws

They are, for instance, defined by the free energy $\Psi(\mathsf{L},\beta,\operatorname{grad}_a\beta)$ and pseudopotential $\Phi(D(\mathbf{U}),d\beta/dt,\operatorname{grad}_x d\beta/dt,\operatorname{grad}_x T)$ where $\mathsf{L}=(\mathsf{F}^T\mathsf{F}-\mathsf{I})/2$ is the Green-Lagrange strain

$$s=-\frac{\partial\Psi}{\partial T},\ \Sigma^{nd}=\rho\mathsf{F}\frac{\partial\Psi}{\partial\mathsf{L}}\mathsf{F}^T,\ B^{nd}=\rho\frac{\partial\Psi}{\partial\beta},\ \mathbf{H}^{nd}=\rho\mathsf{F}\frac{\partial\Psi}{\partial\operatorname{grad}_a\beta},$$

$$\Sigma^d=\frac{\partial\Phi}{\partial(D(\mathbf{U}))},\ B^d=\frac{\partial\Phi}{\partial(d\beta/dt)},\ \mathbf{H}^d=\frac{\partial\Phi}{\partial(\operatorname{grad}_x d\beta/dt)},$$

$$-\mathbf{Q}=\frac{\partial\Phi}{\partial(\operatorname{grad}_x T)},\tag{B.8}$$

and

$$\Sigma = \Sigma^{nd} + \Sigma^{d},\ B = B^{nd} + B^{d},\ \mathbf{H} = \mathbf{H}^{nd} + \mathbf{H}^{d}.$$

We assume the reference configuration is stress free

$$\frac{\partial \Psi}{\partial \mathsf{L}}(0, \beta, \mathrm{grad}_a \beta) = 0.$$

This assumption makes possible that $\Sigma^{nd}$ is of order $\eta$. In the examples, it is to be checked if the internal forces are of order $\eta$ or $\eta^2$ as required.

The small perturbation interior forces may be introduced with small perturbation free energy and pseudopotential of deformation $\hat{\Psi}(\varepsilon(\hat{\mathbf{u}}), \hat{\beta}, \mathrm{grad}\hat{\beta})$ and $\hat{\Phi}(\mathrm{grad}\hat{\theta}, \partial\hat{\beta}/\partial\mathrm{t}, \mathrm{grad}\partial\hat{\beta}/\partial\mathrm{t})$. In this setting, the interior forces satisfy either Assumption 2.1 or 2.1bis, provided the reaction forces, i.e., the elements of the subdifferential sets, satisfy the same assumption. This property does not result from the constitutive law. It is either an assumption or the consequence of a balance equation.

## B.7 Example 1: The Stefan Problem

We assume that temperature $\theta = T - T_0$ and $\beta$ satisfy Assumption 2.1. The free energy and pseudopotential of dissipation are

$$\Psi(T, \hat{\beta}) = -CT \ln T - \hat{\beta}\frac{L}{T_0}(T - T_0) + I(\hat{\beta}),$$

$$\Phi(\mathrm{grad}\, T, T) = \frac{\lambda}{2T}(\mathrm{grad}\, T)^2.$$

The constitutive laws are

$$B \in -\frac{L}{T_0}(T - T_0) + \partial I(\hat{\beta}),$$

$$s = C(1 + \ln T) = C(1 + \ln T_0) + \frac{C}{T_0}(T - T_0) = s(T_0) + \frac{C}{T_0}(T - T_0),$$

$$\mathbf{Q} = -\frac{\lambda}{T}\mathrm{grad}\, T = -\frac{\lambda}{T_0 + \theta}\mathrm{grad}\, \theta.$$

They give

$$B \in -\frac{L}{T_0}\theta + \partial I(\beta),$$

$$s - s(T_0) = \hat{s} - s(T_0) = \frac{C}{T_0}\hat{\theta},$$

$$\hat{\mathbf{Q}} = -\frac{\lambda}{T_0}\mathrm{grad}\, \hat{\theta}.$$

The equation of motion for the microscopic motion is

$$-B = 0.$$

It results that the two quantities $B = \hat{B}$ and $s - s(T_0)$ satisfy Assumption 2.1.

## B.8 Example 2: The Damage Predictive Theory

We assume $\hat{\beta}$ is of order 0. The free energy and pseudopotential are

$$\begin{aligned}\hat{\Psi}(\varepsilon(\hat{\mathbf{u}}),\hat{\beta},\operatorname{grad}\hat{\beta},\hat{\theta}) &= -\frac{C}{2T_0}\hat{\theta}^2 + \frac{\hat{\beta}}{2}\{\lambda_e(tr\varepsilon(\hat{\mathbf{u}}))^2 + 2\mu_e\varepsilon(\hat{\mathbf{u}}) : \varepsilon(\hat{\mathbf{u}})\} \\ &\quad + w(1-\hat{\beta}) + I(\beta) + \frac{k}{2}(\operatorname{grad}\hat{\beta})^2, \\ \hat{\Phi}(\operatorname{grad}\hat{\theta},\frac{\partial\hat{\beta}}{\partial t}) &= \frac{\lambda}{2T_0}(\operatorname{grad}\hat{\theta})^2 + \frac{c}{2}\left(\frac{\partial\hat{\beta}}{\partial t}\right)^2 + I_-(\frac{\partial\hat{\beta}}{\partial t}).\end{aligned}$$

They give the constitutive laws

$$\begin{aligned}\hat{\Sigma} &= \hat{\beta}\left(\lambda_e(tr\varepsilon(\hat{\mathbf{u}})1 + 2\mu_e\varepsilon(\hat{\mathbf{u}})\right), \\ \hat{B} &\in \frac{1}{2}\{\lambda_e(tr\varepsilon(\hat{\mathbf{u}}))^2 + 2\mu_e\varepsilon(\hat{\mathbf{u}}) : \varepsilon(\hat{\mathbf{u}})\} - w + c\frac{\partial\hat{\beta}}{\partial t} + \partial I(\hat{\beta}) + \partial I_-(\frac{\partial\hat{\beta}}{\partial t}), \\ \hat{\mathbf{H}} &= k\operatorname{grad}\hat{\beta},\ \hat{\mathbf{Q}} = -\frac{\lambda}{T_0}\operatorname{grad}\hat{\theta}, \\ \hat{s} &= \frac{C}{T_0}\hat{\theta}.\end{aligned}$$

We assume that $w$ is of order $\eta^2$, $c$ and $k$ are of order $\eta$. We assume also that the exterior works $A$ and $a$ satisfy Assumption 3bis. They are of order $\eta^2$, (in many engineering problems they are 0). The equation of motion (B.6) show that $B$ is of order $\eta^2$. Thus the small perturbation equations for $\hat{\mathbf{u}}$, $\hat{\beta}$ and $\hat{\theta}$ are

$$\begin{aligned}&C\frac{\partial\hat{\theta}}{\partial t} - \lambda\Delta\hat{\theta} = T_0 R,\ in\,\Omega_a, \\ &c\frac{\partial\hat{\beta}}{\partial t} - k\Delta\hat{\beta} + \partial I(\hat{\beta}) + \partial I_-(\frac{\partial\hat{\beta}}{\partial t}) \ni w - \frac{1}{2}\{\lambda_e(tr\varepsilon(\hat{\mathbf{u}}))^2 \\ &\qquad + 2\mu_e\varepsilon(\hat{\mathbf{u}}) : \varepsilon(\hat{\mathbf{u}})\} + A,\ in\,\Omega_a, \\ &\operatorname{div}\left(\hat{\beta}\left(\lambda_e(tr\varepsilon(\hat{u})1 + 2\mu_e\varepsilon(\hat{\mathbf{u}})\right)\right) + \mathbf{f} = 0,\ in\ \Omega_a,\end{aligned}$$

$$\lambda \frac{\partial \hat{\theta}}{\partial N} = T_0 \pi, \; k \frac{\partial \hat{\beta}}{\partial N} = a, \; \sigma \mathbf{N} = \mathbf{g}, \; in \; \partial \Omega_a,$$

$$\hat{\theta}(\mathbf{x}, 0) = \theta^0(\mathbf{x}), \; \hat{\beta}(\mathbf{x}, 0) = \beta^0(\mathbf{x}), \; in \; \Omega_a.$$

Let us note that the small perturbation assumption may be applied partially. For instance, in this example, the quantity $-CT \ln T$ may be kept in the free energy and not be replaced by $-C\hat{\theta}^2/2T_0$ (see the free energy of Chap. 6). It is also possible to keep in the equations some quantities of order $\eta^2$. For instance in damage theory, the dissipative quantities

$$\frac{1}{T}\left(\hat{\Sigma}^d : \varepsilon\left(\frac{\partial \hat{\mathbf{u}}}{\partial t}\right) + \frac{\partial \hat{\beta}}{\partial t} \hat{B}^d\right).$$

may be kept assuming linear viscosity, [38]. To keep this quantity implies clever and sophisticated mathematics to prove mathematical coherency of the model. The necessity to escape the small perturbation assumption leads to difficulties but it is a clear challenge.

### B.8.1 Some Numbers for the Equation for the Microscopic Motion

Let us give some values of the parameters for concrete damage, [120, 175]

$$w = 10^{-4}\,\text{MPa}, \; c = 10^{-2}\,\text{MPa s}, \; k = 10^{-4}\,\text{MPa m}^2, \; \lambda_e \simeq \mu_e \simeq 10^4\,\text{MPa}.$$

Let us recall, the Lamé parameters $\lambda_e$ and $\mu_e$ are of order $\eta^0 = 1$. Thus choosing $\eta = 10^{-4}$, we have

$$\frac{w}{\mu_e} = A_w \eta^2 \; with \; A_w = 1,$$

$$\frac{c}{\mu_e} = A_c \eta \; with \; A_c = 10^{-2},$$

$$\frac{k}{\mu_e} = A_k \eta \; with \; A_k = 10^{-4}.$$

In this setting, $\eta$ represents the intensity of the small deformation. The numerical values show that elasticity and cohesion of concrete, defining damage source

$$w - \frac{1}{2}\{\lambda_e (tr\varepsilon(\hat{\mathbf{u}}))^2 + 2\mu_e \varepsilon(\hat{\mathbf{u}}) : \varepsilon(\hat{\mathbf{u}})\},$$

with Lamé parameters and cohesion $w$, govern the evolution. The microscopic motion viscosity and local interaction are not as important.

## B.9 A Practical Rule

From the previous results, one may note that the small perturbation equation may be obtained rapidly by dealing with products $XY$ which appear in an equation, where only the terms of order $\eta^p$ are kept, with the following rule:

- Assuming $X$ of order $x$, $Y$ of order $y$.
- If $x+y>p$, the product is neglected.
- If $x+y=p$, the product is replaced by $\hat{X}\hat{Y}$.

## B.10 Advantages and Disadvantages of the Small Perturbation Assumption

The small perturbation assumption is mainly used in solid mechanics. In civil engineering most of the theories use it. The advantages of the small perturbation assumptions are:

1. The equations are written on a fixed and known domain, $\Omega_a$, the reference configuration.
2. The mechanical quantities have an immediate and simple physical meaning as those defined on the actual configuration. They may be easily involved in engineering projects.
3. Mathematical results are available giving the opportunity to have efficient numerical methods.
4. The engineering and practical implications of the predictive theories are often useful and instructive. The everyday success of the elasticity theory is so important that it is often forgotten that it is a small perturbation theory.
5. Let us also mention that the theories without the small perturbation assumption are far from being well established and accepted for engineering purposes.

A disadvantage is to add a new subjective assumption, which means more difference between nature and the predictive theories. The small perturbation assumption is a new cause of maladjustment of a theory to predict the evolution of structures.

A careful and useful way for the engineer to deal with predictive theories is to be convinced that they are only models which give schematic information. In case the information is not good or useful, the theory is used either in an unadapted or in an extreme situation. We think that a theory must not be used in extreme situations. In case an extreme situation occurs, the theory has to be quit and replaced by a new theory taking into account the new phenomenon and quantities which are responsible for the extreme situation. In this careful point of view, the constant comparison of theoretical results with experiments is the guide to judge a predictive theory.

Another disadvantage of the small perturbation assumption is that the rules are vague and do not have a sound mathematical basis. This is due to the basic idea which is to find approximation of the large perturbation solutions which are not defined because the large perturbation theories, for instance the constitutive laws, are far from being known. It is very daring to approximate something which is not well known.

Let us also note that the small perturbation theory is not a linearization process because of the internal constraints which make compulsory to deal with non-smooth equations where the notion of linearization is not pertinent.

# References

1. M. Achenbach, 1989, A model for an alloy with shape
2. J. Aguirre-Puente, M. Frémond, 1975, Frost propagation in wet porous media, in Applications of Methods of Functional Analysis to Problems in Mechanics, P. Germain, B. Nayroles eds, Springer, Berlin, Heidelberg.
3. J. Aguirre-Puente, M. Frémond, 1976, Frost and water propagation in porous media, Second Conference on Soil Water Problems in Cold Regions, Edmonton.
4. O. Allix, P. Ladevéze, A. Corigliano, 1995, Damage analysis of interlaminar fracture specimens, Compos. Struct., 31, 61–74.
5. L. Ambrosio, N. Fusco, D. Pallara, 2000, Special functions of bounded variations and free discontinuity problems, Oxford University Press, Oxford.
6. H. Amor, J. J. Marigo, C. Maurini, 2009, Regularized formulation of the variational brittle fracture with unilateral contact: numerical experiments, J. Mech. Phys. Solids, 57, 1209–1229.
7. K. T. Andrews, A. Klarbring, M. Shillor, S. Wright, 1997, A dynamic contact problem with friction and wear, Int. J. Eng. Sci., 35, 1291–1309.
8. S. S. Antman, 1995, Nonlinear problems of elasticity, Springer Verlag, New York.
9. A. Anthoine, P. Pegon, 1996, Numerical analysis and modelling of the damage and softening of brick masonry, in Numerical analysis and modelling of composite materials, J. W. Ball ed, Blacke Academic and Professional, London.
10. P. Argoul, K. Benzarti, F. Freddi, M. Frémond, Ti Hoa Tam Nguycn, 2011, A damage model to predict the durability of bonded assemblies. Part II: parameter identification and preliminary results for accelerated ageing tests, doi:10.1016/j.conbuildmat.2009.12.014, Constr. and Build. Mater., 25, 2, 556–567.
11. L. Ascione, A. Grimaldi, F. Maceri, 1981, Modeling and analysis of beams on tensionless foundations, Int. J. for Model. Simul.
12. F. Ascione, M. Frémond, 2009, Phase change with voids and bubbles, Vietnam J. Mech., VAST, 31, 3, 263–278.
13. M. Aso, M. Frémond, N. Kenmochi, 2004, Phase change problems with temperature dependent constraints for velocity of volume fractions, J. of Nonlinear Anal. Ser. A, Theor. and Methods, 60, 1003–1023.
14. M. Aso, M. Frémond, N. Kenmochi, 2004, Quasi-variational evolution problems for irreversible phase change, Math. Sci. and Appl., 20, 517–535.
15. F. Ascione, 2009, Mechanical behavior of FRP adhesive joints: A theoretical model, Composites Part B, 40, 116–124.
16. F. Ascione, 2009, Ultimate behaviour of adhesively bonded FRP lap joints, Composites Part B, 40, 107–115.

M. Frémond, *Phase Change in Mechanics*, Lecture Notes of the Unione Matematica Italiana 13, DOI 10.1007/978-3-642-24609-8, 

17. F. Ascione, G. Mancusi, 2010, Axial/bending coupled analysis for FRP adhesive lap joints, Mech. of Adv. Mater. and Struct., 17, 85–98.
18. H. Attouch, G. Buttazzo, G. Michaille, 2004, Variational analysis in Sobolev and BV spaces, Application to PDE and optimization, MPS/SIAM Series in Optimization.
19. F. Auricchio, L. Petrini, 2004, A three-dimensional model describing stress-temperature induced solid phase transformations. Part I: solution algorithm and boundary value problems, Int. J. Num. Methods Eng., 61, 807–836.
20. F. Auricchio, L. Petrini, 2004, A three-dimensional model describing stress-temperature induced solid phase transformations. Part II: thermomechanical coupling and hybrid composite applications, Int. J. Num. Methods Eng., 61, 716–737.
21. H. Bailly, D. Menessier, C. Prunier, R. Dautray, 1996, Le combustible nucléaire des réacteurs à eau sous pression et des réacteurs à neutrons rapides : conception et comportement, Collection du Commissariat à l'Énergie Atomique, Série Synthèses, Eyrolles, Paris.
22. X. Balandraud, E. Ernst, E. Soós, 1999, Phénomènes rhéologiques dans les alliages à mémoire de forme, C. R. Acad. Sci, Paris, II, 327, 1, 33–39.
23. J. M. Ball, R. D. James, 1992, Theory for the microstructure of martensite and applications, Proc. Int. Conf. on Martensitic Transformations, C. M. Wayman, J. Perkings eds, Monterey.
24. A. Barbu, G. Martin, 1993, Radiation effects in metals and alloys, Solid State Phenom., 30 and 31, 179–228.
25. G. I. Barenblatt, 1962, The mathematical theory of equilibrium cracks in brittle fracture, Adv. Appl. Mech., 7, 55–129.
26. E. Benvenuti, G. Borino, A. Tralli, 2002, A Thermodynamically consistent nonlocal formulations for damaging materials, Eur. J. of Mech. A/Solids, 21, 535–553.
27. K. Benzarti, M. Frémond, P. Argoul, Ti Hoa Tam Nguyen, 2009, Durability of bonded assemblies. A predictive theory coupling bulk and interfacial damage mechanisms, Eur. J. of Environmental and Civil Eng., 13, 9, 1141–1151.
28. K. Benzarti, F. Freddi, M. Frémond, 2011, A damage model to predict the durability of bonded assemblies. Part I: debonding behaviour of FRP strengthened concrete structures, doi:10.1016/j.conbuildmat.2009.10.018, Constr. and Build. Mater., 25, 2, 547–555.
29. C. Berriet, C. Lexcellent, B. Raniecki, A. Chrysochoos, 1992, Pseudoelastic behaviour analysis by infrared thermography and resistivity measurements of polycristalline shape memory alloys, ICOMAT 92, Monterey.
30. M. Berveiller, E. Patoor, 1997, Micromechanical modelling of the thermomechanical behaviour of shape memory alloys, Mechanics of solids with phase change, M. Berveiller, F. Fischer eds, Springer Verlag.
31. P. Bisegna, F. Maceri, 1998, Delamination of active layers in piezoelectric laminates, IUTAM Symposium Variations de domaines et frontières libres en mécanique des solides, Kluwer, Amsterdam.
32. D. Blanchard, M. Frémond, 1984, The Stefan problem. Computing without the free boundary, Int. J. Num. Methods Eng., 20, 757–771.
33. D. Blanchard, M. Frémond, 1985, Soils Frost Heaving and Thaw Settlement, Ground Freezing 85, S. Kinosita and M. Fukuda eds, 209–216, Balkema, Rotterdam.
34. C. Bonacina, G. Comini, A. Fasano, M. Primicerio, 1973, Numerical solution of phase-change problems, Int. J. Heat Mass Transfer 16, 1825–1832.
35. E. Bonetti, 2002, Global solution to a nonlinear phase transition model with dissipation, Adv. Math. Sci. Appl. 12, 355–376.
36. E. Bonetti, 2003, Global solvability of a dissipative Frémond model for shape memory alloys. Part I: mathematical formulation and uniqueness, Quart. Appl. Math., 61, 759–781.
37. E. Bonetti, 2004, Global solvability of a dissipative Frémond model for shape memory alloys. Part II: existence, Quart. Appl. Math., 62, 53–76.
38. E. Bonetti, G. Bonfanti, 2003, Existence and uniqueness of the solution to a 3D thermoviscoelastic system, Electron. J. Diff. Eqn., 50, 1–15.
39. E. Bonetti, G. Bonfanti, 2005, Asymptotic analysis for vanishing acceleration in a thermoviscoelastic system, Quad. Semin. Mat. Brescia, 16, Abstract and Appl. Anal., 105–120.

40. E. Bonetti, P. Colli, M. Frémond, 2003, A phase field model with thermal memory governed by the entropy balance, Math. Models and Methods in Appl. Sci., 13, 231–256.
41. E. Bonetti, M. Frémond, 2003, A phase transition model with the entropy balance, Math. Methods in the Appl. Sci., 26, 539–556.
42. E. Bonetti, M. Frémond, 2004, Damage theory: microscopic effects of vanishing macroscopic motions, Comput. and Appl. Math., 22, 3, 1–21.
43. E. Bonetti, M. Frémond, 2010, Analytical results on a model for damaging in domains and interfaces, doi: 10.1051/cocv/2010033, pubished on line 18 August 2010, Control, Optim. and Calc. of Var.
44. E. Bonetti, M. Frémond, Ch. Lexcellent, 2004, Modelling shape memory alloys, J. de Phys. IV France, 115, 383–390.
45. E. Bonetti, M. Frémond, Ch. Lexcellent, 2007, Hydrogen storage: modelling and analytical results, Appl. Math. and Optim., 55, 1, 1–59.
46. E. Bonetti, G. Schimperna, 2004, Local existence to Frémond's model for damage in elastic materials, Cont. Mech. Therm., 16, 319–335.
47. G. Bonfanti, M. Frémond, F. Luterotti, 2000, Global solution to a nonlinear system for irreversible phase changes, Adv. Math. Sci. Appl. 10, 1–24.
48. G. Bonfanti, M. Frémond, F. Luterotti, 2001, Local solutions to the full model of phase transitions with dissipation, Adv. Math. Sci. Appl. 11, 791–810.
49. G. Bonfanti, M. Frémond, F. Luterotti, 2004, Existence and uniqueness results to a phase transition model based on microscopic acceleration and movements, Nonlinear Anal.: Real World Appl., 5, 1, 123–140.
50. G. Borino, B. Failla, F. Parrinello, 2003, A Symmetric nonlocal damage theory, Int. J. Solids and Struct., 40, 3621–3645.
51. A. Bossavit, 1976, Définition et calcul d'une perméabilité équivalente pour l'acier saturé. Un problème de Stefan à petit paramètre, Bulletin de la Direction des Études et Recherches, Série C, 2, 45–58.
52. G. Bouchitté, A. Mielke, T. Roubíček, 2007 A complete damage problem at small strains, Z. Angew. Math. Phys. 60, 2009, 205–236 (published eletronically Nov. 2007).
53. B. Bourdin, G. A. Francfort, J. J. Marigo, 2000, Numerical experiments in revisited brittle fracture, J. Mech. Phys. Solids, 48, 797–826.
54. A. Braides, 1998, Approximation of free-discontinuity problems, Springer-Verlag, Berlin.
55. C. M. Brauner, B. Nicolaenco, M Frémond, 1986, Homographic approximations of free boundary problems characterized by variational inequalities. Sci. Comput., Adv. Math. Suppl. Stud., 10, 119–151.
56. M. Brokate, J. Sprekels, 1996, Hysteresis and phase transitions, Appl. Math. Sci. 121, Springer, New York.
57. M. D. Bui, 1978, Mécanique de la rupture fragile, Masson, Paris.
58. L. Cangemi, 1997, Frottement et adhérence : modèle, traitement numérique et application à l'interface fibre/matrice, Thèse de l'Université d'Aix-Marseille.
59. P. Carrara, D. Ferretti, F. Freddi, G. Rosati, 2011, Shear tests of carbon fiber plates bonded to concrete with control of snap-back, Eng. Fract. Mech., 78, 2663–2678.
60. P. Casal, 1972, La théorie du second gradient et la capillarité, C. R. Acad. Sci. Paris, A, 274, II, 1571–1574.
61. A. M. Caucci, M. Frémond, 2005, Collisions and phase change, Free boundary problems: theory and application, Coimbra.
62. A. M. Caucci, M. Frémond, 2009, Thermal effects of collisions: does rain turn into ice when it falls on a frozen ground? J. of Mech. of Mater. and Struct., 4, 2, 225–244, http://pjm.math.berkeley.edu/jomms/2009/4-2/index.xhtml
63. N. Chemetov, 1998, Uniqueness results for the full Frémond model of shape memory alloys, Z. Anal. Anwend., 17, 4, 877–892.
64. A. Chrysochoos, H. Pham, O. Maisonneuve, 1993, Une analyse expérimentale du comportement d'un alliage à mémoire de forme de type Cu-Zn-Al, C. R. Acad. Sci., Paris, 316, II, 1031–1036.

9783642246081